Rafael Cayuela Valencia

**The Future of the Chemical Industry
by 2050**

Related Titles

García-Martínez, J., Serrano-Torregrosa, E. (eds.)

The Chemical Element

The Industry and the Business

2011
Hardcover
ISBN: 978-3-527-32880-2

Pollak, P.

Fine Chemicals

The Industry and the Business

2011
Hardcover
ISBN: 978-0-470-62767-9

Bamfield, P.

Research and Development in the Chemical and Pharmaceutical Industry

2006
Hardcover
ISBN: 978-3-527-31775-2

Budde, F., Felcht, U.-H., Frankemölle, H. (eds.)

Value Creation

Strategies for the Chemical Industry

2006
Hardcover
ISBN: 978-3-527-31266-5

Rafael Cayuela Valencia

The Future of the Chemical Industry by 2050

WILEY-VCH Verlag GmbH & Co. KGaA

The Author

Rafael Cayuela Valencia
Holzmoosrütistr. 6
8820 Wädenswil
Zurich
Switzerland

All books published by **Wiley-VCH** are carefully produced. Nevertheless, authors, editors, and publisher do not warrant the information contained in these books, including this book, to be free of errors. Readers are advised to keep in mind that statements, data, illustrations, procedural details or other items may inadvertently be inaccurate.

Library of Congress Card No.: applied for

British Library Cataloguing-in-Publication Data
A catalogue record for this book is available from the British Library.

Bibliographic information published by the Deutsche Nationalbibliothek
The Deutsche Nationalbibliothek lists this publication in the Deutsche Nationalbibliografie; detailed bibliographic data are available on the Internet at <http://dnb.d-nb.de>.

© 2013 Wiley-VCH Verlag GmbH & Co. KGaA, Boschstr. 12, 69469 Weinheim, Germany

All rights reserved (including those of translation into other languages). No part of this book may be reproduced in any form – by photoprinting, microfilm, or any other means – nor transmitted or translated into a machine language without written permission from the publishers. Registered names, trademarks, etc. used in this book, even when not specifically marked as such, are not to be considered unprotected by law.

Typesetting Laserwords Private Limited, Chennai, India
Printing and Binding Markono Print Media Pte Ltd, Singapore
Cover Design Grafik-Design Schulz, Fußgönheim

Print ISBN: 978-3-527-33257-1
ePDF ISBN: 978-3-527-65702-5
ePub ISBN: 978-3-527-65701-8
mobi ISBN: 978-3-527-65700-1
oBook ISBN: 978-3-527-65699-8

Contents

Preface *IX*
Acknowledgments *XIII*

Introduction *1*
Methodology *4*

1 **Global Megatrends by 2050** *11*
1.1 Social Megatrends *16*
1.1.1 Population Growth *16*
1.1.2 Demographics *18*
1.1.2.1 Area and Age Distribution *18*
1.1.2.2 Change in Age Distribution *18*
1.1.3 Urbanization *25*
1.1.3.1 Megacities *27*
1.2 Economic Megatrends *29*
1.2.1 Foreign Direct Investment(FDI) *40*
1.3 Political Megatrend *42*
1.3.1 Trend – A New International Order *43*
1.3.1.1 Sub Trend – the Emergence of the BRIC Economies *43*
1.3.1.2 Sub Trend – Corporate Mega Economies – (CME) *46*
1.3.1.3 Sub trend – Social Networks *50*
1.3.2 Trend – An increasing role of Governments *54*
1.4 Energy Megatrends *61*
1.4.1 Recent Energy Transitions *63*
1.4.2 Key Lessons from Recent Energy Mix Transitions *69*
1.4.3 Energy Life Cycle *69*
1.4.4 Energy Success Criteria *69*
1.4.5 Shocks Are a Valuable Source of Information *70*
1.4.6 Transitions Occur in "Life" *70*
1.4.7 The Golden Rule – Economics Dictate Energy Transitions *70*
1.4.8 Transitions Always Occur, the Question Is When: The Oil Peak *71*
1.4.9 The Oil Peak – M.King Hubbert *74*

1.4.10	OPEC – Energy projections to 2030	75
1.4.11	Recent Developments	80
1.4.11.1	Nuclear Energy – The Aftermath of Fukushima	80
1.4.11.2	Shale Gas the "Game Changer" – Natural Gas the Energy of the Future	85
1.5	Climate Change	99
1.5.1	Business Case – EU Tire Labeling – CO_2 Emissions Reduction in the Tire and Automotive Industry	113
1.6	Wild Cards	129
1.6.1	Political	131
1.6.2	Social	131
1.6.3	Technological	132
1.6.4	Transportation	132
1.7	Accelerators – Information Technology and "Singularity"	132
	Appendix: Climate Change	139
2	**The World by 2050**	**141**
2.1	"A Much Larger, Wealthier, Healthier, and Sustainable World"	141
2.1.1	Methodology	144
2.2	Status of the World – 2010	145
2.3	The World in 2050	146
2.3.1	BAU Scenario	146
2.3.2	Sustainable Scenario	155
	Appendix: Roadmaps to a World of 4000 g of CO_2 per Capita per Day	161
3	**The Chemical Industry in 2010**	**163**
3.1	Chemical Industry: Economic Relevance	163
3.2	Chemical Industry: Technological Relevance	166
3.3	Industry Relevance: Profitability	168
3.4	Feedstocks and Energy	171
3.5	Major Sectors and Products of the Chemical Industry	173
3.6	Industry Structure and Companies	173
3.7	Safety	184
3.8	Background	186
3.8.1	Recent History of the Chemical Industry Excluding Pharmaceuticals	186
3.8.1.1	1750–1850 Industrial Revolution and Inorganic Chemistry	187
3.8.1.2	1850 Synthetic Dyes from Coal for Textiles, and Chlorine Bleach	187
3.8.1.3	1870 Celluloid	187
3.8.1.4	1880 Rayon from Wood Fibers	188
3.8.1.5	1900 Electrolysis of Brine (Chlorine)	188
3.8.1.6	1913 Synthetic Fertilizers	189
3.8.1.7	1910–1920 Steam Cracker (Ethylene, Propylene and Butadiene)	190

3.8.1.8	1920–1930 – Styrene Cracking (Ethyl-benzene and Styrene) Cracking *190*	
3.8.1.9	Polyamide Nylon (DuPont) *190*	
3.8.1.10	1930s – Synthetic Rubber *190*	
3.8.1.11	1950s – Plastics Demand Explodes *191*	
3.8.1.12	1960s Internationalization *194*	
3.8.1.13	2010–2050 – The Chemical Industry Leads the Revolution against Climate Change *194*	
3.9	Conclusion *194*	
3.10	Summary – Industry Major Features and Upcoming Megatrends *195*	
3.11	Major Features of the Chemical Industry *197*	
3.11.1	Summary: Global Major Megatrends *199*	
	Bibliography *200*	
4	**Impact Assessment of the Global Megatrends on the Chemical Industry** *201*	
4.1	Introduction *201*	
4.2	Megatrends with the Highest Impact into the Chemical Industry (Global & Area Level) *207*	
4.3	Megatrends with the Highest Impact in the Industry (Area Level) – (Figure 4.4) *208*	
4.4	Megatrends with the Highest Impact into the Different Features of the Industry *210*	
4.5	Major Results for the Chemical Industry Globally *212*	
4.6	Major Results for the Chemical Industry in the ADV Economies *214*	
4.7	Major Results for the Chemical Industry in the BRIC Economies *216*	
4.8	Major Results for the Chemical Industry in the REST Economies *218*	
5	**The Chemical Industry by 2050** *221*	
5.1	Introduction *221*	
5.2	Feature 1: The Relevance of the Chemical Industry *225*	
5.2.1	Economic Relevance *226*	
5.2.1.1	Chemicals and Pharmaceuticals by 2050 – per Capita Demand in $US *229*	
5.2.2	Technological Relevance *236*	
5.2.2.1	The Chemical Industry – Long Term Cycles *236*	
5.2.3	Profitability *241*	
5.3	Feature 2: Inputs – Feedstocks *245*	
5.3.1	BAU Scenario for Feedstocks by 2050 *247*	
5.3.2	Feedstock Simulation II by 2050: "Shale Gas I – Ethane + 20% Globally" *250*	
5.3.3	Simulation II by 2050: "Shale Gas II – Ethane at Maximum Capacity Globally" – (Unreal) *252*	
5.3.4	Conclusion and Feedstock Alternatives *254*	

5.4	Feature 3: Outputs – Products	256
5.4.1	Global Ethylene Market by 2050 – BAU Scenario	258
5.5	Feature 4: Climate Change – Greenhouse Emissions – CO_2 Emissions	264
5.5.1	Historical and Future Scenarios on World CO_2 Emissions	268
5.5.2	Summary – Global Emission Trading Systems in Operation	277
5.5.2.1	Chemical Industry – Greenhouse Emissions Abatement in 2005	281
5.6	Feature 5: Industry Structure	283
5.6.1	Markets – Largest World Markets	284
5.6.1.1	Chemicals	284
5.6.1.2	Pharmaceuticals	284
5.6.2	Per Capita Demand	287
5.6.2.1	Chemicals	287
5.6.2.2	Pharmaceuticals	289
5.6.3	Companies – Changes in Global Sales Rankings and Company Structures	291
5.7	Feature 6: Social Awareness	294
	Appendix – Climate Change	296

6   **Conclusion** *307*

Appendix *315*

Index *319*

Preface

The aim of this book is to produce a comprehensive overview of how our world and the chemical industry could look by 2050, by completing a detailed and structured assessment of how the upcoming megatrends could influence our world and the chemical industry during the next decades.

During the last century the world has seen the largest and longest period of wealth creation in human history, with our world living standards reaching their highest levels. During the next decades, that trend is not only going to continue but also accelerate. The world will go through another period of unprecedented economic and social wealth creation. In a world with more than nine billion people and an expected gross domestic product (GDP) four times larger than today, the world by 2050 will host the wealthiest society in human history.

During the next decades the world will be in the midst of one of its largest and most radical transformations. Our societies, economies, political systems, and even the way we work, communicate, and live may be altered. New economic superpowers, like the BRIC (Brazil, Russia, India and China) will become reality, with China and India becoming the two largest economies in the world. New industries and new and large economic players will appear, while many others will quickly loss relevance. The potential improvements in human progress and wealth creation, with millions of people leaving poverty forever, are simply outstanding and fascinating. The world per capita GDP will treble from current levels up to $30 000 by 2050 (2009 based) while the world average life expectancy will move from 67 years in 2010 to 75 in 2050. The potential progress in all fields of human life will be simply enormous. The change will be gigantic, the speed frantic and the need for change will be unquestionable. However, all this positive progress will not come for free and without caveats.

During this journey, our world will be confronted with some of the largest challenges humanity has ever seen. Energy scarcity and climate change might be the most obvious and radical, but not the only ones. Strong global demand across all industries, significant changes in the way we live, communicate, and organize our societies, at national and international level, will also pose significant challenges. Life expectancy and medical progress will reach unseen levels across the world, a factor that might have the potential to alter all aspects of our life to levels never seen or even thought of before.

In a world with more than nine billion people and the host of the wealthiest and most advanced society in human history, our world energy demand under the "business as usual" (BAU) scenario will not only be enormous, but might also be impossible, and even when possible it might not be sustainable. The world average CO_2 emissions per capita per day will move from 12,000 grams a day in 2010 to 28,000 grams by 2050. However, in order to avoid climate change, our world will need to learn to live with just 4,000 grams of CO_2 per capita per day; equivalent to a 30 km drive per day (based on a car with CO_2 emissions of 130 g per km). Under the "BAU" scenario the world energy demand could triple from 12 million tonnes of oil equivalent in 2010 to 35 million tonnes by 2050. Crude oil demand could also more than double from 87 million barrels a day in 2010 to 247 million barrels a day by 2050.

The enormous growth experienced during the last century, in economic terms, energy demand, and population, among others, suggests that similar or higher levels of growth during the next decades might not be attainable or sustainable this time. The world will need to face a gigantic energy and emissions diet, cutting CO_2 emissions to the current levels of a citizen in India.

The chemical industry, as the key enabler of human progress and a building block of many other industries, will again be called into duty. In a world with a GDP of more than $US 280 trillion (2009 dollar) the chemical industry not only has the potential to quadruple by 2050 but, more importantly, it will again have the opportunity to unleash its full potential. During the next decades we will see how China consolidates its role as the largest chemical maker in the world, how the Indian chemical industry will blossom to levels never seen before and how the pharmaceutical industry in the BRIC economies will simple explode with a humungous 1595% growth. The per capita demand for pharmaceuticals in the BRIC economies is expected to grow from $30 in 2010 to $444 by 2050.

The chemical industry is not new to this sort of challenge. It has been enabling human progress and innovation through chemicals and pharmaceuticals for centuries. This industry enabled the first and second industrial revolution and, most recently, all the miracles of the last century's economic growth; the longest period of wealth creation in human history. The chemical industry has been improving living standards and quality of life since its origins, and indeed our current life standards would not be possible without it.

The enormous challenges that climate change and the expected vast economic growth will present, will not only affect directly the chemical industry as an industry itself, but will also force the industry to excel in what it does best: technology and innovation. The industry will soon be called into duty again, forcing it to transition again from a manufacturing and operational model into an innovation model.

This book is intended to provide a logical and comprehensive framework to understand and assess the potential impact of upcoming megatrends in the chemical industry. It is not intended to provide critical solutions or answers to very complex topics, but to provide the right level of understanding, fundamentals and concern, pointing the industry towards its major megatrends. In other words, this book is not intended to be the end, but just the beginning. A starting point and an area of reflection; a book that aims to stimulate proper debate across the chemical industry, other industries, and our society.

I believe, our world is poised to enter another period of unprecedented wealth creation, human progress and change. The challenge will be enormous for our society and for the chemical industry, but the rewards will be equally large and satisfactory. By 2050 the world will have not only the privilege to host the wealthiest and most advanced society in human history, but also the enormous responsibility to leave a sustainable world for the generations to come. At this time there are no answers or solutions, sometimes not even questions, but that is expected to change very quickly and soon.

The stakes are high, very high, and our society and the way we live depend on it. The challenges and rewards are enormous, the responsibility gigantic, and the clock is ticking fast. However, it is in adversity where the best of mankind and the chemical industry has always been seen; and it is my firm conviction that the chemical industry, as a catalyst of human progress and innovation, will be at the core of the solution. There is no other way, and that will be the case.

In this current world, severely affected by perhaps the second largest economic recession in our history after the great depression, I hope this book brings a fresh opportunity to start shifting our focus from the short term to the long term, gearing our attention toward much brighter horizons and the industry into more innovative paradigms. During the next decades our world has the potential to host the wealthiest society in human history, pulling hundreds of millions of people out of poverty; however, if the upcoming concerns are not properly and timely addressed, our generation will move from regretting what has been lost during the last economic crisis, into appreciating what we had and perhaps might not be able to have again.

The world will overcome the current financial crisis and the different economic challenges, as always, but what we may have in front of us might be something as unexpected to some as unbearable to others. The world is on the verge of its most fascinating challenge, the potential for human progress and economic growth is tremendous, however, at stake is the way we live and work, not only today but also tomorrow.

A call for attention to the future will soon reach us, perhaps as soon as the world economy comes back on track, and with that call the world will be confronted with a huge potential and enormous challenge. This shift from the financial crisis to the sustainability of our world and a bright economic future will reach us very soon; and with it, the chemical industry will be called into service once more. I am confident the industry will again be at the forefront of the solution, as always. This time I just hope we will not forget it too soon.

Enjoy the challenging journey into a better world through better chemistry.

Rafael Cayuela Valencia

To my father, Rafael Cayuela Menarguez, for his love, support and for keep showing me the light...

Acknowledgments

To my parents, Rafael and Maria Carmen, who taught me that with perseverance, curiosity, honesty, and care for others everything is possible.

This book would not be possible without the personal and professional collaboration and support of a vast number of people; people who have helped me over the years to become who I am. This book is a result of years of reflection, study, and learning about the chemical industry and its different relations and interactions with our society, economy, and world. It is not intended to be an end point that provides ultimate solutions and conclusions but rather a starting point. A book for learning, reflection, and debate; and ultimately a call for action.

Professionally, I would like to thank The Dow Chemical Company, who gave me the incredible opportunity to learn about the chemical industry from one of the best chemicals companies in the world, working with some of the most talented people in the industry. A place from where I have great memories and friends. The large Dow family with its people, heritage, and passion to strive is something I will appreciate for the rest of my life. After one year working for Dow Chemical, The Chemical Industry News Magazine and Chemical Market Intelligence (ICIS) conducted a survey among young fresh employees in the chemical industry which gave the opportunity to my friend and colleague Mr. Vladimir Jacobson and I to express our complete fascination for working in the chemical industry and for Dow Chemical. Dow's global footprint and working environment, its people and its clear commitment to improve and bring value to its employees, shareholders and society, is something we have never stopped appreciating. Ten years on from that interview, we both remain committed to this fascinating industry, passionate about it and willing to vindicate its proper role in society.

Working and learning from some of the greatest business minds in the chemical industry has not only been a great honor but also a fascinating experience. I could spend several pages mentioning some of the supervisors, mentors, and friends at Dow who have guided me through the different steps of my already fascinating career in the chemical industry. People like Mr. Pedro Suarez, Mr. Julio Hernando, Ms. Martina Bianchini, Ms. Isabelle Driessens, Mr. Guido de Witt, Mr. Luis Quismondo, Mr. Rene den Breeker, Mr. Frank Morgan, Mr. Alfonso Escudero, Mr. Markus Wildi, Mr. Niklas Meintrup, Mr. Chris Easdown, Ms. Carol Dudley, Mr. Torsten Kraef, Mr. Romeo Kreimberg, Mr. Juan Luciano, Mr. Celso Goncalves,

Ms. Sarah Opperman, Mr. Ken van der Wende, Mr. Craig Arnold, Mr. Ralf Irmert, Mr. Frans Hordies, Mr. Peter Kaestner, Mr. Marco Levi are only some of those that have helped and guided me. All taught me something special and I had great experiences working with and for them. I truly admire each of them for their dedication, leadership, and genuine desire to create a positive influence beyond themselves. Thanks to all of you for your support, teaching, and this incredible journey. You all know why.

Special thanks also to Mr. Frank Morgan and Mr. Juan Luciano, two outstanding Dow leaders, had the vision to give me this fascinating challenge. As an economist by education, writing about the chemical industry has been one of the most challenging things I have ever done. Considering the enormous respect and admiration the chemical industry provokes in me, writing about its future has been given me many sleepless nights and many days dreaming. Please consider this book as a small contribution to the chemical industry. An industry often misunderstood and always undervalued.

This manual may not have the answers you were looking for, but I hope it will still give you some of the questions you wanted to hear. In that sense I would also like to thank Mr. Tim Nash, Dean of the Northwood University and Mr. Bill Busby from the Dow/Northwood MBA program for their passion in educating and pushing us to our highest potential. Tim's contagious and brilliant mind has been a constant source of inspiration, an example to follow and a real friend.

Outside Dow Chemical, Mr. Sam Vasegui has been without doubt the catalyst of this project, the visionary who guided me and supported me to dare to go beyond my limits and my own expectations. Sam's delicate presence, often disguised one of the most powerful voices and minds I have ever come across. I thank him for his vision and trust.

Personally, I would like to thanks my parents, Mari Carmen and Rafael for their love, strength, and humble and optimistic approach to life. There are no words or acts to thank them enough. Thanks also to my brother Elias Cayuela for its love, unconditional support and inspirational approach to life. Thanks also to my broad family, including my friends Camilo Bel, Francesco Zanchi, David Carrion, Virginia Canovas, Alexander Perez, Carlos Serrano, David Valverde, Irene Molina, Juan Carlos Herrera, Matteo Rosso, Merari Abigail Escalante, Bernadette Unterkircher, Laurence Conrad, Moreno Volpi, Romeo Volpi, Flavio Volpi, Silvia Rosatto and Rafael Rosatto for your time, your patience, your inspiration and love. You all know why and you all deserve to be here.

Introduction

The purpose of this book is to provide, for a certain set of assumptions and megatrends, a solid understanding and overview of how the chemical industry and our world could look by 2050.

Due to the obvious complexity and difficulty in making such a projection we will provide a very simple and robust framework of analysis, with the aim that every reader can be aware of our assumptions and eventually could replicate this analysis with his own assumptions, expectations, and realities.

This book is not intended to provide a concrete and unique view of the industry by 2050, but a frame of solutions and questions, based on a certain set of assumptions and megatrends. The ultimate purpose is to create a solid understanding of the principles, fundamentals, and guidelines under which the chemical industry operates, with the premise that trends and circumstances might change, but the framework would not.

For a group of selected megatrends and different sets of assumptions we will present different scenarios and alternatives for the chemical industry. In some cases we will provide one or several scenarios, however, on other occasions we might end up by providing a set of questions or some open scenarios.

Therefore, this book is not intended to offer a final view of the industry, but more a directional one. It is intended to be a starting point, providing different trends and scenarios. Basic scenarios that are intended to create awareness and to stimulate appropriate debate across the industry and society. The speed of change in the world in which we operate and live these days will make it almost impossible to make an accurate forecast for a 40 year period. However, we feel very confident the framework of analysis that we will use should still be applicable for decades to come and directionally our study may still be valid in the foreseeable decades.

In Chapter 1 we will present some of the most pervasive megatrends that will impact our world and chemical industry during the next decades. We will review trends in several different areas, such as social, economic, political, energy, and climate change. For each of these megatrends we will also review several small trends, their interconnection and their connection with other megatrends.

Chapter 2 will present a projection of how the world could look by 2050, based on the previous megatrends. Two scenarios will be presented. In the first, the so-called business as usual (BAU) scenario, we will see how our world could look by 2050

The Future of the Chemical Industry by 2050, First Edition. Rafael Cayuela Valencia.
© 2013 Wiley-VCH Verlag GmbH & Co. KGaA. Published 2013 by Wiley-VCH Verlag GmbH & Co. KGaA.

if no major human action takes place and no changes are applied to the expected megatrends. This scenario will present staggering improvements in wealth and living standards across the world, quadrupling the world gross domestic product (GDP) and pulling millions of people out of poverty. Unfortunately, and as we will see later, this scenario might not be sustainable from an energy and emissions perspective. In the second scenario, we will present a sustainable world, a world that will still preserve all the positive aspects of the previous scenario but will manage to avoid climate change and the potential for energy and resources scarcity. For both scenarios we will review some of the fundamental pillars for our society with high impact on the chemical industry. In that sense three fundamental pillars have been selected. The social pillar that includes population and economic conditions; the energy pillar that will establish energy demand projections, including oil and gas; and the sustainable pillar that refers to different measurements for climate change.

In Chapter 3 we will introduce the chemical industry, its history, players, products, and recent status in many critical areas. The idea is to provide a comprehensive summary of the chemical industry, its structure and major features, with the assumption that not all readers may necessarily be knowledgeable about this industry. Even for those already in the chemical industry, certain aspects of this chapter may be interesting and revealing, like those related to its history, products, players, or even its total size and characteristics. As a proud and senior member of the industry, I still find it amazing how little some of us know about our own industry, its history, wonders, roots, and its full positive potential in society. Writing this chapter was an enormous source of personal pleasure and enrichment, a chapter where I kept learning about the complex and fascinating history of our industry. This is a chapter that I recommend to anyone working in the chemical industry or with interest in it.

In Chapter 4, we will review the potential influence of the expected megatrends on the chemical industry. For that purpose we will develop a simple framework of analysis, a framework that we will use for this and further assessments within the book.

Using this framework we will start to assess the impact of the upcoming megatrends on the chemical industry and, ultimately, on the chemical companies of the future. We will review some of the major impacts, the key trade-offs and synergies, however, and for obvious reasons, we will not be able to review all possible interactions.

In the belief that the presented and used framework of analysis may also be valid for different companies within the chemical industry, this framework will be made available for our readers. The idea being that the readers can adapt the model to their own companies, beliefs, and assumptions. After all, our projections are based on some pre-selected numbers of megatrends. Under this premise, we believe that by providing the model to our readers, each will be in the best position to judge our assumptions and projections, while having the option to make their own ones.

In Chapter 5 we will provide an overview of the future of the chemical industry by 2050, by reflecting on the potential impact of each of the megatrends on the major features of the chemical industry under the two defined scenarios: BAU and a "sustainable" world. For each scenario we will present qualitative and quantitative statements about the industry, its major companies, the potential areas of growth, as well as areas of concern and focus. For the first scenario (BAU) we will review how large our chemical industry could become, its energy demand, emissions, major areas of growth, and its key players. Sometimes a modified BAU scenario will also be presented, where the author will introduce some minor changes while remaining as conservative and realistic as possible.

For the second, sustainable scenario we will reflect on the challenges and opportunities the chemical industry will be confronted with when operating in a sustainable way with reduced greenhouse emissions, however, under this second scenario, answers and solutions will tend to be more vague and open. Indeed this is an area that could provoke significant amount of research and innovation during the next decades.

In this chapter we will also provide some conclusions and recommendations for the chemical industry of the future. Considering the enormous area the chemical industry covers – from petrochemicals to polymers, industrial gas, crops, fertilizers, pharmaceuticals, and so on – the aim of this chapter is to provide generic and broad recommendations to the industry, highlighting major changes, areas of focus, growth, or concern. In this chapter, as in some of the previous ones, we do not expect to come with clear answers and solutions, but more with certain generic reflections and guidelines. Guidelines that should be placed into the context of different companies and sectors of the chemical industry.

Finally, in the conclusion we will summarize some of the major opportunities and concerns for the chemical industry in its transition to 2050. The chemical industry is poised to experience significant changes in the next decades, from massive growth, to changing demand, technologies, locations, feedstocks, and even the way it operates and interacts with its different stakeholders. However, and despite the massive changes and challenges expected in these areas, it is in its capacity to innovate and to enable other industries and higher living standards that it will experience the major transformation. Innovation will be by far the major area of focus and attention, as the chemical industry will be required to be at the forefront of the solutions that will save our endangered world.

While the chemical industry has learned to mitigate the negative aspects of change – like the associated cost or potential concerns of the new challenges – the reality is that the upcoming decades with their associated challenges will bring an enormous opportunity for the chemical industry to unleash its full potential, create unprecedented levels of wealth and transform itself into the industry of the new century. The industry will be on the brink of its second revolution; its transition from a manufacturing model into an innovation and growth model. We expect an industry more focused on innovation and value creation, versus its cost, operating efficiency and all sort of other concerns. In other words, an industry more focused on the numerator than on the denominator.

Before concluding this introduction we would like to acknowledge that despite the recent developments in the world economy the world is poised to follow its growth course. The so-called "financial meltdown" in the USA with the Lehman

Brothers collapse in November 2008 and the current European sovereign debt crisis have triggered several years of economic slowdown globally, and perhaps one of the longest and deepest economic recessions since the great depression of 1929.

At this time there are still many unresolved issues and concerns on both sides of the Atlantic. High levels of debt and unemployment across the world, especially in the US and Europe, with the potential default of Greece, Portugal, Ireland, and even Italy or Spain, our world still seems economically "ill", digesting all excesses from the last decades. Weak economic data around the world, growing concerns on the robustness of the Chinese economy and on the long term sustainability of the high levels of debt in the US and Europe do not present a very optimistic outlook for the years to come. In the medium term, the outlook seems much more positive, but in the short term the reality is that the current economic recession might have stolen five to ten years of economic growth.

This economic slowdown has certainly shifted the focus away from the medium to the short term, and from some of the most challenging upcoming megatrends, like climate change and energy efficiency, to the current financial and economic issues. However, this shift will be just temporal, and some of the key issues of our world will come to the forefront of the global debate very soon, as soon as the financial crisis starts to get resolved.

As this book focuses on the long term, we will cover none of the current developments, despite the severity and potential short term influence, see Figure 1. The current slowdown may also have delayed some of the upcoming megatrends, but we do not expect it to have changed the structural and long term relevance of those.

Our world is poised for an unprecedented period of growth in the decades to come, growth that will result in the largest and wealthiest population in human history, and this seems still to be the case. Unfortunately, this growth will not come for free, and significant challenges and opportunities lie ahead. Energy and resource scarcity and climate change will test our capacity to innovate and even put some of our most optimistic projections at risk.

The current economic slowdown is having severe influence in the short term, as well as causing structural changes in the medium term, however, we still believe that in the long term the upcoming megatrends will still apply, as they always did.

Methodology

With such a complex, subjective, and broad topic as the future of our world and the chemical industry by 2050, the aim of this book is to provide the most realistic, factual, and logical set of projections. Projections that are easily replicable, trackable, and understood by anyone. Indeed some of our own projections and calculations will be intentionally very basic in nature, so that they can be easily validated, appearing logical, and simple to our readers.

Despite the fact that sometimes we will present scenarios that might be difficult to believe, we will always try to provide enough argumentation and context to these. Some of the upcoming projections may certainly defy our understanding

IMF data mapper®

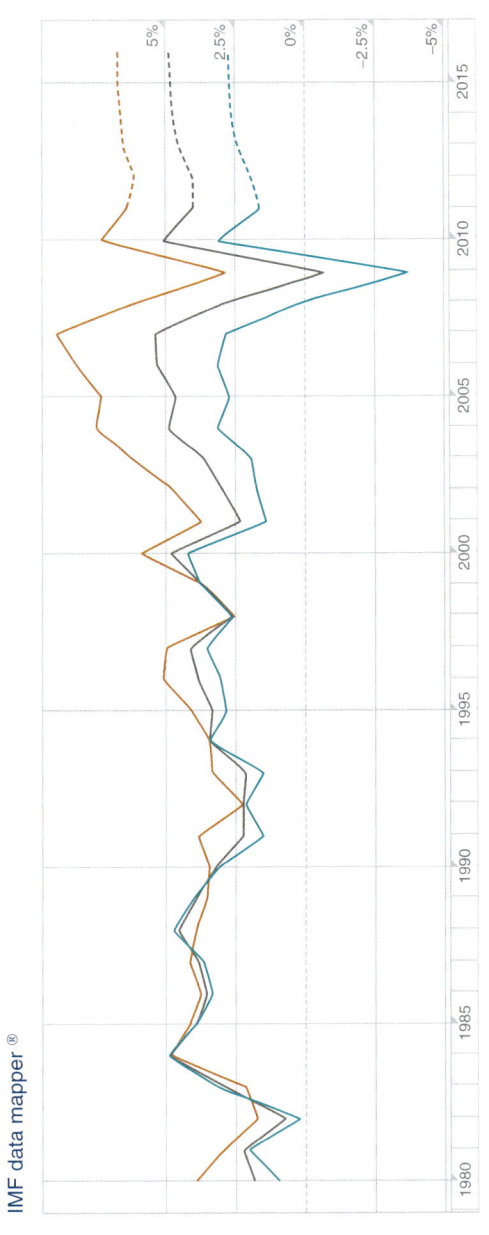

Figure 1 Real GDP growth: black, world economy; blue, advanced economies; brown, emerging and developing economies. World Economic Outlook (September 2011) © IMF, 2010.

and current views of the world, we are aware of and can expect this. However, none of them are expected to defy our common sense and logic. Indeed all scenarios and projections are expected to be easily explainable and ready to be validated with previous and recent historical examples.

The ultimate purpose of this approach is to create a set of conservative and realistic projections that are factual, logical, and usable. Despite that, none of our projections are intended to be considered as 100% accurate, rather more directional. Therefore the purpose of these projections is to highlight some of the major expected trends for the chemical industry, offering sometimes more questions than answers, but certainly enabling the industry to start looking into the right areas and to stimulate objective debates around a very complex topic.

For this purpose, most of the book projections will be based on three major projections: the latest World bank and UN data and projections for 2050 and beyond on several social indicators, like population, demographics, urbanization ratios or life expectancy; and the Economic Projections from PriceWaterhouseCoopers (PWC) on how the world economy will look by 2050.

From PWC we will use their GDP projections, while we will calculate our own GDP per capita values, based on the World Bank population projections. PWC's 2050 GDP projections are based on the 2009 US$ international dollar. In order to better understand PWC's economic projections, see the Box below for their long term economic growth model.

Mr. John Hawksworth, chief UK economist for PWC has granted access and use of the latest GDP projections till 2050, not without warning us that these projections have to be considered as directional and not like completed prophecies.

These three projections will be the bases of our long term projections for 2050.

In this area, and before moving forward, we would like to warn our readers that on some occasions we have been forced to use different sources for some specific data points or certain years. This might create some logical confusion and frustration, when the reader will come across two different numbers for the same year or for the same item, depending on the source. However, please be aware that unfortunately sometimes we have been forced to use different sources, when data or the same source were not available.

We would encourage our readers to focus more on the trends rather than on the specifics, after all we are looking into projections for 2050.

> **Box: PWC's Long-Term Economic Growth Model**
>
> The model used to project long-term economic growth in this paper is described in detail in our earlier series of "The World in 2050" reports. The model is a standard one in the academic research literature in which economic growth is driven by four main factors feeding into an aggregate production function:

- Technological progress, including "catch-up" effects for emerging economies that vary according to their state of institutional development and stability;
- Demographic change, in particular the growth rate of the working age population;
- Investment in plant, machinery, buildings, and other physical assets, which contribute to the long-term growth of the capital stock in the economy; and
- Trends in education levels, which are critical to the quality of the labor force and its ability to make the most of new technologies.

The assumptions used in this model reflect a broad range of research by bodies such as the IMF and the World Bank, as well as leading academic economists. While any such assumptions are subject to many uncertainties, we believe that the baseline economic growth scenario used in this paper is plausible.

Exchange Rate Projections

Purchasing power parity (PPP) exchange rates are assumed to remain constant over time in real terms, while market exchange rates converge gradually over time to these levels in the very long term (due to faster productivity growth in the emerging economies relative to the developed economies).

When making the different projections to 2050, the book will start by using those from the World Bank and PWC as our major base, extrapolating the current value of the selected variables into the future. Sometimes, especially for projections of energy demand emissions under the "sustainable scenario", we will introduce further adjustments and modifications.

For instance, considering our projection for the chemical industry by 2050, we will argue that, if in 2011 the chemical industry amounted to a certain percentage of the World GDP, in 2050 that percentage may be similar. Additionally, and to test the validity and robustness of such an assumption, we will start looking at that percentage in the past, so we have a feeling into the validity of the current ratios. Once that second validation is completed, and in accordance with our principle and desire to remain as conservative and realistic as possible, this projection will be deemed a valid one. This projection will then be subject to further adjustments when considering the emergence of new economies, new technologies, climate change, and so on, but in any case all modifications and assumptions will be explained and documented. We expect this approach will be transparent and straightforward for our readers, while it certainly serves our purpose to create awareness and identify major industry trends, rather than getting very concrete on very complex forecasts based on very complex analysis.

Additionally, and when looking into the future of the world and the chemical industry by 2050, we have segmented the world into three major areas; the advanced (ADV) economies, the "BRIC" (Brazil, Russia, India and China) economies and the REST. In the ADV group we have included the four largest economies: USA, Europe 27, Japan, and Canada. Hereafter, we will refer to Europe as meaning Europe 27. The REST includes all other countries not in the ADV or BRIC groups, which is certainly a large oversimplification, and following updates will start looking deeper into the REST group. However, for this analysis this division will serve our purpose; especially when considering that the ADV and BRIC economies together accounted for almost 70% of the world GDP in 2010 and 75% of the chemical industry.

The REST group is very large with a very heterogeneous set of countries, including such different ones as Switzerland, Norway, Iceland, Argentina, Mozambique, Morocco, Kenya, and Tanzania, just to mention a few. Furthermore, some of the REST members, such as Mexico, Indonesia, Turkey, Vietnam, or South Africa will also be subject to a tremendous growth during the next century. In future updates we will start looking into some of these countries, starting to decompose the REST group into its different members, after all some of these countries will have an enormous growth potential during this new century.

In terms of data and sources, this book has tapped into a multitude of experts, organizations, consultants, and analysis. Considering the broad scope of the presented megatrends and the complexity of the chemical industry, it is almost impossible to list all sources, however, some of the most important ones are listed below.

For all sorts of energy data and analysis, we have made extensive use of data from the 2011 Statistical Review of the World Energy from British Petroleum, as well as data and analysis from the US Energy Information and Administration (EIA) and from the International Energy Agency (IEA). From these bases, and using PWC and World Bank projections as a base, we have done multiple projections on oil, gas, CO_2 emissions and other energy sources on absolute and per capita bases.

From the IEA, as the world's leading organization on energy issues, we have also used some of the latest projections and scenarios for world energy demand and CO_2 emissions by 2030 and 2050. We have devoted special attention to the scenarios relating to the most sustainable scenario, such as 550 and 450 ppm of CO_2 emissions.

For the chemical industry we have used data from different international chemical associations, including the American Chemistry Council (ACC), the European Chemical Industry Council, (CEFIC), the International Council of Chemical Associations (ICCA) and the European Federation of Pharmaceutical Industries and Associations (EFPIA). We have also used chemical related data from several generic consultancy firms like McKinsey or Deloitte and Touch as well as some specific to the chemical industry, like IHS (CMAI), Platts, ICIS, or Dewitt. Additionally we have consulted some others and international organizations like OPEC, IMF, OCDE, the European Commission, Eurostat, World Economic Forum, and so on. For some of the chemical companies data we have used multiple sources, including corporate information from corporate annual reports and web pages, as well as data from specialist magazines and many different industry leaders.

As a result, we will present a set of scenarios and projections for the chemical industry as a whole and for each of its major features. In the first chapters we will assess how the upcoming megatrends will affect the world and the chemical industry, while in the later ones we will start providing concrete and factual scenarios about how the world and the chemical industry could look by 2050.

On some occasions we will present two alternatives, one from a prestigious industry expert or organization, and another based on our own calculations, using the methodology explained above. In other cases we will present only one scenario, either from a well known organization or expert, or from our own study, and on some occasions no projections will be available. For some topics, the scenarios will indeed remain open and with a question mark for future follow ups and evaluations.

As previously mentioned, the purpose of this book is to present a logical and conservative view on how our world and the chemical industry could look by 2050, providing enough data and rationale to take the right considerations and decisions. Most of our projections will be tested with the results learnt from the last decades and centuries, evaluating our performance in these areas during similar periods during the last century.

After all, most of the upcoming projections might be wrong, but at least we should feel confident this book provided the right rationale and argumentation; and the projections are conservative and well documented. Therefore let me wish you a safe and pleasant reading into a better world through better chemistry.

1
Global Megatrends by 2050

> *I never think of the future, it comes soon enough.*
>
> Albert Einstein

During this chapter we will review some of the major global megatrends expected to shape the world during the coming decades.

In that sense we have identified six major megatrends: social, economic, political, energy, climate change, and wild cards. Although most are interconnected we will first review them individually and then consider their potential interrelations. Additionally, information technology has also been recognized as a key accelerator and enabler of change; an accelerator that has the potential to affect all the previously selected megatrends. Figure 1.1 shows the different megatrends and the hierarchy among them.

The forecasted changes in world population and economic growth, will be the major drivers for change in the next decades, not only influencing our world, but also all other megatrends. Meanwhile information technology will have the potential to influence and accelerate all megatrends.

For each of the megatrends we will also explore several subcategories, and their interrelation with each other and with the rest of them, especially with the two major ones, the social and economic megatrends.

For the social megatrend we will review three fundamental aspects, the increase in population, the change in demographics, and the changes in urbanization. With an expected population of nine to eleven billion people, a much more elderly population, higher life expectancy, and more people moving to live in the cities, social megatrends will set the basis for most of the other trends.

For the economic megatrend we will review the economic projections for the world and its largest economic areas, as explained in the methodology. In this sense we will look at the projections for the Gross Domestic Product (GDP) on both an absolute and a per capita basis globally and for the key selected areas. For that purpose we have divided the world into three major areas: the "Advanced" group (ADV) consisting of the four largest economies in the world – USA, Europe (European Union 27), Japan, and Canada, key members of the G8; the BRIC group consisting of Brazil, Russia, India, and China; and all the remaining countries (REST). This is certainly a large oversimplification, and in later updates we will

1 Global Megatrends by 2050

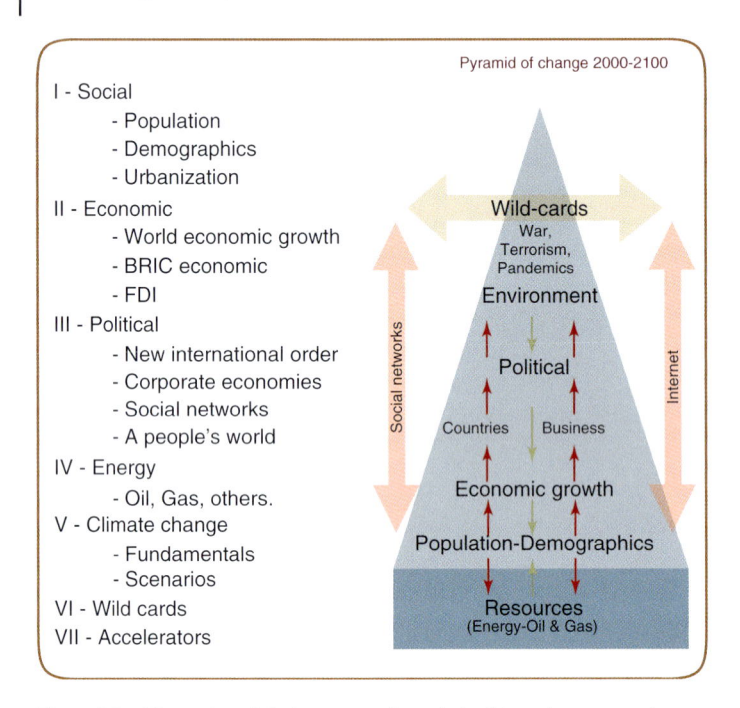

```
I - Social
    - Population
    - Demographics
    - Urbanization
II - Economic
    - World economic growth
    - BRIC economic
    - FDI
III - Political
    - New international order
    - Corporate economies
    - Social networks
    - A people's world
IV - Energy
    - Oil, Gas, others.
V - Climate change
    - Fundamentals
    - Scenarios
VI - Wild cards
VII - Accelerators
```

Pyramid of change 2000-2100
- Wild-cards (War, Terrorism, Pandemics)
- Environment
- Political (Countries, Business)
- Economic growth
- Population-Demographics
- Resources (Energy-Oil & Gas)
- Social networks
- Internet

Figure 1.1 The major global megatrends and the hierarchy among them.

start to look more deeply at the REST group. However, for this analysis this division will serve our purpose; especially when considering that the ADV and BRIC economies together accounted for almost 70% of the World GDP in 2010, and 75% of the chemical industry.

The REST group is a very large group with a very heterogeneous set of members, ranging from countries like Switzerland to Iceland, Angola, Norway, or Argentina. Indeed some of its members, such as Mexico, Indonesia, Turkey, Vietnam, Korea, or South Africa, among many others, are also expected to have a tremendous growth potential in this century.

Finally, we will also look at the projected movements in foreign direct investments (FDI) for the predefined areas. In a world of 9 billion people, the world GDP is expected to grow from the current 63 trillion in 2010 to $280 trillion (purchasing power parity (PPP) based on dollar 2009), according to the consultancy firm PriceWaterhouseCoopers (PWC). *https://spreadsheets.google.com/ccc?key=0AonYZs4MzlZbdC1fandTcXJ0OG9WYW5mZ1NOT1VaTHc&hl=en*.

This massive economic growth implies an equally massive redistribution of wealth on absolute and per capita bases, where the so-called BRIC economies will be the largest beneficiaries. China is expected to become the largest economy in the world and India the third, while the USA will be the second and Europe the fourth.

The GDP per capita of the BRIC countries will increase by a factor of 4 to 16, depending on the country, while the REST countries will triple theirs. However, and despite this formidable growth, the ADV economies will still have much higher GDP per capita than the BRICs or the REST, double that of the BRICs and quadruple that of the REST.

This projected strong global growth will enable the creation of a huge global middle class, unique in human history. Indeed by 2050, more than 50% of the world population, that is to say a staggering 4.5 billion people, will be considered as middle class, with a GDP per capita similar to those that we can see in Europe or the USA in 2011.

This large economic growth in the BRIC economies has attracted significant amounts of FDI, during the last decades. This trend will continue for decades to come. However, the ADV economies, and especially the USA, with still increasing populations will keep their dominant position as the largest recipients of FDI.

In the political megatrends we will review multiple parallel trends. On the one hand, the projected global economic growth, especially in the BRIC economies, will serve to enhance their relevance and political power. In this section we will review how this change in the world balance of power could alter some of the existing international organizations and how some new organizations might be required. On the other hand this massive economic growth will continue to foster the creation of large corporations, hereafter referred to as corporate mega-economies (CME). In 2007, 50% of the largest world economies were companies, if we measure companies by revenue. Globalization has offered companies the opportunity to grow much faster than their country, creating CMEs, companies that can be much larger than countries, while growing much faster than them.

In a completely different area, the fast development of social networks since 2004 has created new ways of communication between citizens, citizens and business, and citizens and governments. Social media have enabled millions of people across the world to get connected, allowing them to share, learn, and discuss about all sorts of topics globally. This new form of communication has grown tremendously fast in recent years, especially in the BRIC economies, reaching more than 3 billion users globally in 2011. Social networks have created a new and very powerful way for citizens to make their voices heard, allowing them to share their views and opinions on any issue, including governments and companies. They have contributed to the globalization of people's ideas and thoughts, giving back to the citizens the power to influence in the short term and real time. They have probably changed the way social stakeholders communicate, and this change may no longer be reversible. Social networks may not only have the power to improve the quality of democracies, but also to accelerate the transition of more countries into more democratic models.

Finally, on the political side, Governments' roles are expected to remain very prominent globally. With the current economic recession, after the collapse of Lehman Brothers in 2008 and the current European sovereign debt issues during 2011 and 2012, governments have been forced to take an even larger role. Initially stimulating their economies with large stimulus packages and high intervention into the economy, and currently, especially in Europe, with much broader government intervention. This includes controlling government expenditure, accelerating reforms in the banking and labor system, promoting industry consolidation and M&A (mergers and acquisitions), while controlling the movement of labor, capital, and goods.

Most of the largest economies have put together large stimulus packages, regulations, and action plans to stimulate the economy. The realization that Governments might need to have a role when their economies slow down, will actually match the fact that companies are becoming larger and larger. In a scenario where companies are not only much bigger than countries, but actually keep growing faster than them, the need to better understand and monitor companies' activities will require very strong governments. That, in combination with the fact that governments will still have a large share of the economy, makes us believe that Governments will continue to have a very dominant role in our societies.

For the energy megatrend we will review the key lessons from the last energy transitions, learning about how, when, and why they occurred. Looking into the future we can still envision a world based on fossil fuels, however, the tremendous expected growth in energy demand, in combination with the need to reduce greenhouse emissions, makes the need for additional and cleaner energy sources more obvious.

In a world of 9 billion people in 2050 compared with almost 7 billion in 2010, and an estimated GDP per capita of more than $30 675 (2009 dollar) versus $9219 in 2010, global crude oil demand under the business as usual "BAU" scenario for 2050 could move from the current 87 to 247 million barrels a day.

In other words, by 2050 crude oil demand could treble versus 2010 demand, at observed energy efficiency ratios. Despite the fact that this figure might be completely unattainable based on today's understanding of crude oil reserves, and also illogical based on the expected need to transition from fossil fuel to cleaner energies due to climate change, the reality is that, looking at history, this would not be the first time that crude oil demand has doubled or trebled during a similar period of time.

Most recently, and during the last 50 years, world crude oil demand almost trebled from 30 million barrels in 1960 to 85 million barrels in 2010; although we should acknowledge that the energy ratio, measured in terms of barrels per day versus absolute GDP in trillion US dollars, went down from 5 in 1960 to almost 1 in 2010. So, theoretically, we could also argue that if the world were able to achieve again such an improvement in its efficiency ratio, crude oil demand could stay at around 60 to 90 million barrels a day by 2050.

All these topics will be reviewed and discussed at length in the section on energy megatrends, including the long term feasibility of crude oil and natural gas, their comparative economics versus other sources, and the potential impact into climate change and the chemical industry.

In the section on the climate change megatrend, we will present the basic background and fundamentals behind it, its major contributors, and potential scenarios. The complexity behind climate change is tremendously high, not only because it is a global issue that requires coordinated action from all countries and industries, but also because the potential solutions imply severe reductions in greenhouse emissions in sectors where they are critical, not only for the economy and society today but also for the future projected economic growth. In 2010 the World per capita CO_2 emissions were 4.4 mt (metric ton) annually or 12 000 g of CO_2 per day, with large differences across countries. The USA with 19.2 mt of CO_2 per capita was among the highest emissions, China with 4.6 mt and India with 1.2 mt were among the lowest.

According to studies from the International Energy Agency (IEA), if the world would like to keep the projected economic growth for the next decades while avoiding the potential negative impacts of climate change, the world carbon productivity will need to be increased drastically. In other words, the World per capita CO_2 emissions should be reduced from the current 4.4 mt annually or 12 000 g a day in 2010 to 1.5 mt annually or 4000 g per day by 2050. To further complicate this issue, the World will need to accomplish this massive reduction in an environment where, according to our own projections for the "BAU" scenario, the World per capita CO_2 emissions could reach 10 Mt or 28 000 g per day by 2050.

We will present a case study, where we could observe the enormous challenges our society will have to reduce greenhouse emissions, even when government, industries and society are committed to achieve it. The reality is that this cannot be achieved in the short term due to lack of suitable technology or still massive growth. This example will show that carbon productivity (the amount of GDP produced per unit of carbon equivalents) of 10 times vs. today's levels is very complex to attain, especially in the short term. All these aspects, and more, will be reviewed at length in the section on the climate change megatrend.

In the final section on the accelerators, we will review how the exponential growth in computer power and the potential arrival of the so-called "singularity", or convergence of human brainpower and computational power, could accelerate most of the previously introduced megatrends. By 2025, computer power is expected to surpass human brainpower, and by 2045, computer power is expected to become much higher that the total brainpower of all humans. The exponential growth in computer power will serve to accelerate technological developments in many fields – from robotics, to medicine, energy, artificial intelligence, biotechnology, or chemistry.

"Singularity" could have the potential to further accelerate most of the presented megatrends. To illustrate that, let us consider perhaps one of the most intuitive and impacting consequences of the application of higher computational power into our world: longer life expectancy.

Higher computational power will have tremendous impact on machinery, robots, biotechnology, forecasting modeling, medicine, and many other disciplines. Realizing all the enhancements in all these disciplines could result in extension of life expectancy to much higher levels that those contemplated in our current forecast. The United Nations has estimated that by 2050, life expectancy will increase from the current 67 to 75 years.

According to some key members of this "singularity" way of thinking, human life could be extended to well above by 2050, and eventually to 150 years by 2100. Obviously, if such a scenario becomes true, or even if it perhaps applies to a part of the world population, the consequences for our world will be massive. To the point that these scenarios will not only affect most of our population, demographics, or economic scenarios, but will also actually influence the way we live, work, and understand life itself.

For obvious reasons at this stage we have decided not to consider information technology and singularity as megatrends but as accelerators. However, we think both information technology and singularity have tremendous potential to become megatrends and are areas that should be carefully monitored and followed up in the next decades.

Finally, and for illustrative purposes, we will present some of the most radical scenarios, like global pandemics, nuclear wars, and other similar events, not because we expect or desire them but just to share the risk and uncertainties embedded in the exercise of making long term forecasts. Looking into such a long period of time is indeed a very complex exercise with plenty of uncertainties, which is why we will consider the most realistic and logical scenarios and megatrends. Certainly we recognize the potential for more radical scenarios but we will not consider these at this time.

1.1
Social Megatrends

> *"A 9 billion people World, with much larger, wealthier, healthier, and older population moving into cities."*

> We must learn to live together as brothers or perish together as fools.
> Martin Luther King Jr.

Social megatrends are in our opinion one of the main and most far-reaching of the megatrends we will review. Social trends will be one of the drivers of change for all other megatrends. In that sense, social megatrends will be explored from three major angles: changes in the absolute number of population; changes in age and geographical distribution of population (demographics); and changes in the ways of living (urbanization).

1.1.1
Population Growth

The unprecedented period of wealth, peace, and prosperity experienced during the last century – world GDP grew from US$ 2.5 trillion in 1900 to US$ 32 trillion in the year 2000 – set the conditions for an incredible population growth. Indeed, the world population increased almost fourfold in the same period, from 1.6 billion people in 1900 to 6 billion in the year 2000 (Figure 1.2).

Although economists will always argue which factor comes first, population or economic growth; most will agree on their positive correlation. Depending on which factor we consider as the leading one, we could analyze and measure growth in an absolute (total GDP) or qualitative (GDP per capita) way, but in the next years both factors will come together (Figure 1.4).

1.1 Social Megatrends 17

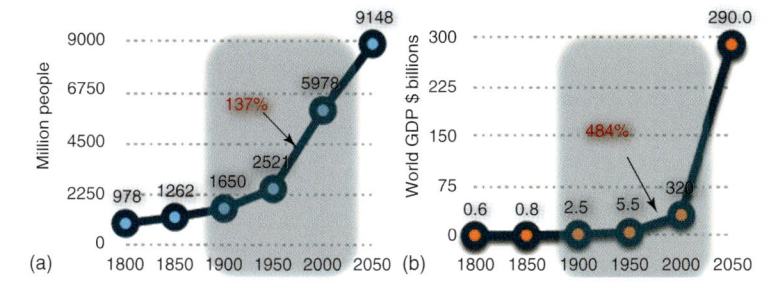

Figure 1.2 (a) World population. Source: Population Division of the Department of Economic and Social Affairs of the united Nations Secretariat. "World Population Prospects. The 2007 Revision" *http://esa.un.org/unpd/ppp/index.htm* (b) World GDP. Source: Historical Statistics of the World Economy 1–2003 AD. Mr. Angus Maddison and PriceWaterHouse "The World by 2050"

Although we will be reviewing economic trends in more detail in Section 1.2, we should acknowledge the fact that our world has been blessed with formidable and unprecedented economic growth during the last century, growth that has enabled our population to more than treble during that period, from 1.6 billion in 1900 to 6 billion in 2000 and 6.8 billion in 2010 (Figure 1.2).

According to the United Nations "normal" scenario, the world population will continue growing at a similar pace in the next decades, reaching 9 billion. The UN also consider "low" and "high" population scenarios with 10.7 and 7.4 billion people, respectively, by 2050, however, for this book and further analysis, we will stay with the UN base scenario of 9.1 billion people by 2050 (Figure 1.3).

A 9.1 billion people projection by 2050 and a 50% growth might seem a very impressive growth and large number, however, if we look at our most recent history, we can observe that our world population has already grown at this pace before in the 1900–1950 period, or even at a much higher rate between 1950 and 2000, when the world population grew by a staggering 137%, from 2.5 to 6 billion (Figure 1.3).

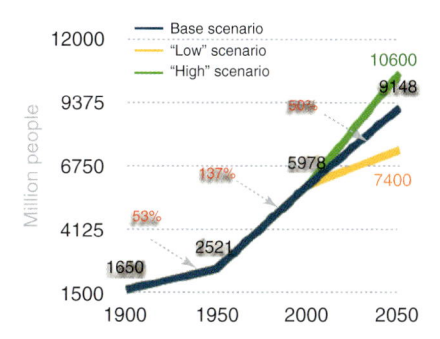

Figure 1.3 UN world population scenarios. Blue line, base scenario; yellow, "low" scenario; green, high scenario. (Source: Population Division of the Department of Economic and Social Affairs of the united Nations Secretariat. "World Population Prospects. The 2007 Revision" *http://esa.un.org/unpd/ppp/index.htm*)

Therefore, and despite the large size and growth rate projected by the UN, the milestone of 9 billion people by 2050 seems quite attainable and realistic. Indeed we could even argue the high-end scenario could become a more realistic one in the next decades, especially when considering that the world population had already reached the 7 billion milestone in 2010, and that the world economy will continue growing fast during the next decades.

In any case the combination of population and economic projections will have a tremendous impact on our world and all the other megatrends. The massive increase in population and economic growth will create formidable challenges and opportunities for those countries, industries, and companies able to understand and profit from them.

1.1.2
Demographics

1.1.2.1 Area and Age Distribution

An additional key aspect of this population growth is its demographics, understood here as age and geographical distribution. To study this aspect, as previously explained, we divided the world into three major areas: ADV, BRIC, and the REST. Within the REST, sometimes we will refer to the NEXT group, or the NEXT 11, as another sub-group of countries with very high growth potential; countries that according to Goldman Sachs could be among the world's largest economies by 2050. In this group we can find the following countries: Indonesia, South Korea, the Philippines, Vietnam, Bangladesh, Turkey, Egypt, Iran, Pakistan, Mexico, and Nigeria.

According to the UN, World population will reach 9 billion in 2050, however, this growth will not be distributed evenly around the different areas. While in 2007 the population of the ADV economies grew at 0.4% – with some economies like Japan having negative growth – the BRIC grew at 1%, and the REST at 2.6% (see Figure 1.4).

The population in these three groups will continue growing; however, both ADV and BRIC countries will see their participation in the total world population decrease significantly, from 15 to 12%, and from 43 to 35%, respectively (Figure 1.5). On the other hand, the REST will continue growing fearlessly at annual rates well above 2%, increasing their share of the world population from 41 to 53%, while the ADV and BRIC economies will see their population growing more slowly, with a much older population, especially in the ADV economies (Figure 1.5). Indeed, the real impact of this new distribution of the world population cannot be fully understood without a review of the demographic changes of all three groups.

1.1.2.2 Change in Age Distribution

The different changes in demographics that each of the different areas will experience in the upcoming years is a critical aspect of this megatrend. According to the UN, the change in population distribution will be certainly accompanied by a change in life expectancy and the pyramids of population of the different regions (Figure 1.6).

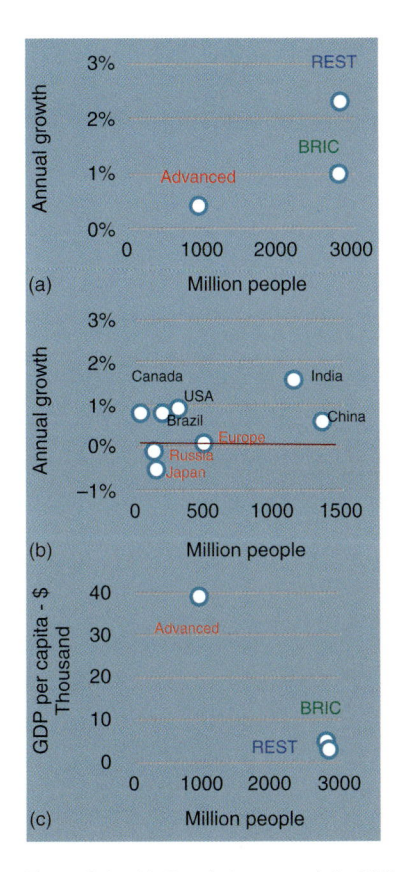

Figure 1.4 (a) Population growth in 2007 for ADV, BRIC and REST economies. (b) Population growth in 2007 for selected countries. (c) GDP per capita in 2007 versus number of people. Source: Population Division of the Department of Economic and Social Affairs of the united Nations Secretariat. "World Population Prospects. The 2007 Revision" http://esa.un.org/unpd/ppp/index.htm

Indeed, even today we can see significant differences in life expectancy, life average, and pyramid composition between the three different groups and also within their members: countries with the highest GDP per capita (ADV group) have higher life expectancy (around 80 years) but also higher average age (45) (Figure 1.7). At the same time their populations tend to be older (17% over 65 in 2007) and grow more slowly than the BRIC and REST economies (0.4% versus 1% and 2.6%, respectively) (Figure 1.4).

Therefore, older populations with slow or negative population growth can be expected in the ADV economies, while younger and rapidly growing populations can be expected in the REST economies.

The BRIC economies will continue to slowly converge to the patterns of the ADV economies, that is, older populations with slower population growth, adding significant pressure to the world population.

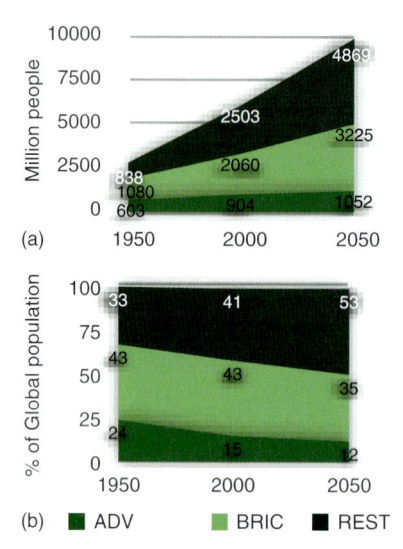

Figure 1.5 World population 1950 to 2050. Dark green relates to ADV economies, mid-green to BRIC and light green to REST. (a) Total population, (b) as a percentage of the total global population. (Source: United Nations (UN), Population Division, 2007. World Population Prospects: The 2007 Revisions. http://esa.un.org/unpp/)

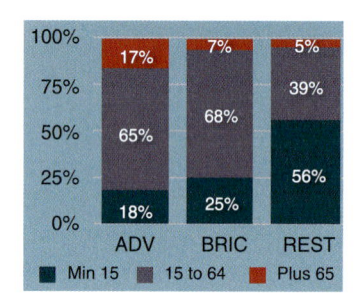

Figure 1.6 Population pyramid in 2007 for the ADV, BRIC and REST economies. Source: Population Division of the Department of Economic and Social Affairs of the united Nations Secretariat. "World Population Prospects. The 2007 Revision" http://esa.un.org/unpd/ppp/index.htm

On a global basis the most important aspect of the change in demographics and age distribution is the fact that, according to the United Nations, the world population is expected to increase its *life expectancy from 67 years in 2010 to 75 years in 2050* (Figure 1.8).

The direct correlation between GDP per capita and life expectancy in combination with the expected increase in wealth and GDP per capita by 2050 and developments in medicine and health will make people live longer (Figure 1.7).

Indeed, a recent article in Time Magazine highlighted the potential for humans to increase their life expectancy to 120 years, or almost to become "immortal", by 2045 thanks to the expected progress in medicine and biology.

1.1 Social Megatrends | 21

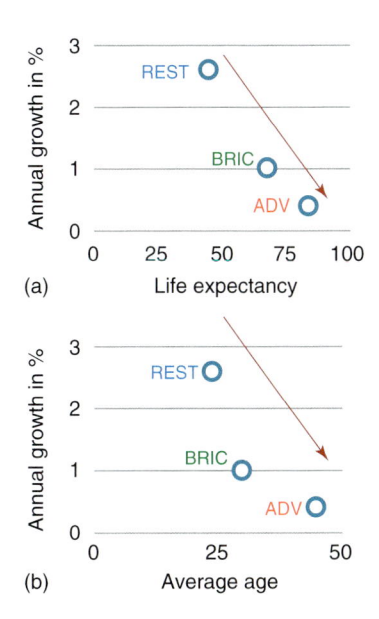

Figure 1.7 (a) Life expectancy versus annual population growth (%). (b) Average age versus annual population growth (%). Source: Population Division of the Department of Economic and Social Affairs of the united Nations Secretariat. "World Population Prospects. The 2007 Revision" *http://esa.un.org/unpd/ppp/index.htm*

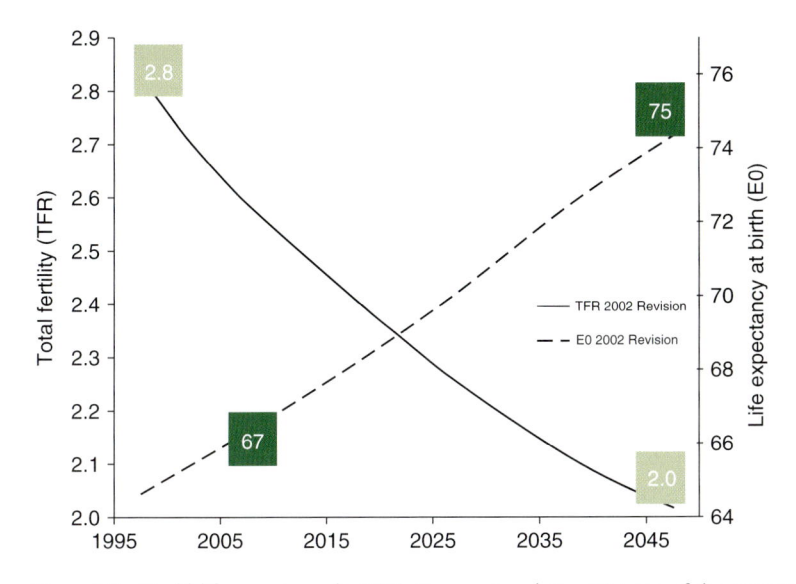

Figure 1.8 World life expectancy by 2050. Source: Population Division of the Department of Economic and Social Affairs of the united Nations Secretariat. "World Population Prospects. The 2007 Revision" *http://esa.un.org/unpd/ppp/index.htm*

For obvious reasons, we will not consider this extreme option for the 2050 scenario, however, we should be aware of these potential trends, especially when looking beyond 2050. On the other hand, and although some of these scenarios might seem extreme, the reality is that world life expectancy almost doubled during the 20th century, from 33 years in 1940 to 65 in 2000 (Figure 1.9 and 1.10). So why could we not expect a similar improvement by 2050, perhaps having an average world life expectancy of 100 years?

Considering the technological advances this generation has witnessed and the speed at which changes have occurred during the last decades, no scenario should be completely underestimated. The next 40 years will be full of advances in human progress, some truly unthinkable, and these, in combination with higher levels of education and wealth, have the potential to extend human life significantly, changing the way we live and work forever.

Indeed on a global basis our population pyramid will become older. According to world bank data, by 2050, 16% of the world's population will be older than 65 years, implying that the number of older people will treble in absolute numbers and double on a world percentile basis (Table 1.1).

In 2010, the World had approximately 500 million people older than 65. By 2050, the World will have more than 1.4 billion people older than 65. Considering the fact that, in 2007, 17% of the ADV population and 7% of the BRIC population were older than 65, compared with just 5% of the REST population, we could expect a much older population in the ADV and BRIC economies.

Perhaps more worrisome seems to be that the number of people below 15 will keep constant on an absolute basis, despite the increase in population. Indeed, in 2010, 1.8 billion people were below 15 years old on a global basis, while by 2050 that figure is expected to remain constant. As fertility rates keep declining in the BRIC and REST economies and life expectancy keeps increasing around the world, our world will become older.

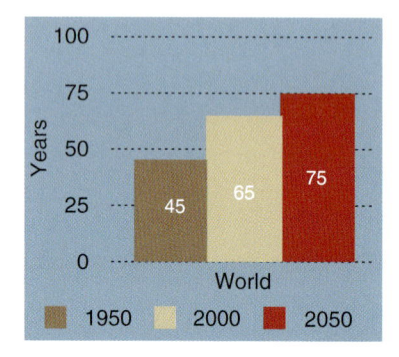

Figure 1.9 World life expectancy. (Source: Population Division of the Department of Economic and Social Affairs of the United Nations Secretariat, 2007. World Population Prospects: The 2006 Revision Dataset on CD-ROM. New York: United Nations. Available on-line at *http://www.un.org/esa/population/ordering.htm*).

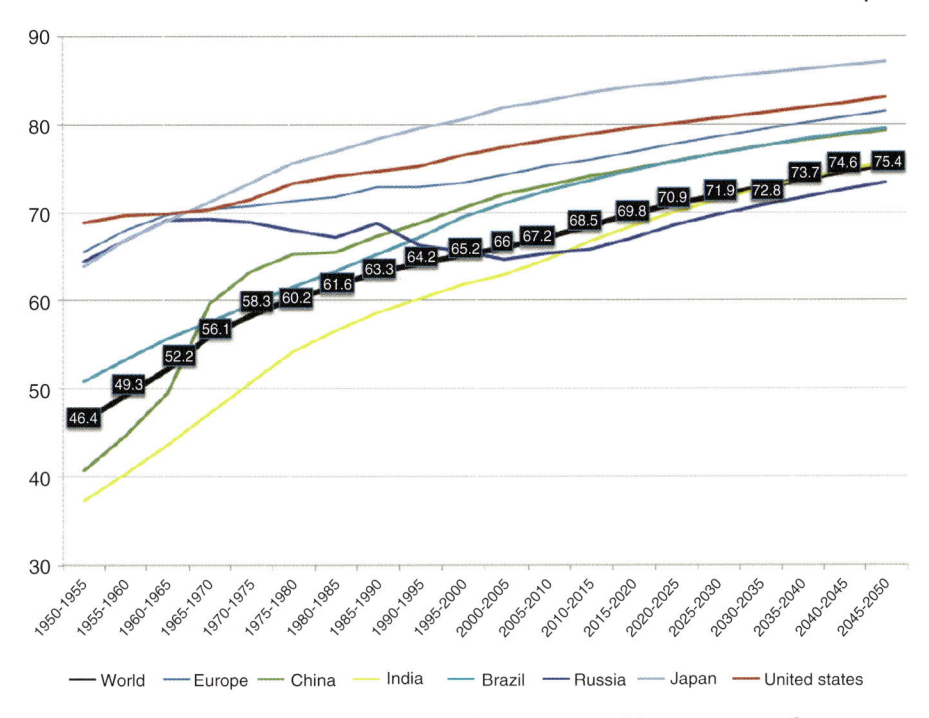

Figure 1.10 World life expectancy. (Source: Population Division of the Department of Economic and Social Affairs of the United Nations Secretariat, 2007. World Population Prospects: The 2006 Revision. Dataset on CD-ROM. New York: United Nations. Available on-line at http://www.un.org/esa/population/ordering.htm.)

Table 1.1 World demographics from 2010 to 2050.

World Demographics (in billion people)	2010 (in billion people)	2010 (in % world population)	2050 (in billion people)	2050 (in % world population)
Min of 15 years	1.830	27	1.887	21
Btw 15 to 65 years	4.487	66	5.837	64
Above 65 years	0.521	8	1.422	16
Total	6.839		9.148	

Source: World Bank Population Projections.
http://data.worldbank.org/data-catalog/population-projection-tables

That fact will be especially clear in the ADV economies where already the ratio of people above 65 is three times larger than in the REST economies.

By 2050 we could envision a world not only with a much larger population, but also a much older, wealthier, and healthier one. As the World population becomes wealthier and life expectancy increases, fertility tends to decrease, as we saw in Figure 1.8.

As the World per capita GDP moves from US$ 9219 in 2010 to US$ 30 675 by 2050, all the areas will experience the positive impact of a wealthier world.

The population in the REST group is expected to grow very fast, amid this fast increase in wealth. Indeed the population in the REST group could grow from 3 billion in 2010 to almost 5 billion in 2050, a staggering 61% increase (Table 1.2).

Among the BRIC group, different countries will face different realities. China and Russia will experience declining populations, especially in the case of Russia, while India and Brazil will be confronted with larger populations. In the case of India the increase will be massive, almost 40%, giving an additional 440 million people (Table 1.2).

In the ADV economies, tremendously high levels of GDP per capita, and decreasing fertility will affect their population size and demographics. Shrinking populations in Japan and Europe will not only create significant resources and labor issues but will certainly contrast with the REST groups where populations will keep growing. Japan and Europe will face two major issues, on the one hand shrinking labor forces and on the other hand a higher number of older people and pensioners.

Table 1.2 Population in millions.

Population	2010	2050	Delta	%
USA	309	397	89	29
Europe 27	502	506	4	0
Japan	127	106	22	−17
Canada	34	43	9	27
ADV Group	973	1052	79	8
China	1339	1273	−65	−5
India	1170	1610	440	38
Russia	142	124	−18	−12
Brazil	195	219	24	12
BRIC Group	2846	3226	380	13
REST Group	3020	4869	1849	61
WORLD	6839	9148	2308	34

Source: World Bank Population Projections.
http://data.worldbank.org/data-catalog/population-projection-tables

Europe and Japan will face large pressure to keep up with their pension systems and living standards. Under these scenarios we could foresee incentives to increase fertility and their respective labor forces. Policies to increase the number of women in the labor market, delay the retirement age, and more flexible immigration policies will be required. After all, these countries will be among the richest in the world, so their capacity to attract labor and talent should be large. Alternatively, large increases in productivity and specialization will be needed to sustain the expected growth with shrinking populations. All this remains to be seen, however, under either of the two scenarios the chemical industry will face significant technological challenge and large business opportunities.

In contrast, the United States and Canada, will continue enjoying population growth. Especially remarkable is the case of the USA that, thanks to its large "Latino" and immigrant populations, will be able to counterbalance the typical correlation between higher wealth and lower fertility rates, and therefore populations (Table 1.2).

Under this scenario we expect a world with much larger population, moving from 6.8 billion in 2010 to the 9.1 billion in 2050, a 34% increase. However, that population growth will not be equally distributed, the REST group will grow extremely fast with a 61% increase during the next 40 years, while the ADV and BRIC economies will have a much more moderate increase with a 13% rise during the same period. In the ADV group, Japan and Europe will face slow or negative population growth at the same time as older populations, while the USA and Canada still enjoy fast growing populations. Among the BRIC group, India will keep growing extremely fast, expecting to add more than 400 million people during the next 40 years. China and Russia will have declining populations which will bring some additional economic complexities, especially to Russia with a much older population than China. In any case, and under this scenario, we expect a much larger World population, growing in most of the countries and especially in India and the REST economies (Table 1.2).

1.1.3
Urbanization

The last of the three aspects analyzed for the social megatrend is the distribution of the population between rural and urban areas. As stated before, population will keep growing globally in the coming decades, however, that growth will not be homogeneous, with higher ratios in the BRIC and REST economies, and declining populations in Japan and Europe. In that context, according to the UN, the world urbanization ratio will increase from 47% in 2000 to 69% in 2050 (Figure 1.11).

This change will be extremely important on a global basis, but more specifically for the BRICs, where the urbanization ratio will be almost doubled in the next 50 years, from 37 to 67% of their total population (Figure 1.12). This significant increase will be particularly true in China and India, where both countries will indeed double their urbanization ratios in this period. India has the lowest urbanization ratio of the BRICs. The REST group will also experience a significant increase in urbanization ratios, from 45 to 68% by 2050 (Figures 1.12 and 1.13).

26 | *1 Global Megatrends by 2050*

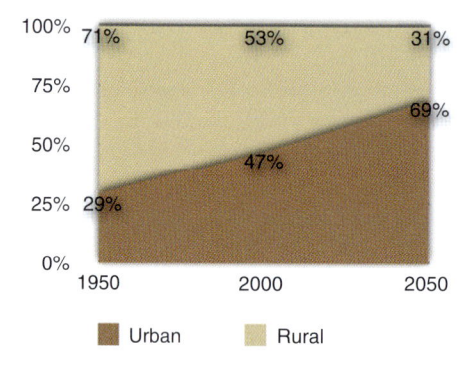

Figure 1.11 World urban population (%). Source: Population Division of the Department of Economic and Social Affairs of the united Nations Secretariat. "World Population Prospects. The 2007 Revision" *http://esa.un.org/unpd/ppp/index.htm*

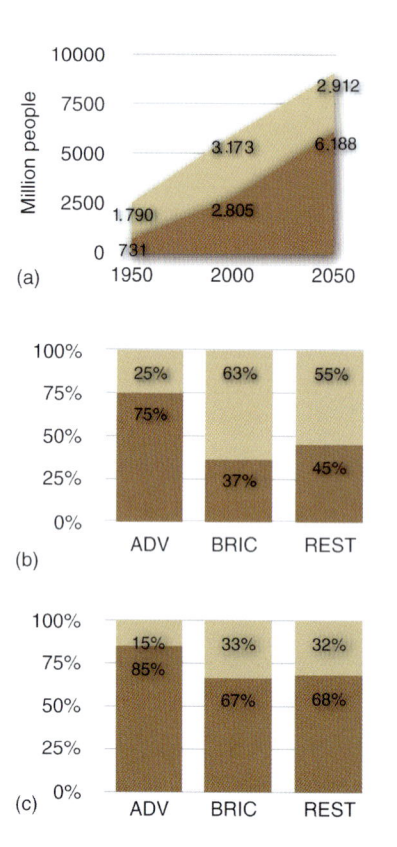

Figure 1.12 (a) World urban population (%). (b) Urban population (%) in 2000. (c) Urban population (%) in 2050. Source: Population Division of the Department of Economic and Social Affairs of the united Nations Secretariat. "World Population Prospects. The 2007 Revision" *http://esa.un.org/unpd/ppp/index.htm*

1.1 Social Megatrends | 27

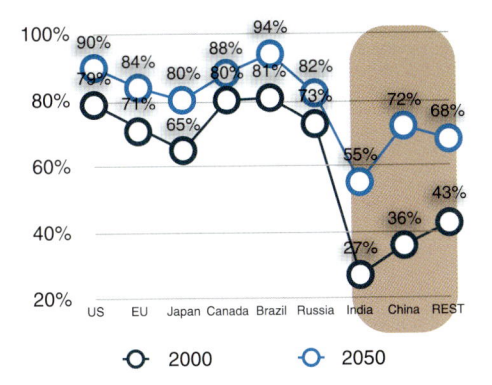

Figure 1.13 Urban population (%) 2000–2050. Source: Population Division of the Department of Economic and Social Affairs of the united Nations Secretariat. "World Population Prospects. The 2007 Revision" *http://esa.un.org/unpd/ppp/index.htm*

These large changes in urbanization ratios will generate very interesting challenges and opportunities globally. A much more urbanized world will create great opportunities and demand for real estate and materials, however, it will pose significant challenges in logistics, climate change, and human health, among many others. It remains to be seen if the benefits derived from the economies of scale will offset the potential negative aspect of much larger cities.

1.1.3.1 Megacities

The increase in the urbanization ratios globally, but especially in China, India and the REST economies will accelerate the creation of so-called "megacities," those with more than 10 million people. Traditionally, megacities have been associated with the ADV economies, however, that has changed dramatically during the last decades and will increase further in future.

In 1950 there were two megacities in the world, both in the ADV economies, Tokyo and New York. In 2010, 22 megacities were identified globally – five in the ADV economies, Tokyo, New York, Los Angeles, Osaka, and Paris; eight in the BRIC economies, with Delhi, Sao Paulo, Mumbai, Shanghai, and Calcutta at the top of the list and nine in the REST economies, with Ciudad de Mexico, Dhaka, Karachi, and Buenos Aires at the top of this ranking (Figure 1.14).

As the world population and economy keep growing, especially in the REST and BRIC, the number of megacities in these areas will continue to rise rapidly, much faster than in the ADV economies. If already in 2010 we can observe that the numbers of megacities in the BRIC and REST have already overtaken those in the ADV economies, that trend will continue in the years until 2050.

According to the latest projection from the UN Department of Economic and Social Affairs, Population Division on World Urbanization Prospects, there will be up to 30 megacities by 2025. These will account for almost 500 million people, 5% of the world population, with Tokyo being the largest with almost 40 million. By 2025, The UN expect a continuation of the observed trend, and by 2025 there will

28 | 1 Global Megatrends by 2050

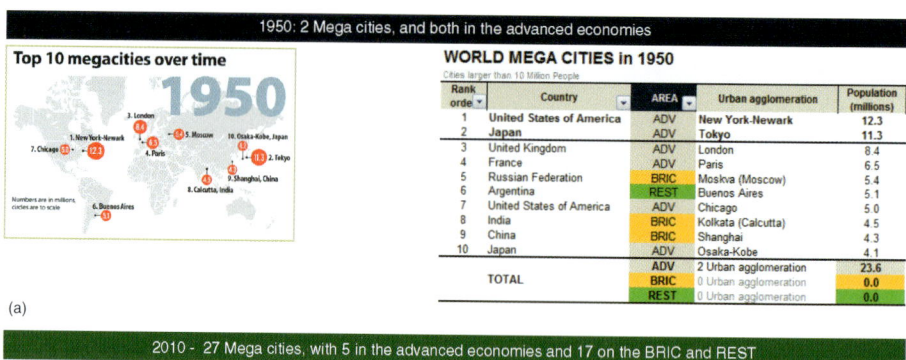

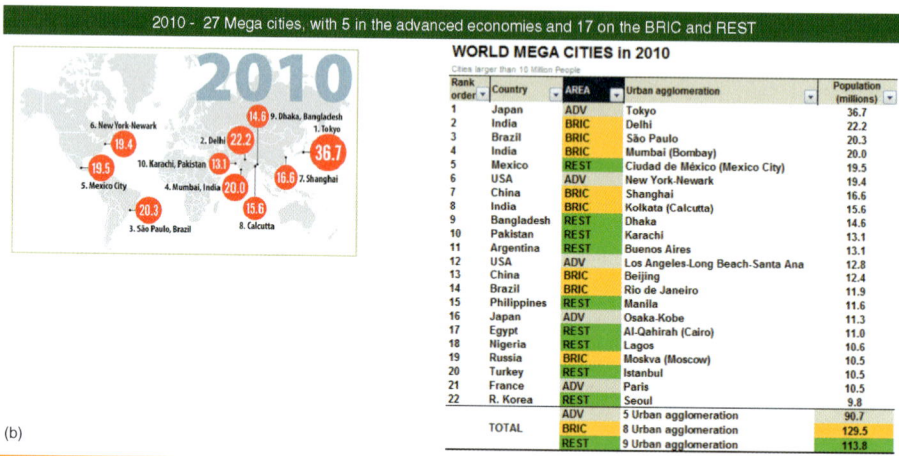

Figure 1.14 World megacities (a) in 1950, (b) in 2010, (c) in 2025. (Source: United Nations, Department of Economic and Social Affairs, Population Division (2010). World Urbanization Prospects: The 2009 Revision. CD-ROM Edition – Data in digital form (POP/DB/WUP/Rev.2009).).

be ever larger megacities in the REST and BRIC economies with 13 in the REST and 11 in the BRIC, double the number of megacities in the ADV economies.

If we consider that the urbanization ratio will rise to 69% globally and the world population will continue growing very fast during the next decades, many more megacities, especially in the REST economies, are expected.

This growth in the number of megacities will provide a significant amount of business opportunities as well as challenges never seen before. Issues such as how to organize and run the daily supply and demand to these huge megacities, and how to manage potential environmental, ecological, and health concerns, especially those related to pollution and waste, and many other size-related issues will require significant attention. The world certainly has experience of running megacities since they have been around for 50 years, however, some of the new cities will be double or even triple the current size, with more than 20–30 million people. By 2025 the top 10 largest cities in the world will have more than 20 million people, with 2 of them in the 30 million league (Tokyo and Delhi). Of these 10 megacities, only 2 will be located in the ADV economies, Tokyo and New York, while the rest will be located in the BRIC and REST.

Some people will argue that these cities will be needed to accommodate the projected increases in population, especially in the REST and BRIC economies, more specifically in India. India is expected to have more than 440 million people in the next 40 years. However, perhaps it is worth remembering that despite the upcoming projected growth from 7 to 9 billion people, the world population density is expected to be around 64 inhabitants per km^2 in 2050, still half the levels observed in Europe today or one quarter those observed in the most populated countries, so there is still plenty of room to grow. Megacities are the simple consequence of millions of people trying to improve their quality of life.

The world population and urbanization ratio is expected to keep growing very fast during the next decades, and with it the number of megacities around the world. As economies become more modern and populations keep growing, individuals will continue moving into the cities to improve their living standards. The transition into cities will be extremely obvious in the REST economies and some of the BRIC economies like India. Large megacities will present both interesting challenges and opportunities for the next generations as well as for the chemical industry.

1.2
Economic Megatrends

"A world with the largest economy and wealthiest society in Human History"

Nobody likes the man who brings bad news.

Sophocles, Antigone

Disclaimer: all economic projections are based on 2009 dollars.

Despite the World economic crisis in 2008 and 2009, the recent sovereign debt issues in Europe and the subsequent slowdown of the European and the World economy, the world is still poised for further decades of strong economic growth. That growth will be similar to the one experienced by our world during the first and second halves of the last century, although its major actors and benefactors will be different this time.

During the next decades, the world economy will continue growing very fast and, according to a 2006 study from PWC, it is expected to reach a staggering GDP of US$ 280 trillion (2009 dollars) by 2050, an impressive 345% increase from 2010. A growth rate that, despite its size, is nothing our world has not seen or experienced before. During the first half of the last century indeed our world grew by a huge 271%, and during the second half by a tremendous and unprecedentedly high 687%. Therefore a 345% growth in the next 40 years might not be so unreasonable after all. Please see Figure 1.15 for additional details and information.

In 2001, Jim O'Neil, Chief Economist of Goldman Sachs, forecast a spectacular growth for the World economy, and most specifically for the BRIC economies, coining this now famous acronym. In his two famous papers,[1] and the "Dreaming BRIC – The path to 2050"[2] he revealed his theory on the massive economic growth for the World and the BRIC. He argued that the strong economic growth observed in these highly populated economies during the last decades is here to stay for the coming decades, creating some of the largest economies in the world.

At a similar time PWC started to look into the World economy by 2050, confirming not only the belief it had a potential for massive growth during the next decades, but estimating the World GDP by 2050 at US$ 280,000 trillion (2009 dollars). PWC estimated and projected forward for each country its potential GDP in PPP terms and 2009 dollars, based on a Cobb–Douglas production function augmented to include human capital. This is a standard type of economic model that is widely used in long-term growth studies of this kind. For the population projections, PWC used World Bank projections, as will we. For the rest of the book, we will use these economic PWC projections, remembering that these are based on 2009 dollars and on a PPP basis. This may lead on some occasions to two different values for GDP, for example, in 2010, where there is some difference between the actual figure and that estimated by PWC in 2006. In any case all the upcoming projections should be considered as trends and best available estimations rather than as exact values. Due to the time length of the study, arguing whether the world GDP by 2050 will be US$ 280 or 320 trillion, or the world population will be 9.1 or 9.7 billion is very difficult, so that is why we recommend to take all projections as directional and for illustration.

According to PWC projections for 2050, China is expected to be the largest world economy, followed by India, USA, and the European Union (EU 27) (Figure 1.16). If Europe is not considered, Brazil will be the fourth largest economy after the USA. The tremendous growth of these BRIC economies will reshuffle the World GDP and the international balance of power (Figure 1.15). In 2010 the ADV

1) Goldman Sachs – Global Economics Paper No: 66 – "Building Better Global Economic BRICs" – Mr. Jim O'Neill – Chief Economist Goldman Sachs – 30th November 2001.
2) Goldman Sachs – Global Economics Paper No: 95 – "Dreaming BRIC the path to 2050" – Mr. Dominic Wilson Roopa Purushothaman – 1st October 2003.

1.2 Economic Megatrends | 31

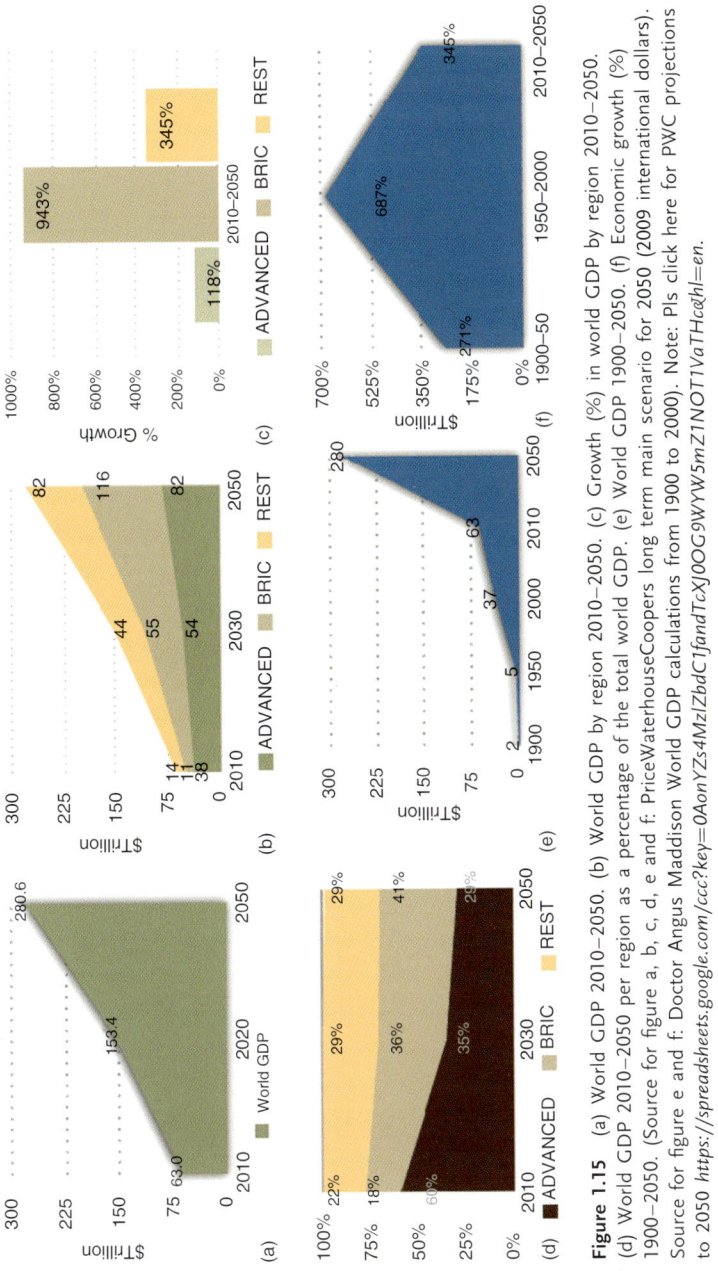

Figure 1.15 (a) World GDP 2010–2050. (b) World GDP by region 2010–2050. (c) Growth (%) in world GDP by region 2010–2050. (d) World GDP 2010–2050 per region as a percentage of the total world GDP. (e) World GDP 1900–2050. (f) Economic growth (%) 1900–2050. (Source for figure a, b, c, d, e and f: PriceWaterhouseCoopers long term main scenario for 2050 (2009 international dollars). Source for figure e and f: Doctor Angus Maddison World GDP calculations from 1900 to 2000). Note: Pls click here for PWC projections to 2050 https://spreadsheets.google.com/ccc?key=0AonYZs4MzIZbdC1fandTcXjOOG9WYW5mZ1NOT1VaTHc&hl=en.

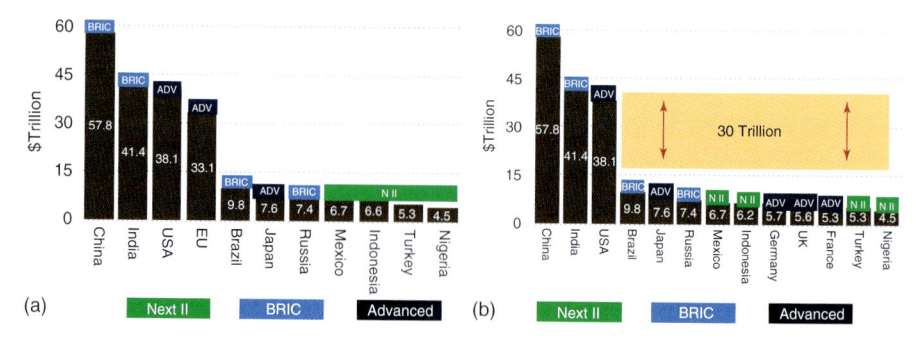

Figure 1.16 (a) Top economies (including EU) by GDP in 2050. (b) Top 10 economies by GDP in 2050. Source for (a) and (b): PWC Long term main scenario for 2050 (2009 dollars).

economies represented 47% of the world economy. However, by 2050 they will represent only 29% of the world economy as they will lose weight against the BRIC. Indeed, the BRIC economies will move from 25% in 2010 to almost 41% by 2050. The REST economies, despite their tremendous increase in population, will not increase their share of the world economy. Within the REST economies there are many different types of economies, however, as a group the increase in GDP in some of the fastest economies will be compensated by the tremendous increase in population this group will have. However, we should be aware that some of the fastest growing economies in the world during the next decades will be within this group. Countries like Indonesia, Mexico, Turkey, or Nigeria will be among the largest economies in the world.

In 2010, the two largest economies in the world were the EU 27 and USA, with US$ 16 and 14.5 trillion, respectively. China was the third largest with US$ 5.8 trillion. Japan with US$ 5.5 trillion, Brazil with US$ 2.1 trillion, India with US$ 1.7 trillion, Canada with US$ 1.6 trillion, and Russia with US$ 1.5 trillion complete this select group. Noticeable is the large difference between the two largest economies and the others in this list (Figure 1.17).

Among the ADV economies and globally, the EU 27 was the largest economy in 2010 with US$ 16.3 trillion; Germany with US$ 3.3 trillion was the largest

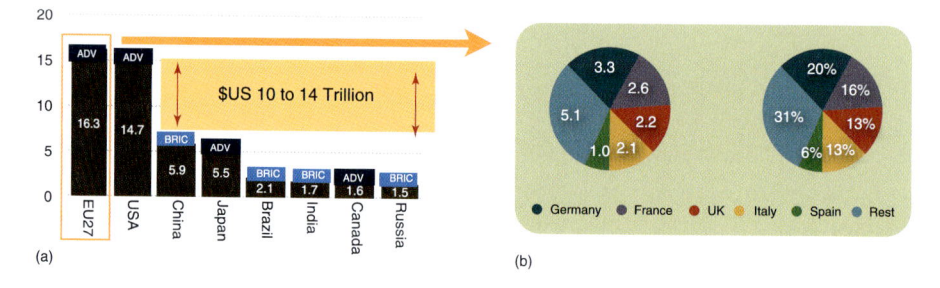

Figure 1.17 (a) GDP in 2010 for the largest economies. (b) Europe 27 largest economies in 2010. Source for (a) and (b): PWC Long term main scenario for 2050 (2009 dollars).

economy within this group followed by France with US$ 2.6 trillion and the UK with US$ 2.2 trillion. In other words, the largest economy in the EU, Germany, was much smaller than the USA, China, and Japan. The UK, France, and Italy with approximately US$ 2 trillion economies were similar to Brazil and India. Spain, the fifth largest European economy with US$ 1 trillion was smaller than Canada or any of the BRIC (Figure 1.18).

Considering the BRIC economies it can be seen that China's GDP is larger than the combined GDP of the other three. Comparing the BRIC with the ADV, China and India today are still far away from the USA and EU 27 in terms of absolute GDP, but that difference is much less when comparing Japan or Canada with the BRIC.

However, as China and India start growing during the next decades the gap will start reducing, in some cases even reversing. At the same time, if the EU remains together as a group, rather than as individual countries, this gives significant influence and power to the union in global terms. However, individually the members could risk becoming very small in comparison to the BRIC if the union breaks down.

Now, looking into the future and according to PWC projections for the World economy by 2050, see Figures 1.16, 1.19–1.21, the BRIC are expected to become among the largest economies in the world, surpassing the US and the EU 27. By 2050 China is expected to be the largest economy in the world with $58 trillion, 20% of the world economy. India will be second with $41 trillion, 14%, while the USA and EU 27 will follow with US$ 38 and 33 trillion, respectively. China, India, and the USA will count for 50% of the world economy, and up to 60% if we include the EU. If we consider the European countries individually rather than as a group, Brazil will become the 4th largest economy with US$ 9.7 trillion, 4%. Japan will become the fifth largest economy with US$ 7.6 trillion, followed closely by Russia, Mexico, and Indonesia all with GDP figures close to $7 billion. Germany will become the ninth largest economy, while remaining as the largest European economy, followed very closely by the UK (10th) and France (11th), all three with very similar GDPs, around US$ 5 trillion.

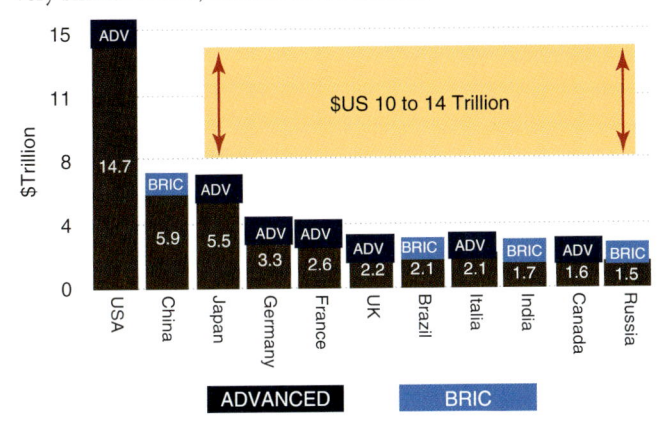

Figure 1.18 GDP in 2010 by top countries. Source: PWC Long term main scenario for 2050 (2009 dollars).

34 | *1 Global Megatrends by 2050*

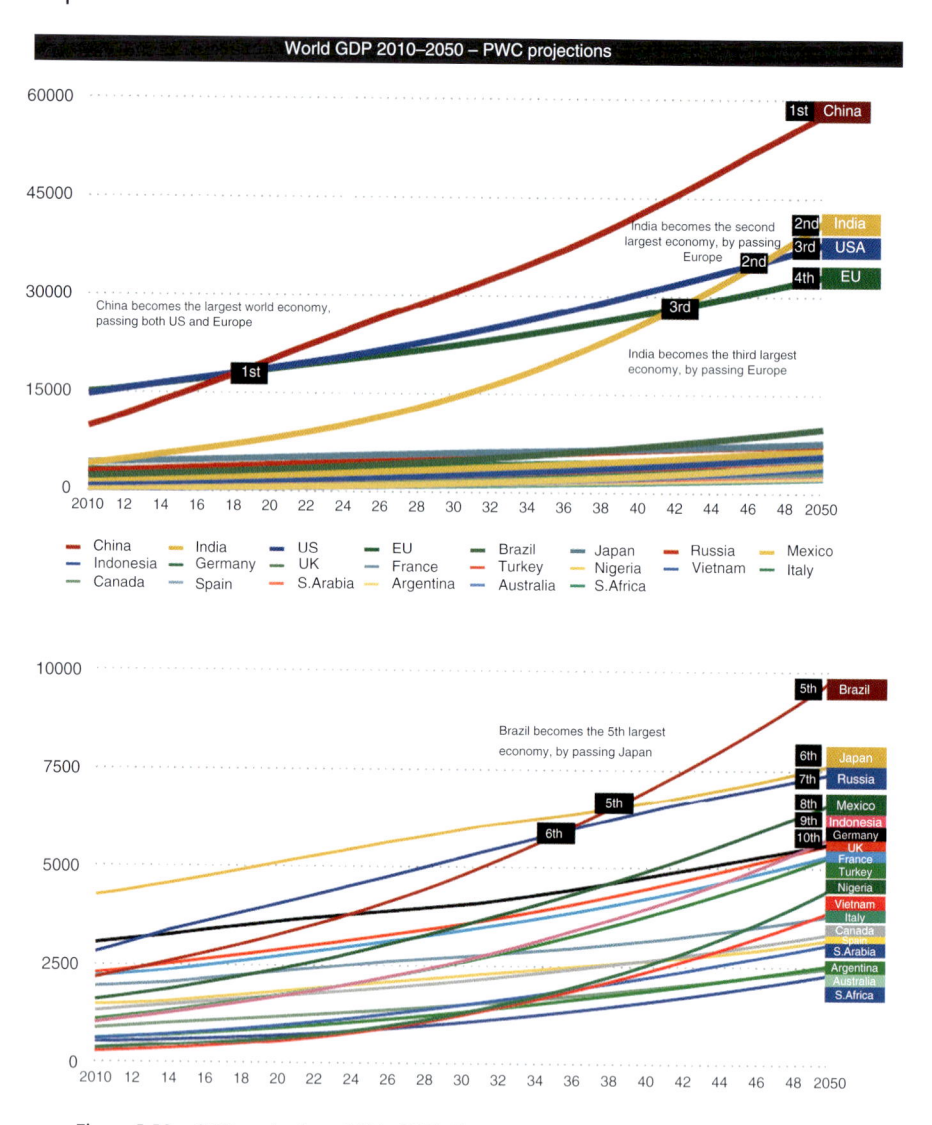

Figure 1.19 GDP projections 2010–2050 (ADV vs BRIC economies). Source: PWC Long term main scenario for 2050 (2009 dollars https://spreadsheets.google.com/ccc?key=0AonYZs4MzlZbdC1fandTcXJ0OG9WYW5mZ1NOT1VaTHcqhl=en).

Beyond the fact that the BRIC economies will become the largest economies in the world, perhaps you might be surprised by some of the other countries that may be among the top 20 world economies. Economies like Mexico (7th), Indonesia (8th), Turkey (12th), Nigeria (14th), Vietnam (15th), South Arabia (18th), Argentina (19th), and Australia (20th) will become powerful economies globally, and clear examples of the growth potential beyond the BRIC and within the REST economies (Figure 1.21).

1.2 Economic Megatrends | 35

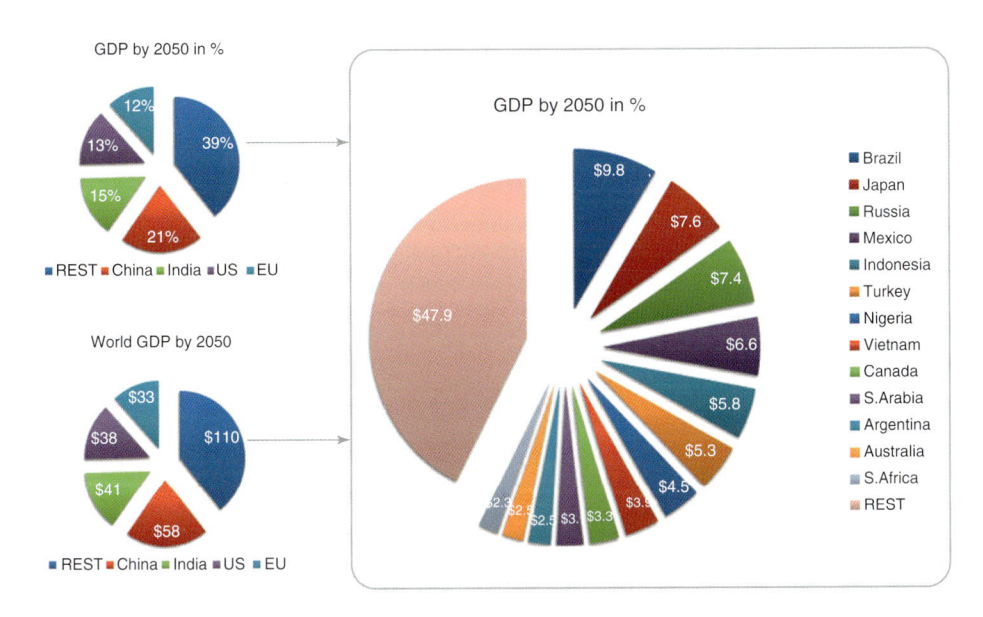

Figure 1.20 World GDP by 2050: top economies including European Union 27 as a group. PWC projections. Source: PWC Long term main scenario for 2050 (2009 dollars).

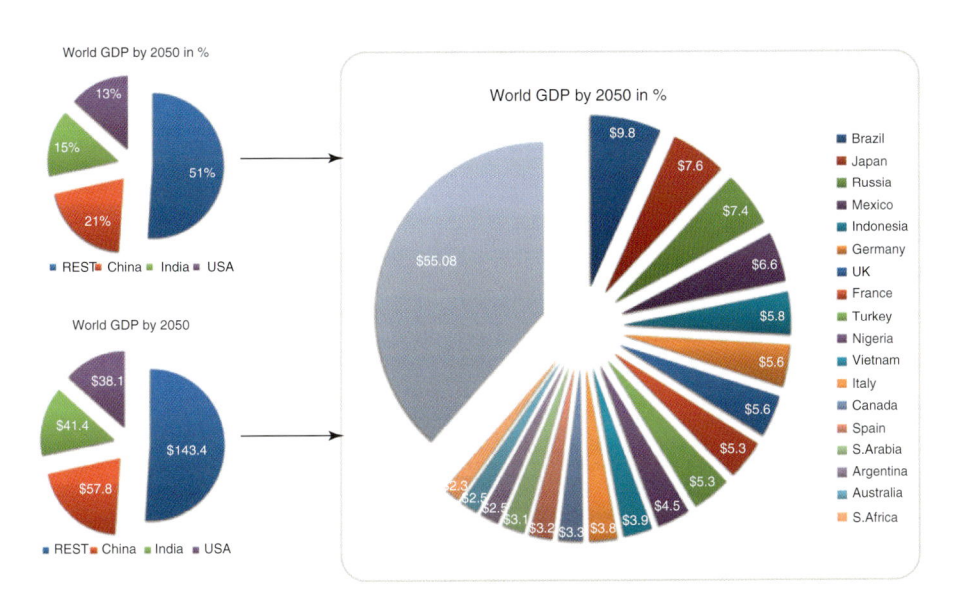

Figure 1.21 World GDP by 2050: top economies excluding European Union 27 as a group. PWC projections. Source: PWC Long term main scenario for 2050 (2009 dollars).

Therefore, China, India, and the USA will become by far the largest economies in the world by 2050 with almost 50% of the world economy. If we include the EU, these four economies will account for almost 60%. In that sense the EU will have a clear role in balancing the incredible power that just these three economies will have, otherwise the world will be managed by a powerful G3 or eventually G4, if we include Europe.

All these projections not only confirm the tremendous growth expected in the BRIC and the REST economies but also the realization of gradual change in the international balance of power and economic order.

If we consider the EU as a country, it will be among the top 10 economies, with the four members of the BRIC, three members of the ADV and three members of the REST. Now, if we consider the countries of the EU individually, the BRIC economies will be among the six largest economies in the world, only accompanied by the USA and Japan, while Germany the largest economy in Europe, might be smaller than any of the BRIC, USA, Japan, Mexico, and Indonesia.

In absolute terms the formidable growth our world economy will experience will influence drastically the international order and, beyond the potential ranking of the different economies in the world economy, the reality is that all economies will experience a tremendous growth. The ADV economies will double their absolute GDP, both as a group and individually. The USA will increase from US$ 15 to 38 trillion, while Europe will also double its GDP from US$ 16 to 33 trillion (Figure 1.22).

Within the BRIC, India is expected to increase its economy by a spectacular 2000% during the next four decades, from US$ 1.6 to 41 trillion in 2050, while China and Brazil are expected to increase by a staggering 1000 and 500%, respectively. China will increase from US$ 5.8 to 58 trillion, and Brazil from US$ 2 to 9.7 trillion.

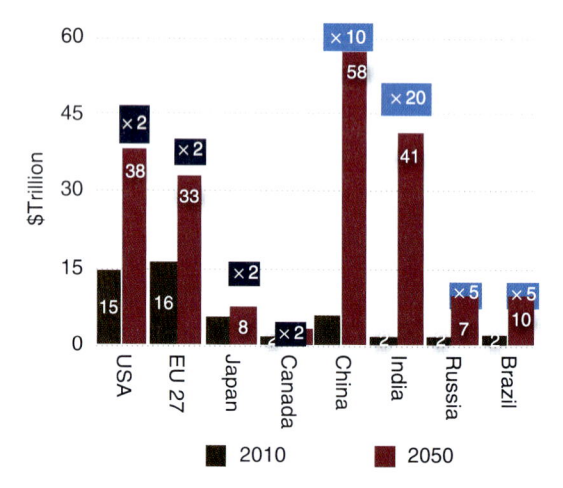

Figure 1.22 GDP evolution 2010–2050. Source: PWC Long term main scenario for 2050 (2009 dollars).

Russia, despite being the BRIC with the slowest growth within the group, will still more than quintuple its economy, from US$ 1.4 to 7.4 trillion (Figure 1.22).

Thus, absolute growth will not be homogeneous across the ADV and BRIC countries, and that growth will place the BRIC economies amid some of the most powerful economies in the world. However, the expected growth will be solid and strong across all countries and areas, with some economies growing at formidable rates.

However when we look at the GDP per capita, the ADV economies will remain as the wealthiest economies, enjoying levels of wealth never seen in human history, and still well above those experienced by the BRIC and most of the REST economies.

The average GDP per capita in the ADV economies is expected to double from $39 028 to 78 673. The BRIC GDP per capita will increase ninefold from an average $3 926 to 36 138; with that of India increasing by a staggering 1700% and China by 1000%. Numbers that highlight and confirm the tremendous growth the BRIC economies are expected to have. The REST will almost quadruple its GDP per capita from $4614 to 16 687. Expectations that, if they are realized, will serve to pull millions of people from poverty, improving living standards around the world to levels never seen before (Figures 1.23 and 1.24).

Globally the World GDP per capita is expected to triple from $9219 to an unprecedented high of $30 675. This average GDP per capita will create the basis for a super wealthy world, a world that will benefit from the highest GDP per capita in human history,

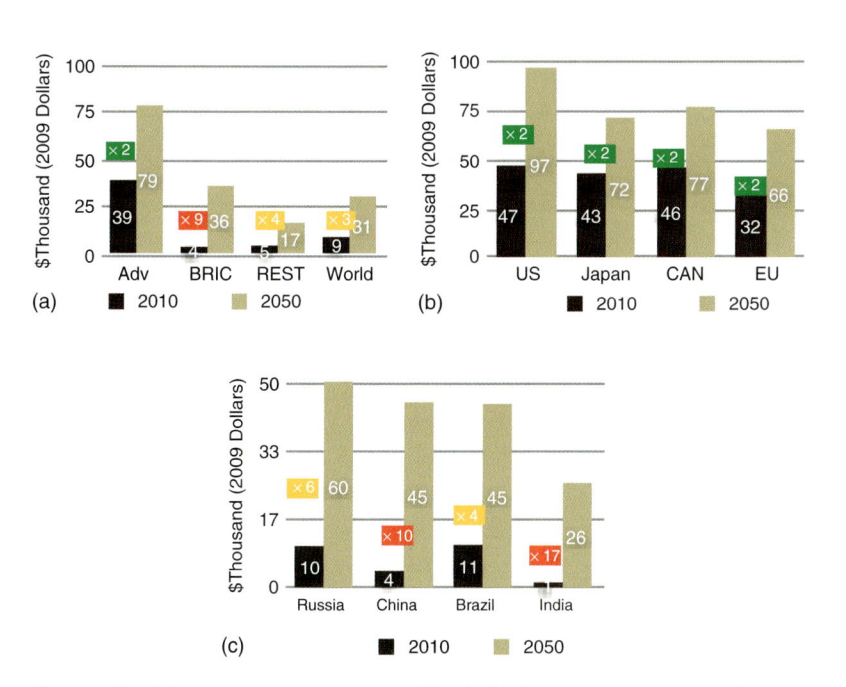

Figure 1.23 GDP per capita evolution: (a) World, (b) ADV economies, (c) BRIC economies. (Source: Calculated by Author based on the 2010 GDP projection for 2050 on GDP from PWC (2009 dollar) and Population from United Nations and Population from the World Bank).

1 Global Megatrends by 2050

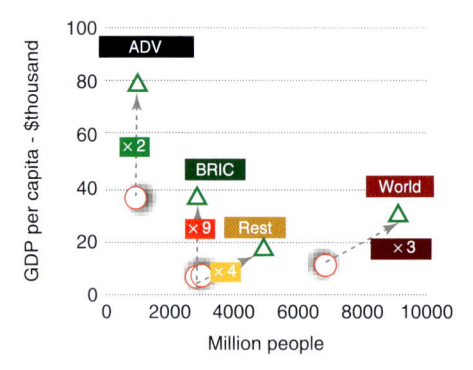

Figure 1.24 GDP per capita versus population. Source: Own elaboration based on the projections for 2050 on GDP from PWC (2009 dollar) and Population from United Nations.

a large middle class society and a generation of super wealthy nations. Despite the fact that disparities among countries will remain high, and in many cases will still increase, the world is expected to become much wealthier than today. Indeed, these large increases in GDP per capita, will create an unprecedented level of wealth; hosting what perhaps will be the wealthiest society in human history.

The ADV economies will have unprecedentedly high levels of GDP per capita, with USA at the front with an expected value of US$ 97 088, followed by Canada with US$ 77 209, Japan with US$ 71 698 and Europe with US$ 65 810. On the other hand the BRIC economies will not only enjoy levels of GDP per capita never seen before but also huge increases during the next decades. India with a forecast increase of 1700%, is expected to have a GDP per capita of US$ 25 761. China will also see a spectacular increase, with 1000% growth, expected to reach US$ 45 428. Among the BRIC, Brazil is expected to have the lowest increase, while enjoying the highest value of the group, US$ 45 000. In contrast Russia is expected to have the largest value of this group, almost US$ 60 000 (Figure 1.23).

To illustrate the magnitude of these increases let us try to imagine that if you live in one of the countries of the ADV economies, say Europe, your GDP per capita will double from today's level. Imagine indeed all the things you could do and buy if you were to double your income from today's levels. If you live in one of the BRICs, your GDP per capita and quality of life would be as good as you can see today in Europe or the USA, while your GDP per capita will be multiplied by nine times, and if you live in the REST group, your GDP per capita is expected to triple during the next decades (Figure 1.24).

If these projections become reality, not only will they serve to pull millions of people around the world out of poverty, but they will also create huge business opportunities globally, but especially in the BRICs and the REST economies. Indeed 50% of the world population will enjoy levels of wealth similar to those of the ADV economies today, creating a huge world middle class. Additionally 10% of the population, mainly in the ADV economies, will become super wealthy, wealth that will benefit our different societies and citizens, assuming a proper distribution of wealth could be ensured.

Some additional results can be extracted from these extraordinary growth projections. We can anticipate that some of the BRIC and REST economies will still have room to grow after 2050 as, after the expected rapid growth in the next decades, they will still be below the levels of the ADV economies. The BRIC economies will enjoy the same levels as does Europe in 2010. India is projected to become the second largest world economy by 2050, growing at 8% annually during the next 40 years, increasing its GDP per capita eightfold to reach $25 761. Albeit with all these impressive projections, India's GDP per capita is still expected to be just one quarter of that in the USA by 2050. *This situation makes us believe that despite the tremendous growth some of the BRIC and REST countries are expected to have during the first half of this century, the second part of the century will be as promising or more so than the first.*

Another interesting fact is the rapid annual growth projected for the next 40 years in many BRIC and some of the fast growing economies within the REST group. Countries like Vietnam, with a projected annual growth of almost 9%, India and Nigeria with 8%, China with 6%, Indonesia, Turkey, South Africa Saudi Arabia, Argentina, and Mexico with almost 5%, Brazil and Russia with 4%, will be some of the fastest growing economies in the world. This rapid growth will make these countries an attractive destination for investment and business opportunities (Figure 1.25).

In contrast, the ADV economies, partly because they started from much higher levels, partly because most of them will experience declining and elderly populations with reductions in their labor force, will have much more modest growth than the above countries. A notable exception will be the USA that, despite being the largest economy and one of the wealthiest economies on a per capita basis in 2010, will have average annual growth of 2.4% during this period, much faster than most of the ADV economies, and almost double the projections for Japan (1%), Germany (1.3%), or Italy (1.4%). The US will remain a very attractive place for investment for the next decades, with a wealthy and growing economy and population (Figure 1.25).

The expected massive economic growth will serve to pull millions of people out of poverty all around the world, creating significant challenges but also huge business opportunities. However, this growth will not be enough to close the wealth gap among the different groups. The ADV and REST economies will double and quadruple their GDP per capita during the next four decades, while the BRIC will increase theirs fivefold, however the ADV economies will still have twice the GDP per capita of the BRICs and three times that of the REST economies.

In any case the World economy is expected to have massive economic growth globally, and specifically in the BRIC economies, growth that will generate massive opportunities and growth for the chemical industry too.

In that sense, we believe population and economic megatrends will have a tremendous impact across all other megatrends, increasing demand globally and stretching all resources and supplies across multiple industries.

40 | *1 Global Megatrends by 2050*

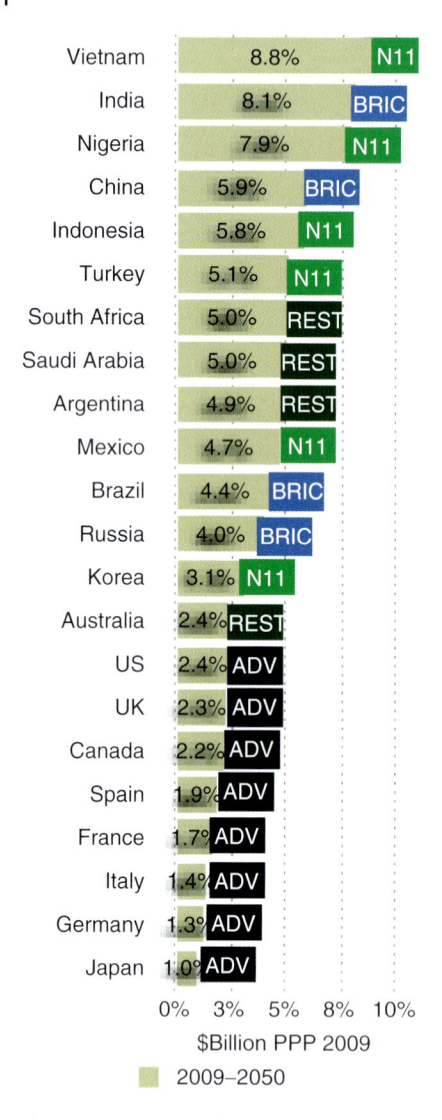

Country	Growth	Group
Vietnam	8.8%	N11
India	8.1%	BRIC
Nigeria	7.9%	N11
China	5.9%	BRIC
Indonesia	5.8%	N11
Turkey	5.1%	N11
South Africa	5.0%	REST
Saudi Arabia	5.0%	REST
Argentina	4.9%	REST
Mexico	4.7%	N11
Brazil	4.4%	BRIC
Russia	4.0%	BRIC
Korea	3.1%	N11
Australia	2.4%	REST
US	2.4%	ADV
UK	2.3%	ADV
Canada	2.2%	ADV
Spain	1.9%	ADV
France	1.7%	ADV
Italy	1.4%	ADV
Germany	1.3%	ADV
Japan	1.0%	ADV

$Billion PPP 2009
2009–2050

Figure 1.25 Projected average annual growth in GDP. (Source: PWC main GDP PPP main scenario; model projections for 2010–2050, based on World; Bank for 2009.)

1.2.1
Foreign Direct Investment (FDI)

Finally, the enormous growth potential of the BRIC economies has generated an unprecedented flow of FDI to the BRICs. According to the World Investment Report (WIR) from 2009, China alone attracted during the last seven years around 10% of the world's FDI, being the second largest receiver of FDI, only preceded by the United States (14%) (Table 1.3).

Table 1.3 World FDI evolution by major countries and areas.

AREAS	1970–79		1980–89		1990–99		2000–07		Annual
Countries	Avg. $	%	Avg. $	%	Avg. $	%	Avg. $	%	Growth 70–08
Europe	9917	41	28719	31	233805	405	374732	39	37
Canada	3135	13	3782	4	20700	4	31140	3	9
USA	3219	13	33681	36	126535	22	133185	14	40
Japan	124	1	181	0	4008	1	6722	1	53
ADV	16395	67	66362	71	385048	66	545779	57	32
Brazil	1270	5	1721	2	13820	2	18579	2	14
China	268	1	3752	4	46004	8	99190	10	370
India	37	0	105	0	1691	0	5004	1	133
Russia		0	0	0	1871	0	12469	1	6
BRIC	1575	6	5578	6	63386	11	135242	14	85
REST	6396	26	21938	23	135867	23	268434	28	41
World	24365		93878		584301		949455		38

Source: United Nation Conference for Trade and Development-UNVCTAD-WR 2007   Unit: US Dolars at current prices in millions

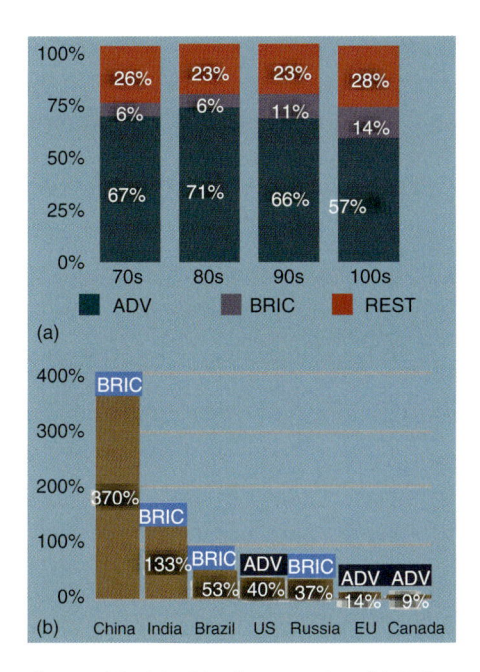

Figure 1.26 (a) FDI inflow overview. (b) FDI annual inflow 1970–2007. (Source: United Nations Conference for Trade and Development.)

These impressive numbers show the strong investor confidence in China's growth potential. Still more impressive is the annual growth rate of FDI experienced by the BRIC economies compared to the other two groups. In the period between 1970 and 2007 the world FDI grew by 38% annually – highlighting the long period of peace and global growth. However, the FDI directed to the BRIC countries doubled this figure with an 85% annual increase. China and India, with a spectacular 370 and 133% annual growth respectively, top the BRIC group (Figure 1.26).

Looking forward and considering the extraordinary amount of growth the BRIC and some of the REST economies are still expected to have, this trend is expected to continue and increase further during the next decades.

1.3
Political Megatrend

"A new international order, with more BRICs, more megacorporations, more government, and much more social accountability."

"A transition to a new Democratic model: the People's Government"

> Change is the law of life. And those who look only to the past or present are certain to miss the future. -
>
> John F. Kennedy

Political megatrends will be one of the most influential, dynamic, and most converging of all the presented megatrends. Within this megatrend, two major trends have been identified. On the one hand the appearance of a new international order, where the BRIC, companies, political unions, and society will play a much larger role, and on the other hand, the acknowledgment that governments will play a much larger role moving forward.

Additionally, some political megatrends are directly related to the previous social and economic megatrends, for example, the increase in economic power of the BRIC, or the formidable growth of multinationals. Other trends will be more related to the recent economic crises in 2008 and 2011, leading to a higher and dominant role for government and the increased involvement of citizens in society and political life. Recent examples could be the "Arab spring," the "Indignados" in Spain or the "Wall Street protesters" in the USA. All these changes will also be accelerated by the disruptive influence of the internet, specifically social networks, in our day to day life. Some of these trends have traditionally been considered as the "softer" ones, versus the "harder" or "factual" ones like economic, population, or energy trends. However, we believe the relevance of these "softer" trends will increase significantly in the decades to come. As the world becomes wealthier and wiser, the "how" will become as important as the "what" and the "who."

The emergence of the BRIC economies will alter the balance of power at an international level, especially in the way international organizations work today. The emergence of super megacompanies, much larger than country economies, will attract a lot of attention from governments, media, and citizens, creating large expectations and duties versus society. The acknowledgment that in a global economy, absolute population, trade, and economic size matter, fostered the creation of several supranational political and economic unions, like the EU or the Association of South East Asian Nations (ASEAN), and several trade or economic unions like the North American Free Trade Agreement (NAFTA) or the Mercado Comun del Sur (MERCOSUR). That acknowledgment, in combination with the most recent realization that in a truly globalized world, global problems, like climate change or global economic recession, require global and coordinated actions, will foster further political and economic integration as well as the creation of larger and more efficient international organizations.

Finally, the sudden and drastic eruption of the social media into our daily life has not only proved its capacity to connect individuals across the world, completing the human aspect of globalization, but has also empowered millions of people around the world, enabling them to discuss, speak, and change things. Governments, companies, and societies cannot afford to neglect the people's will and they will be forced to listen to them, with more democratic countries expected, and more open and transparent ones.

As the world becomes wealthier, older, and wiser, and social media enhance global transparency, new forms of more direct government will be developed, with more direct forms of democracy. More and better democracies might be expected. Indeed these changes will affect the way governments, companies, and the chemical industry work and do business around the world.

1.3.1
Trend – A New International Order

1.3.1.1 Sub Trend – the Emergence of the BRIC Economies

There will be a shift in economic growth and wealth from the ADV to the BRIC economies, where the BRICs will account for the same share of the World economy as the ADV economies by 2050, that is, 30%. This will serve to accelerate a new redistribution of international power. The current international order will be altered, and the existing international organizations will need to be adapted to reflect the new balance of power, or new organizations will need to be created.

Most of the major international organizations (Table 1.4), like the UN, the World Bank, the International Monetary Fund (IMF) or the Organization for Economic Cooperation and Development (OECD) were created after the Second World War, reflecting the balance of power after the war. This is easily observed when looking at simple facts like the location of their headquarters, distribution of obligations, votes, veto powers, and overall distribution of power, presidents, vice presidents and internal working mechanisms. Let us briefly review in the following business case study, the background and status of one of the most powerful international organizations: the IMF.

IMF: Case Study As a "soft" introduction to the topic, let us start by sharing the curious coincidence that the World Bank and the IMF not only have their headquarters across the street in Washington, DC, but actually they are at walking distance from the White House and the US Federal Reserve. It is so impressive

Table 1.4 International organizations.

PURPOSE	ORGANIZATION	AREA	2010
Int. Law, Security, economic development social progress	UN - United Nations Created in 1944 Based in NY, USA Criteria = % of Budget	ADV	70%
		BRIC	4%
		Rest	26%
Economic development and social progress, with special geographical focus.	WORLD BANK Created in 1944 Washington, Dc USA Criteria = % of Budget	ADV	
		BRIC	
		Rest	
	IMF - Int. Mon. Fund Created in 1944 HQ: Washington, Dc USA Criteria = % of Votes	ADV	60%
		BRIC	10%
		Rest	30%

Source: Data from IMF, World Bank, OCDF and own elaboration for 2050.

to notice that some of the most powerful organizations in the world are so close together, despite the fact that they are supposed to represent the whole world.

Beyond curiosities, and more seriously, let us start to review how some of these organizations work, so we can understand what sort of changes and challenges we should expect in the next decades. For this analysis, and as a good representative, we have selected the IMF, where we will review how power, obligations, and voting rights have been allocated.

In terms of power distribution, there is an unwritten agreement by which the managing director of the IMF is always a European, and the president of the World Bank is always an American citizen; this being indeed the case since these organizations were created in the 1940s. In Table 1.5 you can see the unwritten agreement has been fulfilled for the last 66 years, since the inception of the two organizations.

In terms of obligations, budget contributions, and voting rights, these have been calculated using a formula and tend to be directly correlated. This formula considered several economic indicators, like average GDP, openness, economic variability, and international reserves. Based on the results of these calculations, voting rights, and obligations were assigned. In 2010, and after very complex and arduous negotiations, a new formula was agreed, and a new allocation was made. The idea behind this change was to try to better reflect the new economic circumstances and, as a consequence, 54 countries, including the BRICs, had an increase in their voting rights and obligations.

In March 2011 the new allocation was implemented, with the ADV economies getting 60% of the votes, the REST 30% and the BRICs just 10%. The USA kept 17% of the votes, the largest share, and its veto, the only one available. China increased its voting rights to its current level at 3.81%, while the BRIC reached 10% of the votes (Table 1.6).

Indeed, if we consider the expected relative size of the BRICs in the World economy by 2050, we should expect a completely new distribution of power at the

Table 1.5 World Bank Presidents and IMF Directors.

WORLD BANK PRESIDENT		Nationality
Eugene Meyer	1946–1946	USA
John J. McCloy	1947–1949	USA
Eugene R. Black,	1949–1963	USA
George Woods	1963–1968	USA
Robert McNamara	1968–1981	USA
Alden W. Clausen	1981–1986	USA
Barber Conable	1986–1991	USA
Lewis T. Preston	1991–1995	USA
James Wolfensohn	1995–2005	USA
Paul Wolfowitz	2005–2007	USA
IMF DIRECTOR		**Nationality**
Camille Gutt	1946–1951	Belgium
Ivar Rooth	1951–1956	Sweden
Per Jacobsson	1956–1963	Sweden
Paul Schweitzer	1963–1973	France
J. Witteveen	1973–1978	NL
J. Larosière	1978–1987	France
M. Camdessus	1987–2000	France
Horst Köhler	2000–2004	Germany
Rodrigo Rato	2004–2007	Spain
D. Strauss-Kahn	2007–2011	France
C. Lagarde	2011–	France

IMF, somewhere around 30% of the votes for the ADV economies and 30% for the BRICs (Table 1.6).

Despite the IMF's attempt to better reflect the current balance of power, the reality is that the BRIC's voting rights remain at a very low level, 10% of the total, especially when considering that according to the IMF's estimations for 2011, the BRICs accounted already for 25% of the world GDP at PPP. The IMF agreed to

1 Global Megatrends by 2050

Table 1.6 Major international organizations.

IMF	AREA	2011	VETO	2050		
Created in 1944 Washington, DC USA Criteria % of Votes	ADV	60%	No	30?	No	Headquaters move to a BRIC? China?
	USA	16.76%	YES	11%	YES	
	BRIC	10%	No	30?	No	
	China	3.81%	No	12%	YES	
	Rest	30%	No	40?	No	

Source: I MF http://www.imf.org/external/pp/longres.aspx?id=4235

Note: IMF – International Monetary Found, New Voting Rights Formula March 2011 The current quota formula is a weighted average of GDP (weight of 50 percent), openness (30 percent), economic variability (15 percent), and international reserves (5 percent). For this purpose, GDP is measured as a blend of GDP based on market exchange rates (weight of 60 percent) and on PPP exchange rates (40 percent). The formula also includes a "compression factor" that reduces the dispersion in calculated quota shares across members.

continue to review rights and obligations, however, it may take decades before these will match their real economic power. In the future we should expect that the IMF, as well as the other international organizations, will start reorganizing themselves to better reflect the new balance of power, otherwise more radical changes could be expected. So perhaps in a non-distant future we could have a president from outside Europe; the IMF or World Bank headquarters moved to a BRIC country, and/or a new a set of rights and obligations for those institutions more consistent with the actual economic conditions (Table 1.6).

The current economic crisis after the Lehman Brother's collapse in the USA, and now the Sovereign Debt Crisis in Europe, from Ireland to Greece and with potential to spread further into Italy and even Spain, has sparingly helped Emerging Economies and specifically the BRIC to increasing their relevance in the world economy.

Indeed at this time, October 2011, at least three new groups or forums of economic power have been created. The so-called G20, that includes the G8 plus a list of emerging economies, including the BRIC, the G2 that includes China and US and finally the so-called BRIC Group, composed of the BRIC countries.

In a non-distant future it is not difficult to envision the creation of new international organizations, where the new economic order would be better reflected. It would also not be strange to see that old institutions (i.e., OCED) could be replaced by new ones, where, for instance, the G20 and BRIC could meet regularly. Many combinations could be argued, however, two things appear clear to us: first, Governments are going to become even more relevant; and secondly international organizations are going to face significant transformations.

1.3.1.2 Sub Trend – Corporate Mega Economies – (CME)

However, the already complex and radical changes in the balance of power among countries or economic areas from the ADV to the BRIC economies will not come alone, as two other powerful trends will also emerge.

First, the creation of "CMEs" – corporations that will be much bigger than countries for long periods of time. Indeed, and although corporations can go bankrupt, merge, or simply disappear, we have also observed that some of these can keep their size, power, and influence through the decades and even centuries.

Secondly, and most recently, we have seen how the power of social media has served to create a new powerful voice in our societies – power that can talk directly to governments or corporations, exchanging views and opinions all the time and in real time.

These "new" enhanced actors will be reviewed in the next pages, and we should also consider them carefully when looking toward 2050. Questions like what would be the power of social media with a more mature, educated, wiser, and wealthier world population? What role, influence and responsibilities will large mega-economies, or corporations have in society? What new forms of democracy could better capture people's desire to share and decide? Will social media serve to improve democracies? and so on, are all questions that can have a tremendous influence on our world and on how the chemical industry works, sells, consumes, and interacts with society.

The accelerated process of economic growth and globalization experienced during the last century has created a new set of organizations and economic agents: the CMEs. CMEs brought new challenges and opportunities to the economy, attracting much attention.

In 2007 the list of the top 100 economies, considering both countries and companies (measured by GDP/Revenue), included 47 multinationals. Likewise, among the top 150 economies in the world, 100 were companies (Table 1.7).

The formidable size, dynamism, and growth of the new CMEs have created a vast amount of interest and concerns among countries and citizens. These huge CMEs not only keep constantly growing but they are ubiquitous in the economy. The tremendous forces of global competition have forced them to become competitive, resilient, and agile.

Looking at the Fortune 500 largest USA CMEs we can clearly see the fierce competition existing among them to grow and remain competitive. The largest and most dynamic and competitive economy in the world, the US economy, increased 35-fold between 1955 and 2008, while the top 500 companies, in the same period, increased 74-fold, and Dow Chemical, as the largest US Chemical Company, increased by a staggering 131-fold (Table 1.8).

Additionally, of the top ten companies in 1958, including DuPont, the second largest US Chemical Company, only three remain at the top of the list in 2008. Some of them have merged and others have simply not been able to keep up with the strong competition. This illustrates the fierce rivalry existing among companies to survive and to strive forward.

In 1958 the top 500 companies in the US accounted for almost 10% of the US economy. In 2008 the top 500 corporations accounted for 74% of the US economy. The economic relevance of these formidable competitive CMEs is noticeable not only in the US economy but also globally (Tables 1.7 and 1.8).

Table 1.7 World Top 100 Economies in 2007 – national and corporate.

RANK	COUNTRY COMPANY	GDP REVENUES	% WORLD GDP	GROUP	RANK	COUNTRY COMPANY	GDP REVENUES	% WORLD GDP	GROUP
Rank 1	United States	13,794	22.9%	Advanced	Rank 51	Malaysia	165	0.3%	Rest
Rank 2	Japan	4,346	7.22%	Advanced	Rank 52	Chile	161	0.3%	Rest
Rank 3	Germany	3,259	5.41%	Advanced	Rank 53	Ford Motor	160	0.3%	Advanced
Rank 4	China	3,249	5.39%	BRIC	Rank 54	Romania	159	0.3%	Advanced
Rank 5	United Kingdom	2,756	4.58%	Advanced	Rank 55	ING Group	158	0.3%	Advanced
Rank 6	France	2,515	4.18%	Advanced	Rank 56	Israel	154	0.3%	Rest
Rank 7	Italy	2,068	3.43%	Advanced	Rank 57	Singapore	153	0.3%	Rest
Rank 8	Spain	1,415	2.35%	Advanced	Rank 58	Citigroup	147	0.2%	Advanced
Rank 9	Canada	1,406	2.34%	Advanced	Rank 59	Pakistan	144	0.2%	Rest
Rank 10	Brazil	1,295	2.15%	BRIC	Rank 60	Philippines	141	0.2%	Rest
Rank 11	Russia	1,224	2.03%	BRIC	Rank 61	AXA	140	0.2%	Advanced
Rank 12	India	1,090	1.81%	BRIC	Rank 62	Hungary	136	0.2%	Advanced
Rank 13	Korea	950	1.58%	Rest	Rank 63	Volkswagen	132	0.2%	Advanced
Rank 14	Australia	890	1.48%	Rest	Rank 64	Sinopec	132	0.2%	BRIC
Rank 15	Mexico	886	1.47%	Rest	Rank 65	Ukraine	131	0.2%	Rest
Rank 16	Netherlands	755	1.25%	Advanced	Rank 66	Crédit Agricole	128	0.2%	Advanced
Rank 17	Turkey	482	0.80%	Advanced	Rank 67	Egypt	128	0.2%	Rest
Rank 18	Belgium	443	0.74%	Advanced	Rank 68	Nigeria	127	0.2%	Rest
Rank 19	Sweden	432	0.72%	Advanced	Rank 69	Algeria	126	0.2%	Rest
Rank 20	Switzerland	414	0.69%	Rest	Rank 70	Allianz	125	0.2%	Advanced
Rank 21	Poland	413	0.69%	Advanced	Rank 71	New Zealand	124	0.2%	Rest
Rank 22	Indonesia	410	0.68%	Rest	Rank 72	Fortis	121	0.2%	Advanced
Rank 23	Taiwan Province of China	376	0.62%	Rest	Rank 73	Bank of America	117	0.2%	Advanced
Rank 24	Saudi Arabia	374	0.62%	Rest	Rank 74	HSBC Holdings	115	0.2%	Advanced
Rank 25	Norway	369	0.61%	Advanced	Rank 75	AIG	113	0.2%	Advanced
Rank 26	Austria	367	0.61%	Advanced	Rank 76	China Nat. Petroleum	111	0.2%	BRIC
Rank 27	Greece	356	0.59%	Advanced	Rank 77	BNP Paribas	109	0.2%	Advanced
Rank 28	Wal-Mart	351	0.58%	Advanced	Rank 78	ENI	109	0.2%	Advanced
Rank 29	Exxon Mobil	347	0.58%	Advanced	Rank 79	UBS	108	0.2%	Advanced
Rank 30	SHELL	319	0.53%	Advanced	Rank 80	Siemens	107	0.2%	Advanced
Rank 31	Denmark	311	0.52%	Advanced	Rank 81	State Grid	107	0.2%	Advanced
Rank 32	Iran	278	0.46%	Rest	Rank 82	Kuwait	103	0.2%	Rest
Rank 33	South Africa	275	0.46%	Rest	Rank 83	Assicurazioni Generali	102	0.2%	Advanced
Rank 34	BP	274	0.46%	Advanced	Rank 84	Peru	102	0.2%	Rest
Rank 35	Ireland	253	0.42%	Advanced	Rank 85	J.P. Morgan Chase	100	0.2%	Advanced
Rank 36	Argentina	248	0.41%	Rest	Rank 86	Carrefour	99	0.2%	Advanced
Rank 37	Finland	236	0.39%	Advanced	Rank 86	Berkshire Hathaway	99	0.2%	Advanced
Rank 38	Venezuela	227	0.38%	Rest	Rank 87	Pemex	97	0.2%	Advanced
Rank 38	Thailand	226	0.38%	Rest	Rank 88	Deutsche Bank	96	0.2%	Advanced
Rank 39	Portugal	220	0.36%	Advanced	Rank 89	Dexia Group	96	0.2%	Advanced
Rank 40	General Motors	207	0.34%	Advanced	Rank 90	Kazakhstan	95	0.2%	Rest
Rank 41	Toyota Motor	205	0.34%	Advanced	Rank 91	Honda	95	0.2%	Advanced
Rank 42	Hong Kong SAR	203	0.34%	Advanced	Rank 92	McKesson	94	0.2%	Advanced
Rank 43	Chevron	201	0.33%	Advanced	Rank 93	Verizon	93	0.2%	Advanced
Rank 44	DaimlerChrysler	190	0.32%	Advanced	Rank 94	Nippon Telephone	92	0.2%	Advanced
Rank 45	United Arab Emirates	190	0.31%	Rest	Rank 95	Hewlett-Packard	92	0.2%	Advanced
Rank 46	ConocoPhillips	172	0.29%	Advanced	Rank 96	International Business Machines	91	0.2%	Advanced
Rank 47	Colombia	172	0.29%	Rest	Rank 97	Valero Energy	91	0.2%	Advanced
Rank 48	Total	168	0.28%	Advanced	Rank 98	Home Depot	91	0.2%	Advanced
Rank 49	General Electric	168	0.28%	Advanced	Rank 99	Nissan Motor	90	0.1%	Advanced
Rank 50	Czech Republic	168	0.28%	Advanced	Rank 100	Samsung Electronics	90	0.1%	0%

	GDP REVENUE	% TOTAL	NUMBER
TOP 100 COMPANNIES	6,361	11%	47
TOP 100 COUNTRIES	57,658	96%	53
TOTAL WORLD	60,217		

	COUNTRIES	FIRMS	% TOP 100 / GROUP
TOP 100 ADVACNCED	43,030	6,119	14%
TOP 100 BRIC	6,858	242	4%
TOP 100 REST	7,770	0	-
TOTAL WORLD	60,217		

Source: Fortune 500, IMF GDP estimate 2007

Units: Billion

Table 1.8 Top 10 Fortune 500 mega corporations – 1955–2008.

TOP 10 Companies in 1955	1955 Revenue ($Million)	Sector	TOP 10 Companies in 2007	2007 Revenue ($Million)	Sector	% TNI Index (Unctad)
1 - General Motor	9823	Automotive	1- Wall Mart	378799	Grocery	26
2 - Exxon Mobile	5661	Oil	2 - Exxon Mobile	372824	Oil	67
3 - U.S. Steel	3251	Steel	3 - Chevron	210783	Oil	57
4-General Electric	2959	Conglemerat	4 - General Motor	182714	Automotive	43
5 - Esmark	2510	Steel	Conoco Philips	178347	Oil	41
6 - Chrysler	2071	Automotive	General Electric	176000	Conglomerate	50
7 - Armour	2056	Food	Ford Motor	172989	Automotive	48
8 - Gulf Oil	1705	Oil	Citigroup	159089	Banking	56
9 - Mobil	1703	Oil	Bank of America	119987	Banking	
9 - DuPont	1697	Chemical	AT&T	118989	Communication	10
Top 10	30477	22 /	Top 10 (55 times)	1690000	9	
Dow Chemical	427		DOW × 131	56000	131 times	52
Top 500	136251	33	Top 500 × 77	10600507	74	
US GDP	402000		US GDP × 35	14195000		

Source: Fortune Magazine Historical Overview 1955–2007 and UNCATD - UN Conference on Trade and Development.
TNI, the Transnational Index, is calculated as the average of the following three ratios: foreign assets to total assets, foreign sales to total sales and foreign employment to total employment.

The crucial need of these new CMEs to remain competitive in order to survive has forced them to take all possible measures to keep growing while remaining agile and fit. These CMEs have become the "new warriors" of the new millennium. Among all the resources available for them to keep competitive, like operational efficiencies, organic growth and external growth (M&A), and so on, globalization has become their key partner and tool.

Indeed looking at the UN Conference for Trade and Development – UNCTAD – Transnational Index (TNI) for the top 10 USA Fortune 500 corporations, clearly these companies are no longer American companies but global ones. Most, including Dow Chemical, have a TNI score higher than 50%, confirming their strong presence in sales, assets, or employees outside the USA and explaining why, for instance, Dow Chemical could grow twice as fast as the US economy.

As the world economy is expected to keep growing in a formidable manner in the next decades, CMEs will keep growing even faster, acquiring sizes and dimensions which are just unthinkable nowadays. In 2007 the largest CME (Wall Mart) ranked just 29 in the World economy; by 2050, with a global economy 10 times bigger than today, CMEs are expected to rank even higher (Table 1.7).

The greater relevance of the CMEs will increase their exposure to society, requiring further levels of governance, transparency, and sustainable development. Therefore, the new CMEs have to be ready for this new challenge, increasing their governmental and social transparency and relations. Tremendous size entails formidable responsibilities and challenges, but also unbelievable opportunities. In a world of wealthier, wiser, and older people, social reputation and social accountability will become even more critical. Under this scenario which forces could balance the growing power of larger CMEs – larger governments, society, or themselves?

For instance, the Dow Chemical Company grew three times faster than the USA economy during the last decades and that might easily happen again. By 2050, BASF could become a US$ 300 billion giant and the Dow Chemical Company, as the second largest chemical company in the world, could become a US$ 200 billion company. Could we imagine chemical companies so large? What kind of obligations, responsibilities and obligations could these huge companies have?

1.3.1.3 Sub trend – Social Networks

Finally, the fast introduction and development of the "internet" and "social networks" has accelerated and changed forever the way people communicate and learn. In 1997 the internet was introduced and since then its use has been increasing year after year, to reach almost 22% of the world population in 2007 (Figure 1.27).

In the same period, and even more impressive, is the degree of penetration of the internet on the ADV economies, reaching a staggering 62% of the population in just 10 years. This large growth in the ADV economies only sets the expectations for the other regions in the years to come.

In 2004 the so-called "social networks" appeared, Facebook being one of the largest and most notorious examples. Since then multiple networks have been created globally, engaging millions of users across the world. Social networks have created a new vehicle for their users to communicate, share data, views, and opinions in real time across the world. The number of citizens engaged, the speed of communication, and their global and ubiquitous nature have made these networks a tremendous source of power.

Another interesting aspect of the fast development of social media is that despite the fact the degree of internet penetration was faster in the ADV economies than

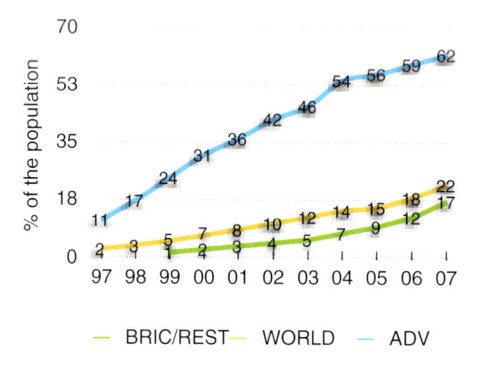

Figure 1.27 Global internet use. Source: International Telecommunication Union (ITU) – UN Agency.

Table 1.9 Top social networks in 2011.

Name	Users (million)	Area
Facebook	800	Global
Tencent	660	BRIC / China
Ozone	480	BRIC/China
Netease	360	BRIC/China
Twitter	360	Global
Windows	330	Global
Weibo	230	BRIC/China
Skype	145	Global
Vkontake	140	BRIC/Russia
LinkedIn	100	Global
Habbo	230	Finland

Source: International Telecommunication Union.

in the BRICs or REST, when we review the list of the largest social networks in terms of members we can see how the BRICs are leading in terms of numbers of networks and millions of users. Indeed, Facebook is currently the largest social network in 2011 with 800 million people, but Tencent in China with its 660 million users is very close. Additionally, there are four other social networks in China, and Chinese social networks have already double the number of users on Facebook globally. Social networks have progressed very fast globally, but it seems they have grown even faster in the BRIC economies (Table 1.9).

To better illustrate the tremendous power and growth of social networks let us review briefly the history and achievements of perhaps the most famous and successful social network in the world, Facebook. Facebook started in 2004 as a small student network at Harvard University, USA, with less than 10 000 members in its first year (Figure 1.28).

Seven years later, Facebook has more than 800 million users around the world, generating more than four billion dollars revenue in 2011. The speed of growth has not only been formidable in terms of its size but also in its global reach. Despite the fact that Facebook started as a student network across the USA, seven years later almost 60% of its members were outside United States, and were not necessarily students. Indeed, among the countries with the largest number of active members we can find countries like Indonesia with 10%, India 9%, UK 7%, Turkey 8%, and Brazil and Mexico with 7% each. It is quite remarkable to notice that although the world average share of internet penetration was around 22% globally, and even

1 Global Megatrends by 2050

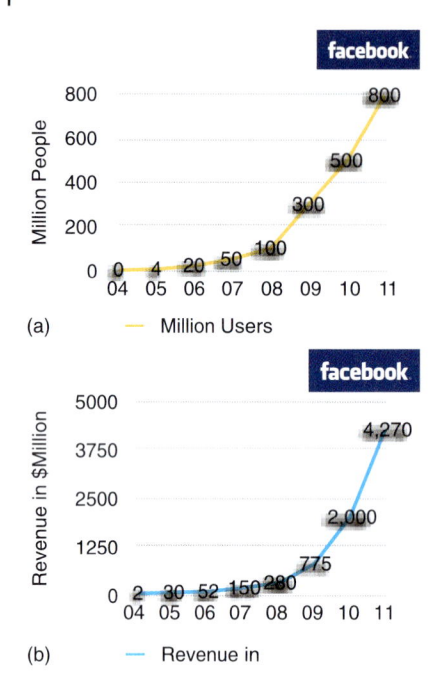

Figure 1.28 (a) Active users of facebook 2004–2011. (b) Facebook revenue 2004–2011. (Source: Wikipedia.)

lower in the BRIC and REST economies with just 17%; among the top nationalities using Facebook we can find BRIC countries (Figure 1.29).

The fact that social networks allow the connection of people from all over the world and of different ages (see Figure 1.29) has contributed to further "flattening" the world – alluding to Thomas Friedman and his book "The World is Flat" – contributing to the creation of a more transparent and horizontal world.

Several recent examples have shown us the increasing relevance of social networks, like during the so-called "Arab Spring" or the "Indignados" in Spain where millions of people used the power of social networks to discuss and share ideas, organizing themselves for demonstrations or concerted actions.

In the past when people wanted to discuss certain topics and organize themselves for action, they used to do it by newspapers, radio, telephone, and so on. Today they can share information 24 hours a day, organizing and gathering thousands of people through a web page in just a "click."

Additionally, when governments have tried to control or stop some of these communications, like during the last revolts in Egypt or Libya, this has proved to be very complicated, not to say almost impossible. In the past it was relatively easy, it was enough to go to the premises of the respective newspaper, radio, or TV station and close it. Now, with the internet, it is more difficult as governments would have to close a web page that may be located in another country, or eventually

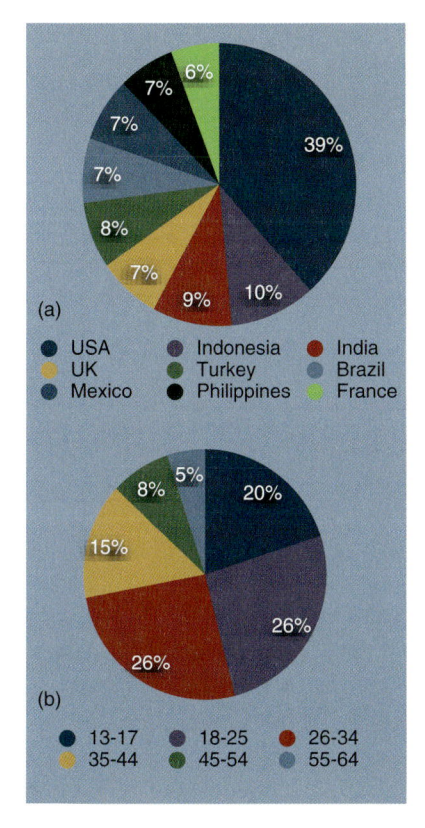

Figure 1.29 (a) Facebook users by country. (b) Facebook users by age. (Source: Wikipedia.)

shutdown the whole internet. That might seem a very drastic action that could not only impact thousands of people but also its own government web pages, pages perhaps needed to run the country. Hence, internet and social networks have not only become very powerful by their capacity to accelerate communications but also for their resilience and difficulty to be controlled.

As social networks keep developing, enhancing their services and the number of members, and users increase their understanding of them, we could envision a world where social networks could become much more relevant than today. The numbers of scenarios are so diverse and the potential so vast that it is very difficult to think today about some of them. However, if we consider the simplest ones, could you imagine a world where citizens could discuss, propose, and vote in real time about the critical things for society? Could you imagine a much more horizontal political system where citizens, politicians, media, companies, and non-governmental organizations (NGOs) could gather in a social network to discuss, work, and decide on the critical subjects on an equal basis? perhaps by 2050?

In a not very distant future, all these things might happen. Indeed, technically most of these "bizarre" ideas might be possible today, however, our societies and their key stakeholders have not realized it or seen the value yet. In some democracies, as in Switzerland, most of the key government decisions are already voted on with several referenda during the year, so that citizens are not only much more aware and engage in political discussions but are also more accountable for their government's actions and their country's progress and evolution.

As social media, society, and democracy progress in that direction during the next decades, we can anticipate more open democracies with a much more direct involvement, not only from individuals but also from the rest of the key social stakeholders, including corporations and public and private organizations. Thus social media could serve to accelerate this transformation, further engaging stakeholders in the key discussions of a particular society.

Social media could be a strong catalyst for political change. On the one hand they could enable significant improvements in the quality of our democracies, fostering not only transparency and openness but also agility and speed of the government's reaction. On the other hand, the capacity to connect people across different countries, geographies, and political regimes, might serve to stimulate further democracies. After all, "Envy is the basis of democracy?" (Bertrand Russell).

1.3.2
Trend – An increasing role of Governments

Finally, on this point we will cover the second aspect of the political trends, the increased role of Governments after Lehman Brothers. Government's role has traditionally been very large in our societies, however, this role has become even larger during the current world economic crisis after the collapse of Lehman Brothers in 2008 in the USA and the Sovereign Debt Crisis in Europe in 2011.

Governments, traditionally, have accounted for 6–35% of their economies, 15–20% being the normal average (Figure 1.30). The current economic crisis forced governments around the world to embark on large economic programs to stimulate the economy, injecting billions into it and increasing substantially their role in the economy.

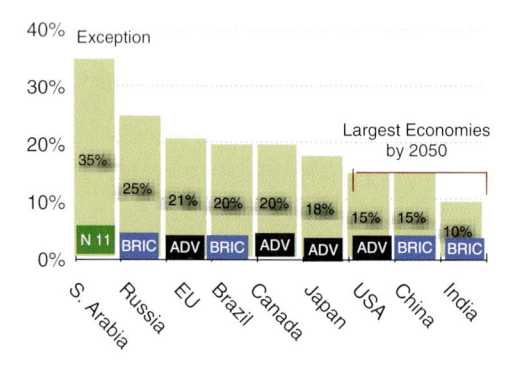

Figure 1.30 Government percentage of GDP. (Source: July 2008, OCDE Annual National Accounts 1970–2006, July 2008.)

1.3 Political Megatrend

Thus we believe government's role will remain very large, as governments will still represent a large part of our economy, and might need to become even larger, as they will not only need to keep investing in and monitoring the economy but also they will need to monitor and balance the upcoming power of the megacorporations.

Governments not only ensure the proper functioning of the economy, society, and the political and legal framework, but in the future they will also need to ensure certain minimum living and quality standards for their citizens and their companies, while balancing and monitoring the tremendous power CMEs will enjoy.

In 2007, 50% of the top 100 economies in the world were companies. With the fast development of the economy in the decades to come that ratio is likely to increase. As we saw earlier, companies, thanks to globalization, tend to grow even faster than countries. So what would happen if 80% of the companies were among the largest economies of the world? Who will balance their power and how will society ensure this power is use for good purposes?

From an economic point of view, the role of the Governments in the economy is already substantial, but it could become even larger, especially if larger economic groups keep being created, as we discussed earlier. Before moving forward let us quickly review the basis of government's role in the economy, so we can better understand how governments influence the economy and how big that role is and could become.

Perhaps one of the most famous and basic economic formula is that for the gross domestic product (GDP) or economic growth (Y). In its simplest version, the GDP of a country or economic area can be calculated by adding the private consumption of goods and services of that country or area, hereafter referred to as C for consumption, plus the gross investment into that economy, I, plus the

FORMULA: GDP - Gross domestic product

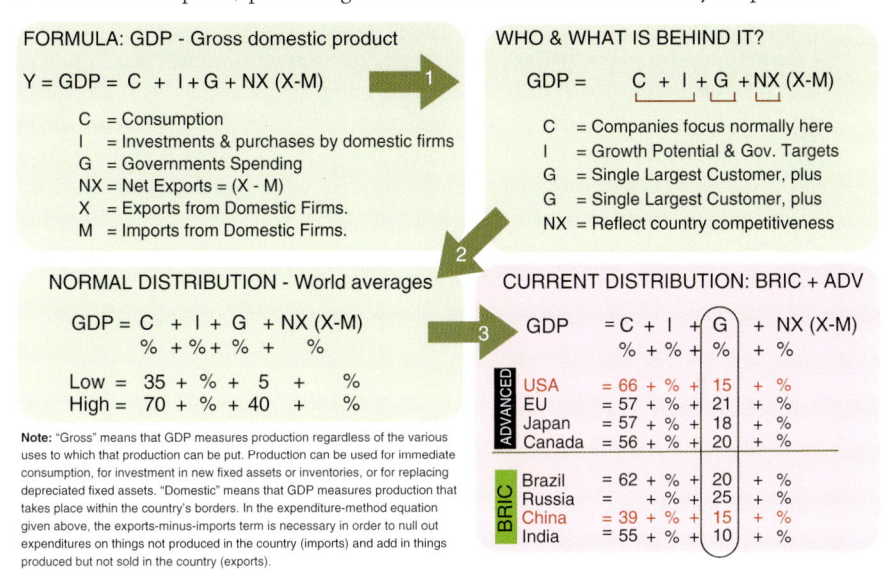

$Y = GDP = C + I + G + NX (X-M)$

- C = Consumption
- I = Investments & purchases by domestic firms
- G = Governments Spending
- NX = Net Exports = (X - M)
- X = Exports from Domestic Firms.
- M = Imports from Domestic Firms.

WHO & WHAT IS BEHIND IT?

$GDP = C + I + G + NX (X-M)$

- C = Companies focus normally here
- I = Growth Potential & Gov. Targets
- G = Single Largest Customer, plus
- G = Single Largest Customer, plus
- NX = Reflect country competitiveness

NORMAL DISTRIBUTION - World averages

$GDP = C + I + G + NX (X-M)$
$\quad\quad\quad \% + \% + \% + \%$

- Low = 35 + % + 5 + %
- High = 70 + % + 40 + %

Note: "Gross" means that GDP measures production regardless of the various uses to which that production can be put. Production can be used for immediate consumption, for investment in new fixed assets or inventories, or for replacing depreciated fixed assets. "Domestic" means that GDP measures production that takes place within the country's borders. In the expenditure-method equation given above, the exports-minus-imports term is necessary in order to null out expenditures on things not produced in the country (imports) and add in things produced but not sold in the country (exports).

CURRENT DISTRIBUTION: BRIC + ADV

	GDP	= C + I + G + NX (X-M)
		% + % + % + %
ADVANCED	USA	= 66 + % + 15 + %
	EU	= 57 + % + 21 + %
	Japan	= 57 + % + 18 + %
	Canada	= 56 + % + 20 + %
BRIC	Brazil	= 62 + % + 20 + %
	Russia	= + % + 25 + %
	China	= 39 + % + 15 + %
	India	= 55 + % + 10 + %

Figure 1.31 GDP – gross domestic product national formula. (Source: July 2008, OCDE Annual National Accounts 1970–2006, July 2008. Comments based on own elaboration or extracted from Wikipedia.)

Government spending in that economy, G, plus the net trade balance or difference between exports, X and imports, M, of that country, X – M (Figure 1.31).

By understanding these elements and how they work, we can understand how the GDP is formed and how an economy works and creates wealth. For instance, considering the consumption we can understand how self-sufficient is the growth of that economy. For instance, when we look into the growth model of the US economy, 66% of the growth is due to its internal consumption, while if we look into the growth model of China, their internal consumption is only 39%, actually the lowest level in all the BRICs and ADV economies. That is why, traditionally, we tend to say that China is growing despite a very low internal consumption, and that it should be stimulated further to keep China's economic growth sustainable in the future.

Also, when we look at Government spending we can see how big is the role of the government in the economy. In this area, we can observe that the US is the country in the ADV group with the lowest Government spending at 15%, the same as China, but much lower than in Europe (21%) or Russia (25%) (Figure 1.30).

When we look at the trade balance we can also learn how competitive a country is. Negative trade balances imply the country imports more than it exports, and highlight a lack of competitive advantage or a certain economic model in which perhaps there is a particular need to import more at a certain time.

Once we have understood the GDP formula, we can then review and quantify the role of the government in a particular economy. Although government's role and influence in the economy and society goes well beyond its spending, it also regulates and organizes societies in many different ways but in this chapter and for our purpose we will just focus on the spending aspect.

From an economic point of view, the facts that governments represent a significant part of the economy, up to 40%, that their spending can be anti-cyclical, as exposed during the current economic crisis, and that they have the capacity to set the strategic agenda for the future of their countries, make us believe that any industry and company should have a clear understanding of the government's role and its aspirations. Industries and companies could be very positively or negatively impacted by government decisions, therefore understanding of the government's role is of critical importance.

After all Governments remain as the single "largest customer" of any given economy. This is a role that has been significantly increased during the recent economic crisis, where governments around the world have not only used their financial muscle to bail out companies but also have established multibillion dollar stimulus packages to stimulate their local and global economies.

The recent developments in our World economy, with probably the single and deepest economic recession since 1929, have proved again the economic relevance of the National Governments. Governments that emerged again in the most critical moments of our economies as the last resource for economic stabilization and ultimate economic growth.

Table 1.10 illustrates the relevance of the different governments in their respective economies, as well as the size of the stimulus that governments around the world

Table 1.10 Governments and stimulus as percentage of GDP in 2009.

Country	GDP in $ Billion	"Government % in the Economy"	Stimulus Packages $ Billion	% GDP	Total "G" in % vs. GDP
EUROPE	18394	21	560	3	24
USA	14264	15	787	6	21
JAPAN	4923	18	125	3	21
CANADA	1510	20	30	2	22
ADV	$39091	18	1502	4	22
BRAZIL	1572	20	28	2	22
RUSSIA	1671	25	90	5	30
INDIA	1209	10	4	0	11
CHINA	4401	15	585	13	28
BRIC	8853	17	707	8	25
MEXICO	1088	12	90	8	20
S. KOREA	947	15	11	1	16
S.ARABIA	487	35	126	26	61
VIETNAM	89	6	6	7	12
SAMPLE	2611	17	233	9	26

Source: Own elaboration based on OCDE Annual National Accounts for 2009 and different information on the government subsidies from different governments.

have been putting in place to "fight" this economic recession. In some cases stimulus packages have reached huge sizes, like in *Saudi Arabia, China, or the USA with 26, 13, or 6% of their GDPs.*

We will not discuss the effectiveness or not of these stimulus packages, although we do not perceive them to be very effective in the medium/long term, however, some major trends can be expected after this crisis:

1) Government's role in the economy will increase even further.
2) The world will have a new "Government economics" after this crisis.
3) Governments and their huge stimulus packages are a tremendous source of business opportunities for companies.
4) As globalization progresses larger political and economic unions can be expected.

Governments are not only the single "largest customer" of any single economy but, as discussed, they have a tremendous capacity to generate resources and increase their size in the economy if needed.

As we saw above, governments can temporarily significantly increase their spending in the economy, especially during times of recession and when most needed. So we can see how, for instance, the Governments of Canada, Brazil, Japan, or the EU, increased their spending by ~2–3% of their economies, the USA and Russia by 5–6%, China by 13%, and Saudi Arabia by almost 26% (Figure 1.32).

It may be worth mentioning the economic theory behind the use of stimulus packages. John Maynard Keynes elaborated his economic theory across several decades in his book, "General Theory of Employment, Interest and Money" published in 1936, in which he established the basis for most of the economic actions implemented after the Second World War and during the current economic crisis.

He established the basic principles to escape from an economic recession. Traditionally, the classical treatment of unemployment was to leave markets to adjust via the balance of supply and demand. In other words, salaries will decline until a point where demand will start to increase again. After the 1929 crisis, when millions of people lost their jobs, Keynes discovered what he called the *"paradox of thrift."* This implies that in times of recession, when workers remain unemployed, consumption remains low and savings high, the economy could enter into a self-reinforcing equilibrium, where the economy stalls, keeping economic growth low and millions of people unemployed and in poorer conditions than before the recession.

Under that situation, and keeping in mind the Growth Economic Formula, governments should proactively intervene in the economy, spending more money in the economy, counterbalancing the lack of consumption and high levels of savings.

As governments might or might not have the resources to stimulate the economy in the short term, governments could incur large deficits or print money to stimulate demand. Interest rates could increase for several reasons but, more

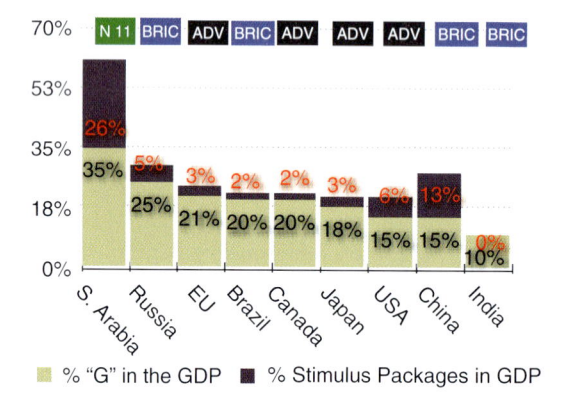

Figure 1.32 Percentage of stimulus package as percentage of GDP 2009. (Source: Own elaboration based on OCDE Annual National Accounts for 2009 and different information on the government subsidies from different governments.)

importantly, consumers who now get jobs or receive money could boost demand, and therefore employment, and ultimately consumption and economic growth. This set of "simple" and "intuitive" correlations defined the basis of Keynes economic thinking. Graphically, economic thoughts were presented under the so-called IS/LM supply model; where IS stands for investment savings and LM for liquidity money (see Figure 1.33).

In this model we can see two major lines: the IS that reflects the equilibrium in any given society between total private investment and total savings, and assumes that what you do not save you consume or invest, and the LM curve that shows the combinations of interest rates and levels of real income for which the money market is in equilibrium. The LM function is the set of equilibrium points between the liquidity preference or Demand for Money function and the Money Supply Function, as determined by banks and central banks.

According to Keynes's model, when governments increase their spending versus their saving ratio, that moves their IS line to the right (IS2), creating a new equilibrium, with a higher GDP or Y growth (Y2) and higher interest rate (i2). In other words, higher government spending could stimulate economic growth. In contrast, interest rates will increase, a fact that will take away some of the private investment from the economy, as investors will prefer to put that money to work in deposits. However, Keynes argued the potential negative impact on economy growth derived from higher interest rates could be offset by the fact that the economy will create larger consumption. This model is a very simple representation of the economic fundamentals used by most governments around the world to resolve the current economic crisis, but certainly provides a basic background to understand why that has happened and why in the future governments may continue using this kind of approach to solve similar situations.

Under these assumptions we believe Government's critical role in the economy will not only stay but will probably increase. Government's capacity to support the economy when it is most needed, in combination with its already large size make us believe companies will need to learn how to work better together.

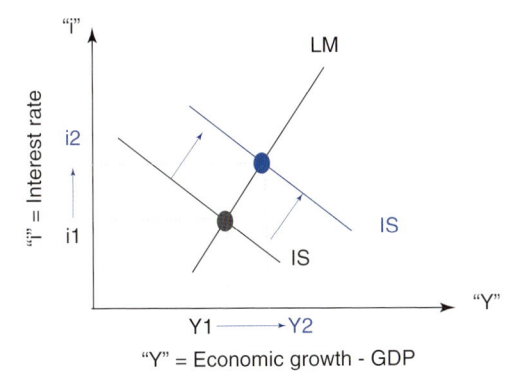

Figure 1.33 IS/LM model. Source: John Maynard Keynes – "General Theory of Employment, Interest and Money"

Looking into the future and combining all these political trends, the question is how all of them will shape the world by 2050.

Certainly we could envision a greater role from governments in society, not only as they try to stimulate and keep economies strong, but also as they try to counterbalance the continuous growth of too large corporations. We can also envision a new economic order where the BRICs and emerging economies will not only have more economic power but also more international and political power and representation. In the next decades we will move towards a tremendous shift in power from the ADV economies to the rest of the world. A shift that will become especially visible across the multiple international organizations, where indeed new ones will also be created. As companies will keep growing, becoming enormously large, much larger than many countries, their rights and obligations will also increase, and they should be conscious of this. Ultimately, in a much connected, intelligent, and democratic world, government and business decisions will become not only more transparent to society but they will be subject to a much higher scrutiny and accountability. Finally, and even more importantly, social media will have the power to catalyze not only more democracies around the world, but actually, better ones.

So, if we aggregate all these changes we can foresee a world where governments, BRICs, and megacorporations will increase their power, but citizens will also have a much stronger voice. Citizens currently express their powers in society by voting for governments and consuming the products of companies. Social media and the internet now allow citizens to express their power by sharing and speaking, affecting governments' and companies' reputations. In a non-distant future citizens will not only have the option to vote more frequently on the most crucial topics in their societies but they will also be able to express their opinions, debate, and share them across the net, influencing millions of citizens around the world.

In this world, companies and governments will need to learn how to work and engage with citizens again, as more than ever they will dictate their success. Please see Figure 1.34 with the new Engagement Rules, where social media and the internet will accelerate the exchange of knowledge, while enhancing citizens' and society's power over governments' and companies' actions and performance.

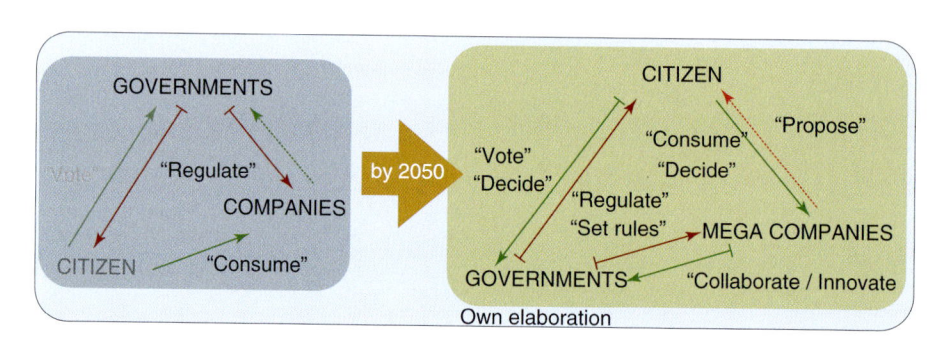

Figure 1.34 New engagement rules. Own elaboration.

In a much more flat and interconnected world more and better democracies will be created and new forms of government may need to appear at local, national, and international levels. At international levels we will see how the BRICs and emerging economies will increase their power and levels of representation, but forums of governments will need to be created. Forums where not only governments but also companies, NGOs, and citizens could have an equal say.

Could we foresee organizations where not only countries but also companies and citizens could discuss together? A potential example of this could be the World Economic Forum where companies, governments, international organizations and key citizens can gather to discuss the status of the world in all its aspects – from political issues to economic or social ones. The World Economic Forum is certainly a very interesting and unique type of organization. Could our world need more of these? How could these work in the future? How should the balance of power be distributed? Who should join? These are questions that will certainly open the door to tremendous controversy and debate, not a very simple idea but it may be necessary.

Political megatrends will bring significant challenges and opportunities for the chemical industry and its companies. By 2050 the industry will still need to improve its way of working, communicating, and engaging with governments, citizens, and with other general stakeholders around the world.

1.4
Energy Megatrends

"Energy transitions occur even when the incumbent is still available, and the new one is more efficient, cheaper, cleaner, and sustainable"

Necessity, who is the mother of invention.

Platon, The Republic

For centuries, energy supply and demand, energy prices, and energy trends have been the subjects of passionate and intense discussions. However, that statement has never been more obvious than during the last century; where fossil fuels, and especially crude oil, enabled the largest period of economic growth and prosperity in human history. New energy sources and technologies, energy availability, uncertainty, and scarcity, have triggered confrontations and even wars among countries.

Moving forward and as our world is poised to face further decades of spectacular economic and population growth, energy demand will keep rising, remaining a key factor for the prosperity of our world and society. Questions like, how much extra energy will our world need to enable the expected growth? or Are our existing energy sources enough to cover that growth? and At what prices? What new energies will be needed? and What could be the impact of this extra energy demand on our environment? are some of the critical questions that will intrigue our society and industry. The understanding of the different energy sources and their alternatives will not simply be a critical subject, but a vital one. As our world will need to address converging and conflicting megatrends, the right energy selection, will determine the long term sustainability of our world and of the chemical industry as we know it today.

The "recent" emergence of climate change as a key megatrend, will also have a tremendous impact on any energy mix decision for our world and our industry. Climate change will expand and complicate the scope of the energy discussion. The debate will move from the traditional short term supply and demand, prices, and long term availability, to whether the world can "afford" that energy from a climate change perspective? In order to answer some of these questions we have divided our answers across three different chapters.

In this chapter we will learn about some of the most recent and critical energy transitions, the reasons for these and their consequences. By reviewing and determining the reasons why, when, and how energy transitions occurred, we will gain insights and understanding on how and when the new transitions might occur. For that purpose we will also review some energy projections to 2030 and 2050 by the Organization of the Petroleum Exporting Countries (OPEC) and the IEA.

In this review we will focus on the major projections for oil and gas, including unconventional sources such as shale gas, as these are the major energy sources for the world and for the chemical industry. At this time we do not intend to discuss too much about the validity of these projections and the different scenarios used by these organizations, but just to introduce them. We also will not discuss here the potential impact of these projections on the chemical industry as we will discuss this in the following chapters. We will also not correlate these now with other important megatrends like economic growth or climate change, as we will discuss this in Chapter 2.

Here, we will just review the latest energy transitions and what we learn from them, while presenting some of the most relevant projections on energy demand. We will also provide an extract of some of the key supply and demand data for the different energies, extracted from the 2011 Statistical Review of World Energy from British Petroleum (BP).

In Chapter 2 we will present and discuss two scenarios on world energy demand based on current demand. In the first scenario, called *Business As Usual* or the unsustainable one, we will project the world energy demand as it could occur if climate change did not exist. In the second, the sustainable scenario, we will present how much energy demand could we afford if the world needs to avoid climate change. The first scenario will be based on our own projections, while the second will be based on the latest scenarios from the IEA.

Finally, in Chapter 5 we will review the impact of these two scenarios on the chemical industry. We will discuss how much energy could the chemical industry need under the BAU scenario, and how much could we actually afford under the sustainable scenario. We will also review the some of the potential impacts of shale gas on the chemical industry.

1.4.1
Recent Energy Transitions

For centuries the world energy mix has been fluctuating following the most basic principles of supply and demand. As different countries were confronted with different energy resources and energy requirements, as well as with different states of economic development, they have gone through different energy transitions. For illustration we will review the most recent energy transitions from some of the largest economies in the world, like the USA, Japan, China, or the UK.

Looking now at the USA and its latest major energy transitions, we can observe at least three different transitions since 1650 (Figure 1.35). From 1650 to 1900, wood and biomass were the preferred energy choices. Wood was widely available, cheap, renewable, and served perfectly to satisfy the energy needs of the country. However, with the arrival of the industrial revolution and the increased energy needs for industry the first large energy transitions occurred. The increasing requirements of the transportation industry, with the arrival of steam boats and trains, the textile industry, the metal industry, and latterly the electrical industry, set the ground for the transition to more efficient energy sources.

The high heat intensity of coal, 30 MJ kg^{-1}, versus wood, 15 MJ kg^{-1}, in combination with its wide availability and competitive price, triggered the first energy transition in the USA. However, this energy transition from wood to coal was much more than a simple energy transition, it implied the transition from biomass to fossil fuels (coal, oil, and gas) or, more importantly, from a renewable source of energy to a non-renewable one (Tables 1.11 and 1.12).

From a heat intensity point of view, as well as for other reasons like cost and availability, this transition made perfect sense. Fossil fuels in general can be regenerated in nature, however, they are considered as a non-renewable source

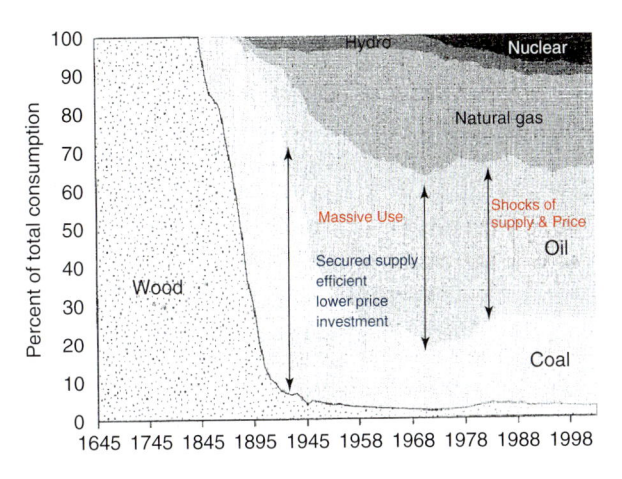

Figure 1.35 Long-term evolution of the US energy mix by percentage of total. (Source: Extract from; "A Thousand Barrel A Second" by Peter Tertzakian. Adapted from USA energy information agency and ARC financial.)

Table 1.11 USA long term evolution energy mix.

USA ENERGY TRANSITIONS		ENERGY	SOURCE	RENEWABLE
1st Transition	1650–1900	WOOD TO COAL	Biomass to Fossil Fuel	Renewable to non-renewable
2nd Transition	1900–1940	COAL TO OIL	Fossil Fuel	Non-renewable
3rd Transition	1940–2010	OIL TO GAS	Fossil Fuel	Non-renewable
		OIL & GAS to RENEWABLE	Non-Fossil Fuel	Non-renewable to renewable
4th Transition	20XX to ?	?	Non-Fossil Fuel	Renewable

Source: Own Elaboration based on an extract from a "Thousand Barrel a Second" by Peter Tertzatian. Adapted from Energy Information agency and ARC Financial.

Table 1.12 Energy–heat intensity.

ENERGY- Heat Intensity		MJ^{-1} per KG	Times Wood
Nuclear	NUCLEAR		25
Fossil Fuel	HYDROGEN	140	9
	GAS - Methane	55	3.5
	GAS - Propane	51	3
	GAS - Ethane	50	3
	PETROLEUM	46	3
	COAL - Anthracite	30	2
	COAL - Ignite	15	1
Biomass	WOOD	15	NA
	PEAT - Dry	15	1
	PEAT - Damp	6	0.5

Source: European Nuclear Society: Coal equivalent.
http://www.euronuclear.org/info/encyclopedia/coalequivalent.htm

of energy, as that process could take millions of years. Despite its non-renewable status, for many other obvious reasons, coal became a key component of the US and global energy mix since the beginning of the 19th century.

The second transition started at the beginning of the 19th century, with the transition from coal to crude oil, both fossil fuels. Crude oil has been known and used by humans for thousands of years. The Greek and Chinese civilizations, 2000 years BC used it already for illumination purposes.

The automotive and transportation revolution at the end of 1900 and the need for higher heat intensity and more easily transportable energy (crude oil is liquid), triggered the second large energy transition in the USA. The higher heat intensity of crude oil, 45 MJ kg^{-1}, versus anthracite and lignite, 30 and 15 MJ kg^{-1}, in combination with availability, competitive prices, and its liquid condition fostered this second transition and the beginning of the world oil dependence.

In the beginning, as with any new energy and technology, crude oil prices were tremendously high. As crude oil demand increased and the extraction technology started to improve, investments in supply followed. As supply started to improve, with large discoveries made in Persia, Saudi Arabia, and Kuwait, crude oil started to become a solid and reliable source of energy. Indeed crude oil during the 1930s and 1940s started to set the basis for a complete and successful energy transition from coal to oil (Figure 1.36).

The higher heat intensity of crude oil versus the incumbent technology, its more efficient and cheaper performance on a cost/energy basis, and its wide availability fostered this transition. Since the 1940s crude oil has become the preferred energy source in the USA and globally. Indeed, in 2010, crude oil accounted for 34% of the world energy demand, followed by coal with 30% and natural gas 24%, according to the BP 2011 World Energy Statistical Report.

During the 1950s and 1960s investments in extraction and additional capacity were made on a global basis, providing an expectation of unlimited supply. On the back of all these investments and this belief, industries, and especially transportation, grew very rapidly, to a point at which demand was gradually and constantly starting to exceed supply. World oil demand went from 5 million barrels a day in 1930 to almost 88 million barrels a day in 2010 (Figure 1.36).

During the first decades the speed of growth in demand was simply extraordinary, almost doubling every 10 years between 1945 and 1975. During that period crude oil demand went from around 7 million barrels a day in 1945 to 15 million in 1955, 30 million in 1965, and 60 million in 1977. Since the mid-1970s and with the different crude oil shocks, growth in demand has slowed, no longer doubling. Indeed, in 2010 world oil demand peaked at almost 88 million barrels a day, a mere 46% increase in 30 years.

The oil shocks of 1973, 1979, and 1980 had tremendous impact on the view of oil as the preferred world energy and the overall oil demand growth. The 1973 and 1979 shocks were supply shocks that increased oil prices massively (Figure 1.37).

The 1973 oil shock, also known as the *Arab Embargo*, implied a massive reduction in the supply of crude oil, tripling the price. The 1979 shock was created on the back first of the Iranian Revolution and then the war between Iran and Iraq in the 1980s. As a consequence, Iran reduced massively its oil supply and oil prices doubled, from $15 to 35 per barrel. Finally, and after a decade where oil prices increased by a staggering 1800%, from $2 in 1970 to $36 in 1980, the world economy entered into recession (Figure 1.36). Suppliers' expectation of an inelastic and ever growing demand triggered a massive reduction in demand, a huge oversupply, and the last of the first demand shocks. Prices collapsed, from $36 in 1980 to $14 in 1986 and 1988.

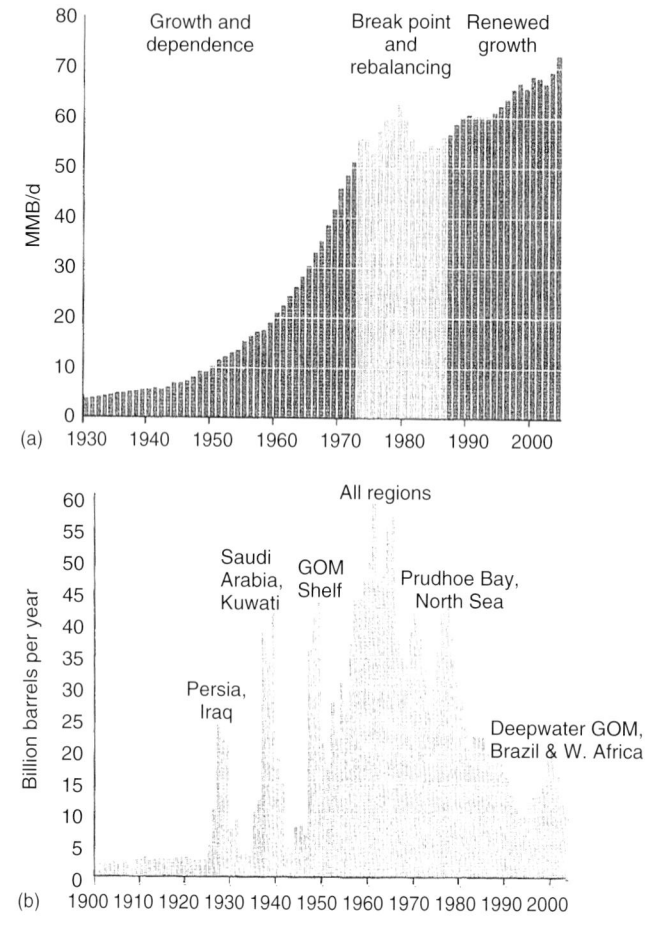

Figure 1.36 (a) Historical world oil demand. (b) Historical new oil discoveries. Long-term evolution of the US energy mix by percentage of total. (Source: Extract from; "A Thousand Barrel A Second" by Peter Tertzakian. Adapted from USA energy information agency and ARC financial.)

As is well known in economic theory, in the short term prices tend to resolve supply and demand imbalances. In the long term more structural changes occurred. Fundamentally, no industry can keep winning or losing money into the long term, and when that occurs, substitution and structural changes start happening. Thus the first two supply shocks triggered a massive search for oil alternatives, reduction in oil consumption, and more efficient ways to use energy. The USA started to reduced its crude oil production and gas started to be widely used as a complement and gradually an alternative to crude oil. The consideration of gas as a new alternative to crude oil, marked the third energy transition in the USA and globally. The crude oil shocks also served to stimulate other energy sources, and the use of nuclear energy, as a sustainable and renewable source, started to accelerate in 1970. However, a

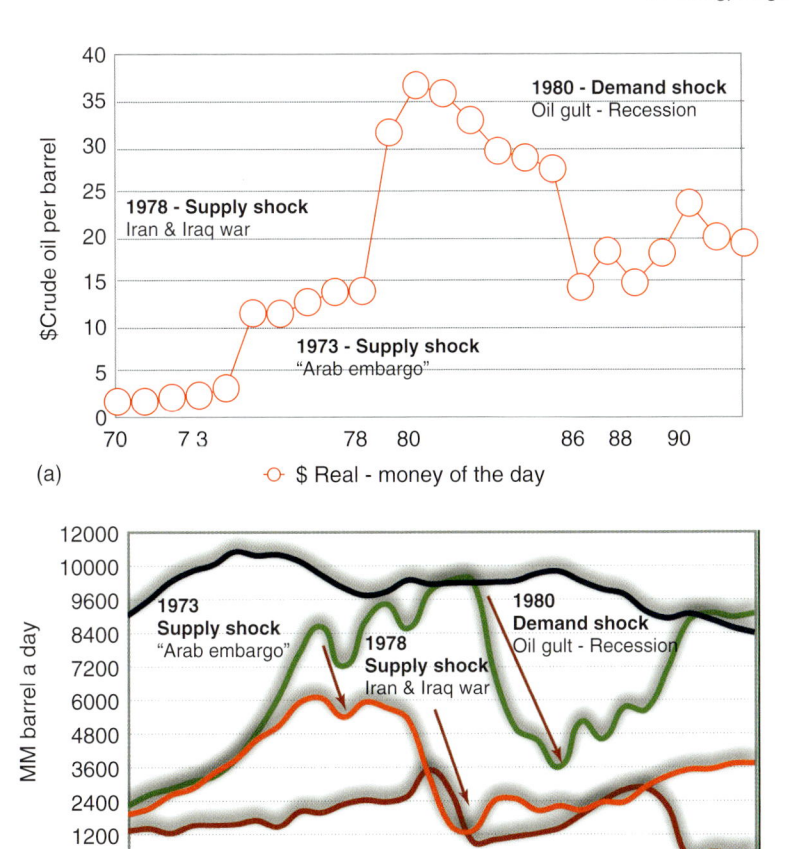

Figure 1.37 (a) Crude oil Brent prices. (b) Crude oil production. Source: Own elaboration and comments based on british petroleum (BP) – Review of the World Energy 2011.

nuclear accident in 1979 at Three Mile Island, Pennsylvania, temporarily tempered that growth in The USA. Russia's nuclear accident at Chernobyl in the mid-1980s, served to balance the strong growth that nuclear energy had enjoyed globally during the 1960s and 1970s.

As we briefly consider other key economies we can observe similar patterns, both in respect to energy mix transitions and in respect to their reaction after the crude oil shocks of the 1970s (Figure 1.38). After each of the crude oil shocks, and with the only exception of China, crude oil demand dropped gradually in the USA, Japan, and the UK. Indeed, after these shocks Japan and the UK never got back to the levels reached before the first oil shock. Both countries started to diversify their energy mix to other energy sources. Nuclear, gas, and coal were the selected alternatives for Japan, while natural gas and nuclear were the choices of the UK. In contrast, China and the USA, despite the crude oil shocks, have kept increasing

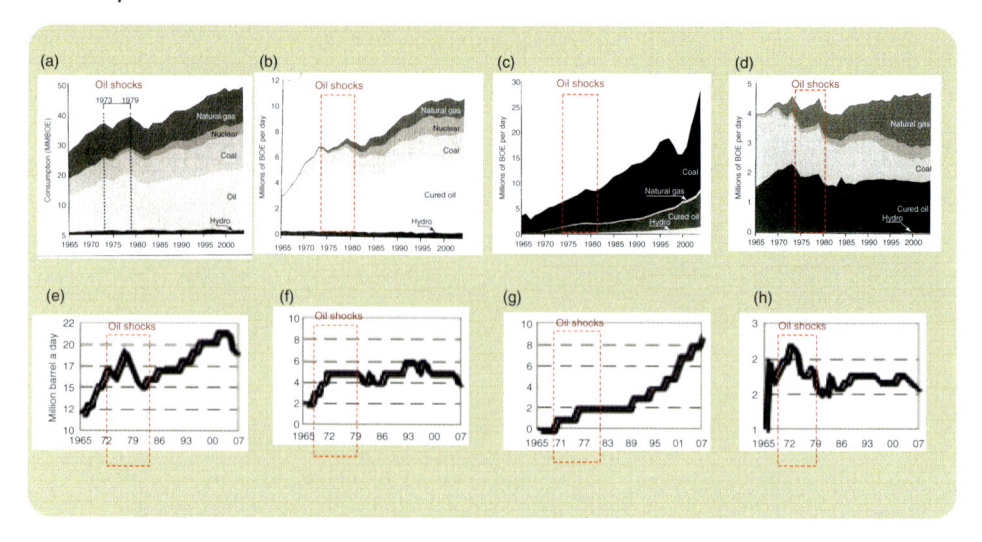

Figure 1.38 Energy mix: (a) USA, (b) Japan, (c) China, (d) UK. (Source: Extract from Peter Tertzakian, "A Thousand Barrels a Second". Adapted from USA Energy Information Agency and ARC Financial.) Oil demand: (e) USA, (f) Japan, (g) China, (h) UK. (Source: British Petroleum Statistical Review 2011.)

their crude oil demand until now, especially in the case of China. Indeed, with a much lower economic development and oil consumption than the rest of the selected countries, demand went from 1 to 10 million barrels a day over the last 40 years, with still potential for further increases.

In terms of the differences in energy mix and transitions across the selected countries and in general, these can be easily explained by historical and economic factors, including the degree of economic development of the different countries, as well as by the availability of these resources (Figure 1.38). As an example, the USA, Japan, and the UK transitioned their energy mix in a very similar way: from wood to coal in the 1800s and from coal to oil and natural gas in the 1900s. China, being at a lower stage of economic development during the last century, and with a much larger supply of coal, have still today not fully completed the transition from coal to crude oil and natural gas. China's demand for crude oil has skyrocketed during the last decades, going from 2 million barrels a day in 1990 to 8 million barrels a day in 2010. However, coal still remains a very large portion of their energy use and nuclear and gas remain very minor. Considering the expected growth potential in China for the following decades and its still very low crude oil demand on a per capita basis; crude oil and energy demand there may still grow spectacularly.

Before we continue to review the latest energy projections, we will summarize some of what has been learnt from the last energy transitions. Lessons that we could use in the future when considering the different energy scenarios by 2050.

1.4.2
Key Lessons from Recent Energy Mix Transitions
1) Energies have life cycles
2) Energies success criteria
3) Shocks are a valuable source of information
4) Transitions occur in life
5) Economics dictate energy: the "golden" rule
6) Transitions always occur, the question is when? – the oil "peak."

1.4.3
Energy Life Cycle

Energies like most commodity products, move through product life cycles – inception, growth, maturity, and decline. In the inception phase, technology is very inefficient, and costs and prices are very high. However, even then the energy has to comply with certain key criteria. If these criteria are met and the demand is there, the energy supply will start to grow. During the growth period, the energy costs decline very rapidly, reaching its trading price. At that competitive price the energy starts to have massive demand as well as massive supply. As demand and supply keep rising over time, prices tend to remain sustainable within reasonable price ranges for a long period. During these periods supply and demand imbalances can occur that can result in price shocks. Shocks that if not properly managed can start significant and lasting consequences for that energy, as we saw during the last oil shocks and we will review later. Ultimately, if the energy source is depleted or prices remain very high for a long period of time, alternatives will start blossoming. New technologies could foster new more efficient and competitive energies, and when that happens the existing energy starts to be replaced by the new one. In some cases the incumbent energy can disappear completely, while in others it might simply see its demand reduced and be used in more competitive applications. That happened with biomass and with coal, and might happen one day with crude oil. This last period is known as the *demise* or *sleeping period* of the energy. These phases can be easily recognized in any product, technology, and energy. Although the length of each of these phases can vary significantly, depending on the particular energy supply and demand of each country, in the end all energies will go through these transitions.

1.4.4
Energy Success Criteria

Three major criteria have been identified for any energy to become successful: availability, economics measured in terms of price versus energy efficiency, and sustainability. The new energies have to be widely available, on a sustainable basis and their heating and economics have to be more efficient than the incumbent ones. These simple principles have been followed for all energy sources during the last energy transitions and they are still expected to apply in the future.

We have all probably observed how some of the newest renewable energies, like solar, wind, and so on, do not yet meet all these criteria, delaying their introduction and massive use in the short term. In most of these cases, economics has been the major impediment to their success. However, the arrival of climate change might start changing the way we calculate these economics, as the abatement of green house emission will need to be factored in. Climate change might have the potential to improve the chances of some of the newest renewable energies, accelerating their introduction and outlook. The impact of climate change is something that should be considered when evaluating the long-term success of some of the newest renewable energies.

1.4.5
Shocks Are a Valuable Source of Information

During the last century we saw at least three major oil price shocks as we recently reviewed. These shocks changed forever the supply and demand behavior for some countries. Japan and the UK kept their crude oil demand stable despite increasing their energy demand over the last decades. Shocks are a tremendous source of valuable information, giving hints to the market about future outlooks and directions. Energy industries specifically, and all industries in general, have to pay significant attention to the root causes and resolution of these shocks, because after these data points may lie the future of that energy or industry.

1.4.6
Transitions Occur in "Life"

Indeed one of the most interesting lessons is the fact that energy transitions occur even when the incumbent energy is still in place. So energy depletion is not the only reason to migrate to an alternative energy. Indeed, energy depletion has never so far been the fundamental reason behind the latest energy transitions. A combination of efficiency, economics, long-term availability, and ecological concerns has been the major reason for transitions. The USA moved from wood to coal during the 19th century, and from coal to crude oil and gas during the last century for efficiency and economic reasons, not because of depletion. In the future, still not as close as we might think, crude oil will also be replaced, and perhaps despite its availability.

1.4.7
The Golden Rule – Economics Dictate Energy Transitions

New energies emerge when they provide higher energy value at the same or lower cost than the incumbent one. That happened in the last transitions and it seems to be a very solid argument when looking at comparative energy economics and feasibility. Please refer to Table 1.12 for the comparative heating value for the different energies.

1.4 Energy Megatrends

The other "golden rule" in any energy transition, is that when an energy emerges its cost and price tend to be at their highest level, as the new energy does not have the proper economies of scale, size, and technology. As those start to improve, and supply and demand increase, the cost and price of that energy tends to start to decline. Normally energy cost and price decline until a point at which entry into the market is possible. This is normally defined as the starting trading level.

It is worth noticing that, in 2008 and 2010, crude oil prices crossed the line of this golden rule. Indeed, in both cases crude oil in real terms reached the $110 a barrel level. That price, however, is similar to the nominal price of crude oil in 1864, when it reached a staggering $111 a barrel, or in other words when its nominal price was more than 10 times its real price (Figure 1.39). When something like that happens industries start wondering how can an energy have the same price in maturity, when the technology is solid and the economies of scale are maximized, as at its inception when the technologies were new and inefficient? Normally when this situation occurs, industries have tremendous incentives to start energy transitions. That is why this "Golden Rule" is another powerful indicator of the sustainability and outlook of any given energy. Economics has always needed to be measured in different contexts and alternatives, including its own energy price dynamics.

1.4.8
Transitions Always Occur, the Question Is When: The Oil Peak

Energy transitions have been occurring for centuries, the question is when and why they happened as well as what is the replacing alternative source. We do not know when fossil fuels, and specifically crude oil, could be replaced, but for logical reasons this should occur at some time. The fundamental question here is whether that time is near and do we have alternatives?

Looking at the figures provided by BP in the last 2011 Statistical Review of the World Energy, it seems that with the 2010 world oil demand level at 88 million

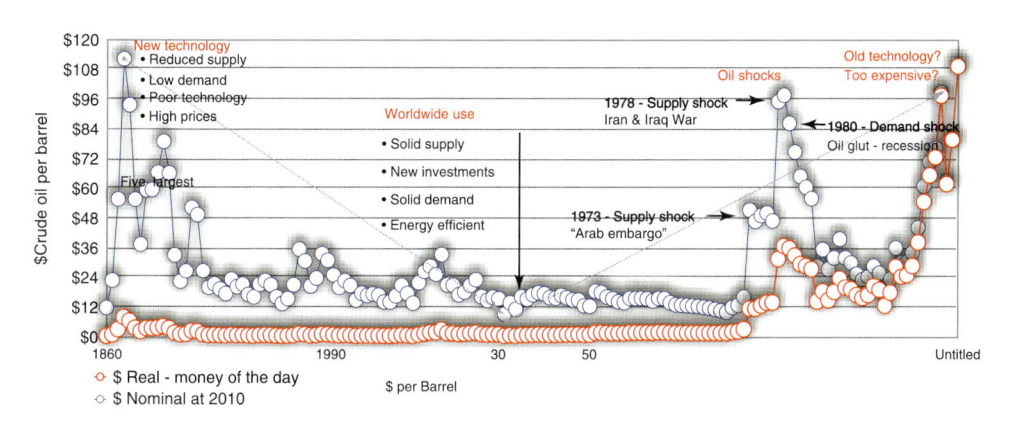

Figure 1.39 Historical crude oil Brent prices. (Source: British Petroleum Statistical Report 2011.)

72 | *1 Global Megatrends by 2050*

barrels a day and the recorded world proved oil reserves the world could consume oil at the 2010 pace for the next 45 years (Figure 1.40). As demand is expected to increase in the coming years, these numbers might need to be reviewed, however, it seems the decline or demise of crude oil does not seem very near. On the contrary, if we look at the annual refining capacity, it seems that during the last decades the increase in crude oil consumption has been larger than the increase in refining capacity, and indeed refining capacity could be a potential bottleneck to the growth ambitions of crude oil.

Therefore, the current concerns seem to be not about oil depletion but more about the available refining capacity to meet that demand in the short term and about the potential price oil will need to have in order to make use of the extra reserves.

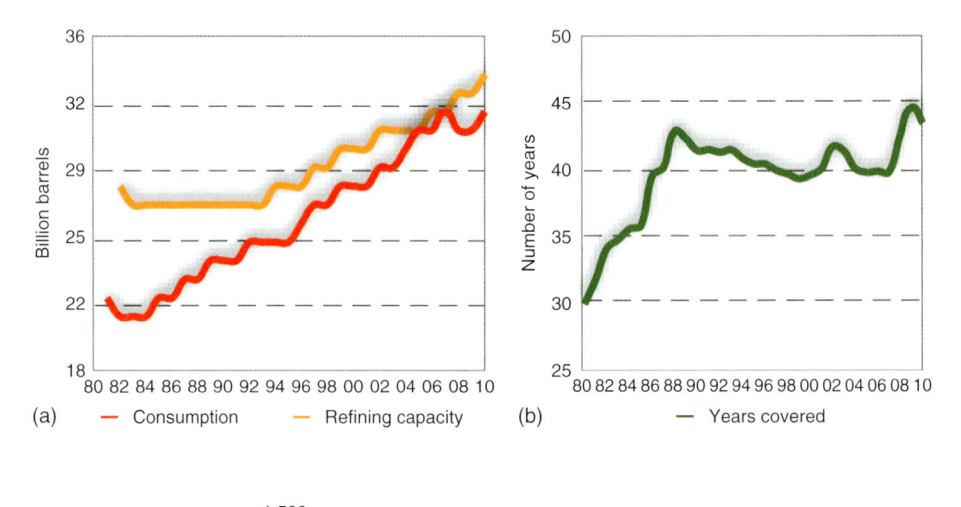

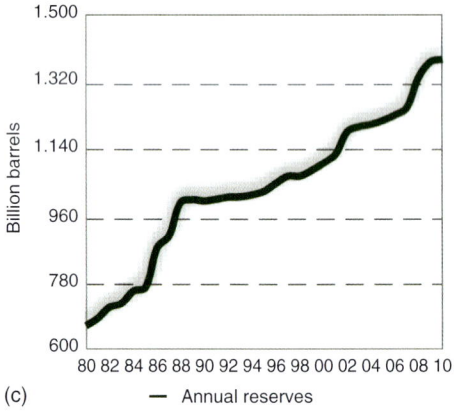

Figure 1.40 (a) World oil annual demand and refining capacity. (b) World oil proved reserves in number of years. (c) world oil proved reserves in billion barrels. (Source: British Petroleum Statistical Report 2011.)

On this second point we would like to illustrate the complexity of that topic by sharing two similar views from the IEA.

In Figure 1.41a, we can see the 2008 production cost of oil depending on the different sources around the world. The first thing to notice is that to fulfill the current levels of demand, around $80 million barrels, that could be done with relatively cheap oil. However, as we move above these levels, production costs start raising significantly.

Indeed there are clear differences in cost for the different oil sources. OPEC, FSU, and other conventional oil producers tend to enjoy very low production costs and, at current prices, also very large margins. However, as we enter the next levels, like deep water oil or Arctic oil, costs start rising sharply.

Therefore, crude oil might be available, but what is the price that will provide suppliers the right incentives to produce it? If a supplier would like to exploit the oil from the Arctic with the same margins that an OPEC supplier has at current oil prices (~$70 per barrel), oil from these regions will need to be priced at almost $160 a barrel. Is that sustainable?

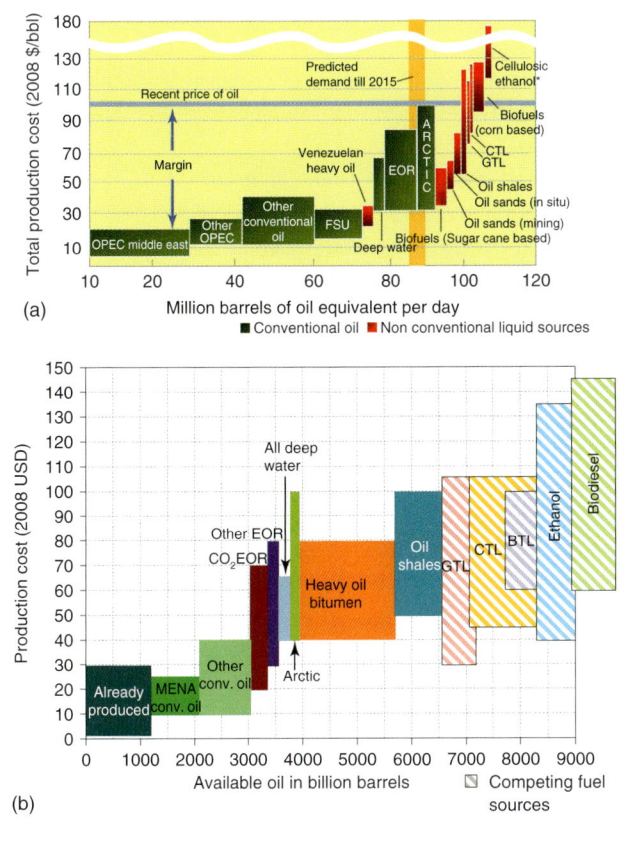

Figure 1.41 (a) Crude oil production cost globally. (b) Oil production cost curve (excluding carbon pricing). Source: International Energy Agency.

Along these lines and in a recent study from the IEA, called "Resources to Reserves 2012", we can see how at production cost levels above $40 we can have alternative sources to oil, so a crude oil from the Arctic at $160 might never need to happen.

Therefore, and to conclude this section on oil, it seems that demand will keep increasing, as will supply. Refining capacity might become a concern in the short term if investments are not made on time, and at the current record high prices oil will soon be above its inception price (Figure 1.39). In the medium to long term, the question is not whether we will have enough reserves to cover similar levels of demand as we have today, but whether the prices of the next barrels of oil will be economically feasible for suppliers and for consumers, bearing in mind that the price levels that the last barrels will need to have will be completely uncompetitive versus alternative solutions. Let us now move to the energy projections from the OPEC, and especially for oil and gas.

1.4.9
The Oil Peak – M. King Hubbert

This review of the latest energy transitions would not be complete without referring to perhaps one of the most famous and controversial energy forecasts from the last century, the so-called "Oil Peak."

In 1956, an oil engineer and geophysicist from Shell Oil, M. King Hubbert made two major interesting predictions in relation to crude oil. First, he predicted that during the period 1965–1970 the US economy would start a decline in oil production after reaching a peak at 10 million barrels, as their natural reserves had been rapidly reduced during the previous decades (Figure 1.42).

Interestingly enough that prediction occurred after the two oil shocks and since the 1980s the crude oil production has been declining over the decades, never reaching again the original peak of 10 million barrels.

The second and more controversial one was when he predicted that by 2010, or when the world reached the mark of 80 million barrels a day, crude oil would reach its production peak. He argued that at the observed and forecast demand levels, the level of proved reserves would not be enough to support that growth, provoking large price increases and ultimate decline in supply. Interestingly enough, in 2010, we surpassed the 80 million barrels a day demand and nothing happened, but we did get some of the highest oil prices ever.

Indeed, and as previously explained, in Q4 2008 and Q4 2011, oil prices reached $98 and $111 per barrel, close to the nominal price at its inception in 1968 at $111 per barrel.

According to Hubbert when that will happen we should see clear incentives from consumers for more efficient alternatives. So this prediction still needs to be fulfilled, oil supply remains stable although prices indeed reached record high levels and crossed the golden rule of energy. Let us see how this prediction develops.

1.4 Energy Megatrends | **75**

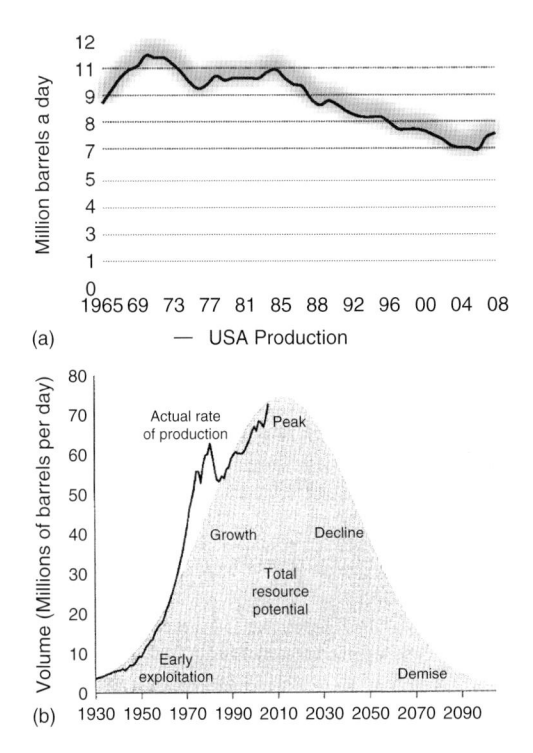

Figure 1.42 Hubbert predictions: (a) USA oil peak. (Source: British Petroleum Statistical Report 2011.) (b) World oil peak. (Source: Extract Peter Tertzakian, "A Thousand Barrels a Second". Adapted from USA Energy Information Agency and ARC Financial.)

1.4.10
OPEC – Energy projections to 2030

After reviewing and summarizing some of the major lessons from our recent history and the last energy transitions, we will review the *long-term world energy demand by 2030 according to OPEC, with special attention on oil and gas* as the major sources of energy for the chemical industry. Finally, in Chapter 2 we will also review in detail what could be the potential energy demand for the world and what our world can actually afford, when we start correlating economic growth, energy demand, and climate change.

According to OPEC the demand for all energy sources will continue to grow significantly in the next 20 years. Energy demand is expected to grow from around 11 million barrels of oil equivalent (MBOE) of energy in 2010 to almost 16 MBOE by 2030 (Figure 1.43).

This increase will be based on the unprecedented growth of the world economy, and most specifically of the BRIC countries, please see Figure 1.44.

Looking at the 2007 world crude oil per capita consumption and taking into account the tremendous growth expected in the BRIC and REST groups, a very

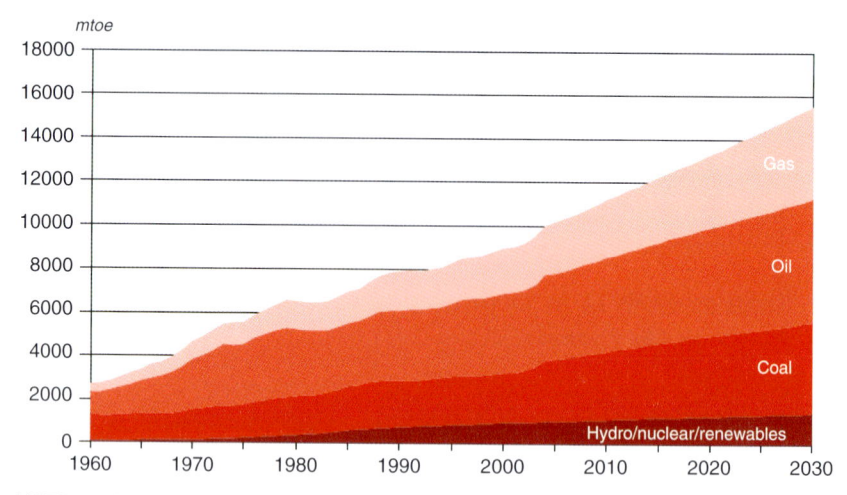

1 Global Megatrends by 2050

Figure 1.43 World energy demand by 2030. Source: Organization of the Petroleum Exporting Countries - World Oil Outlook 2009.

NOTE: MBOE stands for million barrels of equivalent

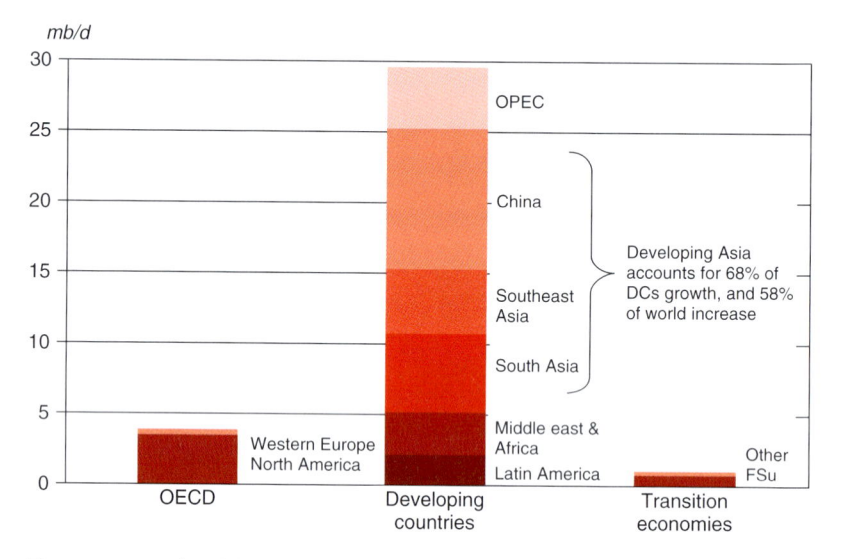

Figure 1.44 Crude oil demand growth by 2030. (Source: Organization of the Petroleum Exporting Countries – World Oil Outlook 2009.)

high oil demand can be anticipated. Indeed, the BRIC economies currently consume one quarter of what the ADV economies consume (Figure 1.45).

The USA and Canada consume around 20 barrels of oil annually per capita. Europe consumes half this amount while China consumes around 4 to 5 barrels.

The expected economic growth in the BRIC economies, in combination with a very low consumption on a per capita basis will accelerate world oil demand.

1.4 Energy Megatrends | 77

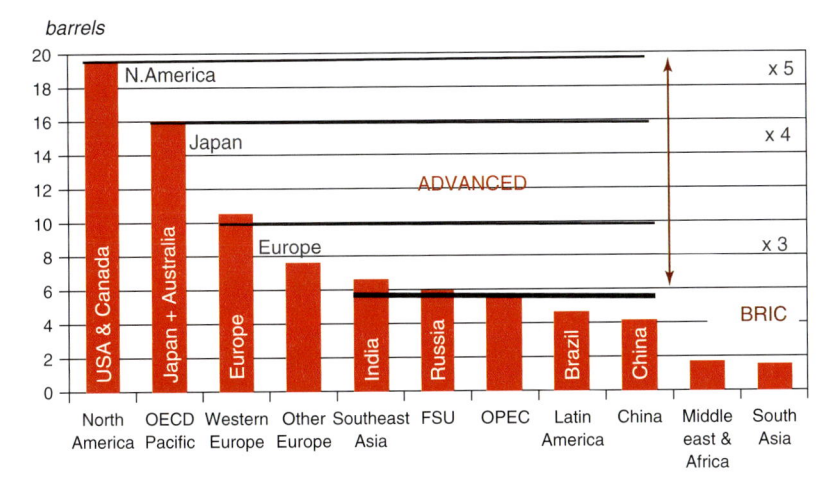

Figure 1.45 Per capita oil consumption. (Source: Organization of the Petroleum Exporting Countries – World Oil Outlook 2009.)

Table 1.13 World crude oil demand 2005–2030.

	2005	2010	2015	2020	2025	2030
North America	25.5	26.1	26.9	27.7	28.4	29.0
Western Europe	15.5	15.6	15.8	15.9	15.9	15.8
OECD Pacific	8.6	8.6	8.6	8.6	8.6	8.5
OECD	49.6	50.3	51.3	52.2	52.9	53.4
Latin America	4.6	5.0	5.5	5.9	6.4	6.8
Middle East & Africa	3.0	3.4	4.0	4.6	5.2	5.9
South Asia	3.1	3.9	5.0	6.1	7.3	8.6
South-East Asia	4.4	5.2	6.1	7.1	8.0	9.0
China	6.5	8.7	10.4	12.3	14.3	16.4
OPEC	7.4	8.2	9.1	9.9	10.8	11.7
DCs	29.0	34.5	40.0	45.9	52.0	58.5
FSU	3.8	4.0	4.2	4.3	4.5	4.6
Other Europe	0.9	0.9	1.0	1.0	1.0	1.1
Transition economies	4.7	4.9	5.2	5.4	5.5	5.7
World	83.3	89.7	96.5	103.5	110.4	117.6

Source: Organization of the Petroleum Exporting Countries – World Oil Outlook 2009.

Indeed the major increases in crude oil demand are expected to come from the BRIC economies, elevating the world crude oil demand from 83.3 million barrels in 2005 to 117 million barrels in 2030, where China alone will add almost 10 million barrels during the next 20 years (Table 1.13).

1 Global Megatrends by 2050

According to OPEC, the sectors where crude oil demand will grow fastest will be transportation and industry. As before, this growth will be sustained by the economic growth of both the BRIC and REST economies (Figure 1.46). Looking at the transportation sector and using the number of cars per 1000 inhabitants as an indicator of future oil demand, the numbers cannot be more compelling (Figure 1.47).

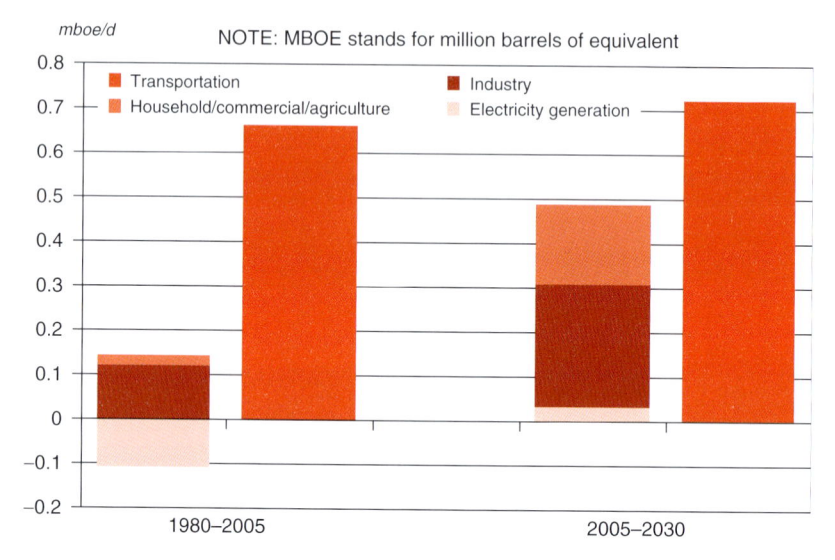

Figure 1.46 World energy demand by 2030. Source: Organization of the Petroleum Exporting Countries – World Oil Outlook 2009.

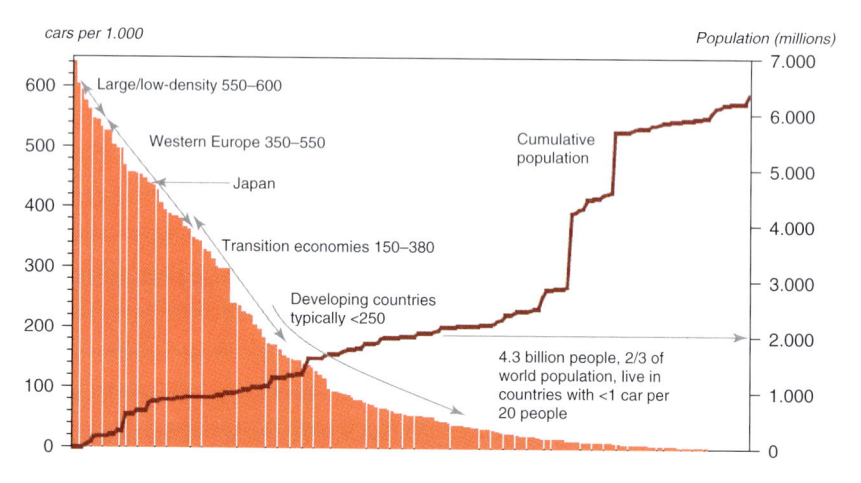

Figure 1.47 Per capita car ownership. (Source: World Road Statistics, International Road Federation (various editions), OPEC Secretariat estimates.)

1.4 Energy Megatrends

In 2007 the ADV Group had a ratio of cars per 1000 inhabitants 7–12 times higher than that in the BRIC economies and 20 times that of the REST group. For instance, the USA has around 700 cars per 1000 people, while China has just 70 cars per 1000 people. This certainly shows a tremendous potential for car consumption from the BRIC and REST economies (see Figure 1.47).

Massive growth in the car and transportation industries is one of the key expected developments fueling the increase in demand for crude oil, especially from the REST/BRIC economies. Electric cars will start to become very popular, especially in the ADV economies, however, they will remain relatively minor on a global basis.

Industry will continue to be the largest driver of energy demand. That was the case during the last 25 years, from 1980 to 2005, and will be the case also for the next 25 years until 2030. As the world becomes wealthier and population continues to rise, demand for energy and for more industry will increase.

All in all, most of the energy demand will come from what the OPEC defined as developing economies, mainly BRICs. The ADV economies will also increase, but with all the improvements in energy maximization, that demand will be much less, see Figure 1.48.

Certainly OPEC projections seems very solid and conservative indeed, as the unprecedented rate of economic growth expected in the world economy will certainly fuel a massive growth in energy consumption. However, that demand will vary significantly depending on the country and industry of origin of the demand. As discussed above, different countries today present different energy models, resulting from a large array of factors: starting from the price and their own availability of the different energy resources, the stage of their economic development, and social preferences.

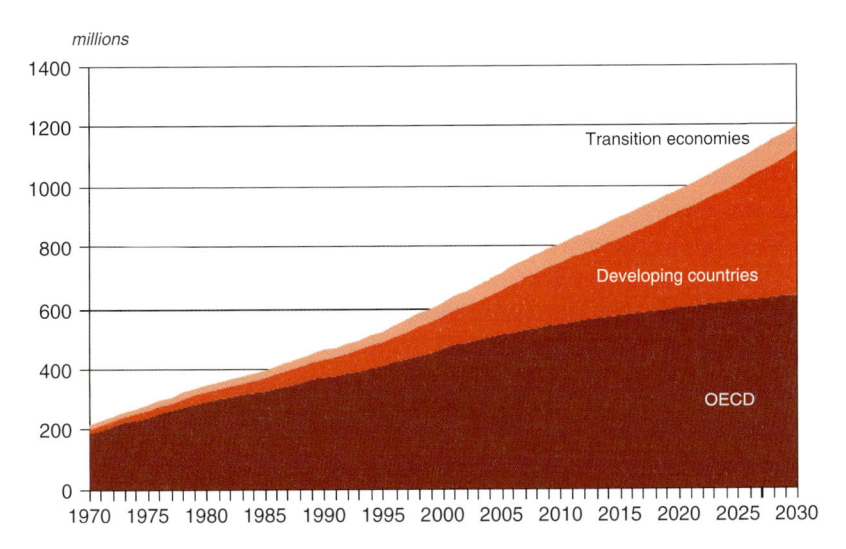

Figure 1.48 Energy demand by area. (Source: Organization of the Petroleum Exporting Countries – World Oil Outlook 2009.)

The ADV economies have already enjoyed significant periods of wealth and growth, and they are currently facing high oil prices in combination with strong social and political pressures for environmentally friendly and clean energy sources. Climate change indeed might have the potential to significantly curb the demand for crude oil and fossil fuel in the years to come.

The ADV economies especially, and the world in general, will have huge incentives to foster massive research on alternative, cheaper, and "greener" sources of energy. Solar, wind, biomass, hydrogen, nuclear energy, or new energy sources will need to take the lead in the energy future. Crude oil, despite high price levels, will certainly reach its production – and more importantly, demand – peak, since all the ADV economies and their transportation industries will have gradually moved to cleaner, cheaper, and more efficient energy sources. The BRIC economies might need to rely for longer on crude oil, emulating the energy mix path experienced by the USA and Europe but with 20–30 years delay. Crude oil will then complete its energy cycle, being gradually substituted by other more efficient, alternative energy sources. Migration will vary across countries and industries depending on its regional supply and demand balance and their stage of development.

In Chapter 2, we will present two scenarios for crude oil demand. One based on the potential crude oil demand if the world continues with BAU, and a second, from the IEA, on how much crude oil and energy demand can the world afford if we want to avoid climate change. For additional information please see Figure 1.49.

1.4.11
Recent Developments

Finally, we would like to reflect on two others aspects that might have an impact on the development of energy supply over the next decades.

First, we would like to consider the potential consequences for the nuclear energy industry after the March 2011 earthquake in Japan and the subsequent nuclear accident at the Fukushima nuclear plant.

Secondly, and especially critical for the chemical industry, we would like to review the recent developments in the natural gas industry, more specifically the development and potential of shale gas in the USA and globally, as well as its potential consequences for the chemical industry.

1.4.11.1 Nuclear Energy – The Aftermath of Fukushima

On Friday, 11 March 2011, Japan suffered its most powerful earthquake ever, and one of the five most powerful earthquakes in the world since 1900, with a magnitude of 9.0 on the Richter scale. The earthquake hit Japan on the pacific coast, triggering a powerful tsunami, with waves that reached 40 m high, and traveled up to 10 km inland in certain areas of Japan.

The tsunami caused several accidents, including some nuclear ones, the most notorious being that in the Fukushima nuclear plant. Several reactors were affected and, according to the Japanese Government, the accident reached a level 7 on the nuclear event scale. This is the highest level of severity according to the

1.4 Energy Megatrends | 81

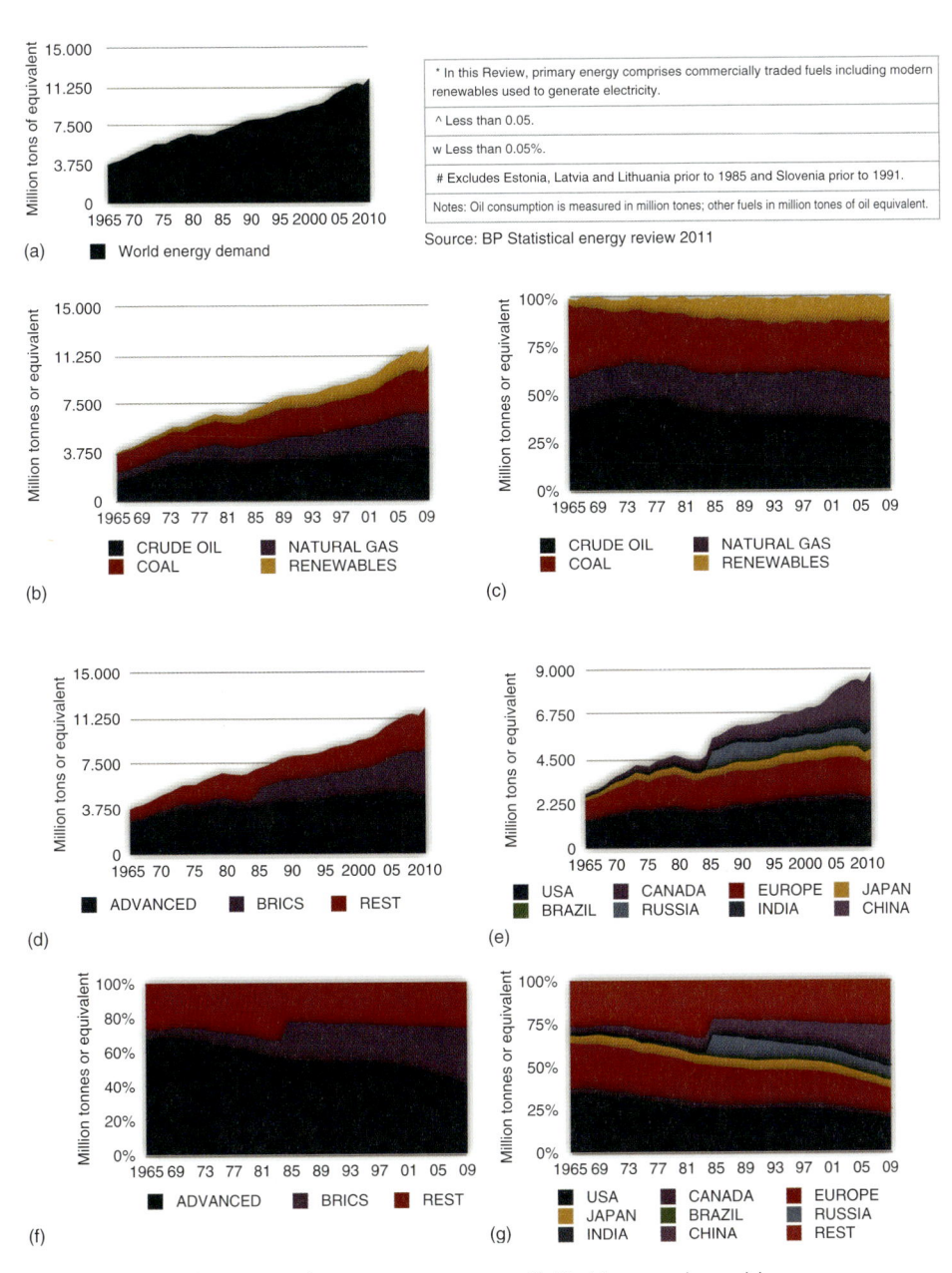

* In this Review, primary energy comprises commercially traded fuels including modern renewables used to generate electricity.
^ Less than 0.05.
w Less than 0.05%.
Excludes Estonia, Latvia and Lithuania prior to 1985 and Slovenia prior to 1991.
Notes: Oil consumption is measured in million tones; other fuels in million tones of oil equivalent.
Source: BP Statistical energy review 2011

Figure 1.49 World energy product mix 1965–2010. (a) World energy demand. (b) World energy demand by energy. (c) World energy demand by energy expressed as a percentage of the total. (d) World energy demand by area. (e) World energy demand by country. (f) World energy demand by area expressed as a percentage of the total. (g) World energy demand by country expressed as a percentage of the total. (Source: BP Statistical Energy Review 2011.)

82 | *1 Global Megatrends by 2050*

International Atomic Nuclear Agency (IANE). Please see the International Nuclear and Radiology Event Scale (Figure 1.50).

The accident triggered the relocation of thousands of residents in peril of being affected by high levels of radiation, radiation that could cause severe damage to the health of the people, and even death in the case of direct exposure. Residents within a 20 km radius were evacuated. In addition, the USA recommended that its citizens evacuate up to 80 km from the plant. The tremendous drama and suffering related to this accident put millions of people in Japan and globally in a televised drama for several months.

The nuclear accidents at Fukushima in Japan, and Chernobyl in Russia, are the only two accidents that have reached level 7 on the Nuclear Event Scale of the IANE. The severity, length, and enormous risk of the accident, in combination with the tremendous difficulties presented to the government and Fukushima's owner to control the accident, triggered a massive debate about the value and sustainability of nuclear energy in the future.

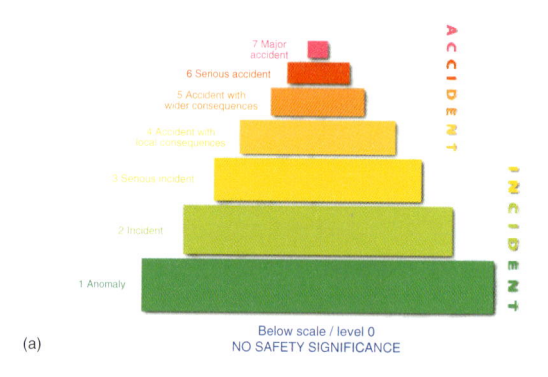

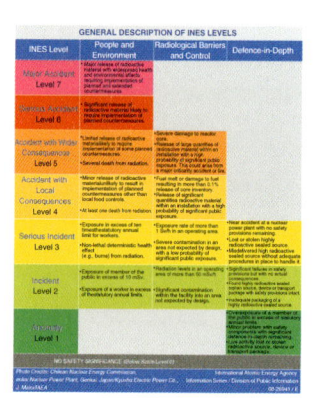

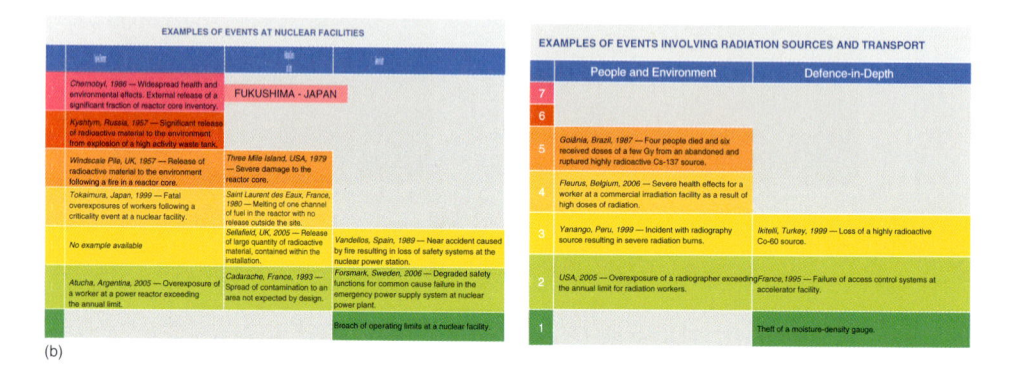

Figure 1.50 (a) International nuclear and radiology event scale. (b) Large nuclear accidents and incidents globally. (Source: *http://www.iaea.org/Publications/Factsheets/English/ines.pdf*).

Despite the fact that the circumstances surrounding the Fukushima nuclear accident were of the most extraordinary nature and almost impossible to be replicated again, the negative perception of nuclear energy seemed to be very solid after Fukushima. During Fukushima's crisis, Japan dealt with a unique combination of all worst case scenarios: dealing with the largest earthquake in Japan's history and one of the five largest earthquakes in human history, a huge tsunami impacting a nuclear facility, large human and material casualties, including severe disruptions to the infrastructure, energy, and communications.

Furthermore, if we look at the list of nuclear accidents, it seems quite clear that nuclear energy has been very safe during the last 50 years, with just six accidents, two level 7 (the highest), one level 6, and three level 5 (Figure 1.50).

However, after the Fukushima accident, as occurred during the period immediately after the Chernobyl accident in Russia and the Three Mile Island accident in the USA, the sentiment around nuclear energy sank to its lowest levels, questioning again the future and outlook of nuclear energy for our world.

Fukushima triggered a wave of concerns on the sustainability and the real value of nuclear energy. Concerns that concluded with eight of the oldest reactors in Germany being shut down and the official declaration from the German and Swiss Governments to move out of nuclear energy by 2020 and 2034, respectively. Italy stated they would have no additional reactors, while China scaled down its large plans to move into nuclear energy by almost 20%.

After Chernobyl, many governments around the world expressed their concern and their plans to reduce their dependence on nuclear energy, however, China and others have, until very recently, still been pushing for additional reactors. Each time we have experienced major accidents nuclear energy growth has slowed in the respective countries, that was the case in the USA, Russia, Japan, and the UK. Only the USA and France, with the absence of major accidents during the last 30 years, now seem to have stable and moderate growth, see Figure 1.51.

Therefore, the question now is what will be the impact of the Fukushima accident on the future of nuclear energy, especially after the subsequent declarations and new intentions toward it from Germany, Switzerland, Italy, and China. OPEC, the IEA, and the IANE, as we will see later, still expect a very bright future for nuclear energy. Indeed, the expected increase in economic growth and energy demand, in combination with the need to reduce the world fossil fuel consumption in order to face climate change, might be a very supportive trend. On the other hand, the recent and sad events of Fukushima might have changed that outlook.

The questions to be answered in the following decades are: Will the recent events change the course of nuclear energy? Can our world meet the expected growth in energy demand without nuclear energy? Will climate change foster the use of nuclear energy as an alternative to fossil fuels? We will answer some of these questions in Chapter 2.

84 | *1 Global Megatrends by 2050*

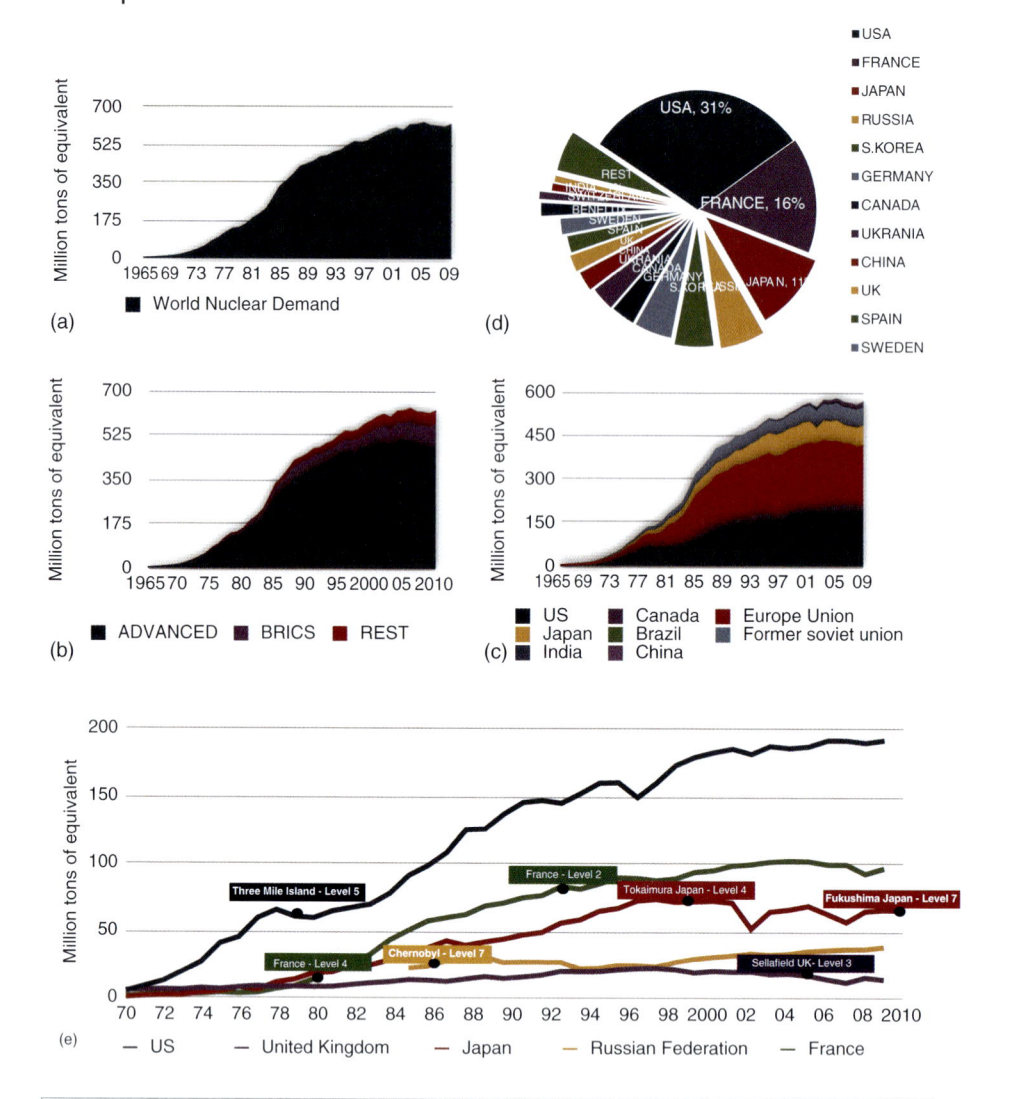

Figure 1.51 World nuclear appendix 1965–2010. (a) World nuclear demand. (b) World nuclear demand by area. (c) World nuclear demand by country. (d) Largest nuclear consumers 2010. (e) Countries with major events (4 to 7) at nuclear facilities since 1970. (Source: BP Statistical Energy Review 2011.)

1.4.11.2 Shale Gas the "Game Changer" – Natural Gas the Energy of the Future

The recent discovery of large and accessible reserves of non-conventional natural gas, also known as *shale gas*, initially in the USA and then globally, has changed forever the supply and demand balance of natural gas.

As mentioned earlier, natural gas already accounts for almost 24% of the world energy demand in 2010, after crude oil with 34% and coal with 30%. Additionally, natural gas accounts for 26% of the energy demand in the ADV economies, 14% in the BRICs and 33% in the REST, see Figure 1.52. However, and as observed with crude oil, the distribution of supply and demand do not match. World natural gas demand is mainly concentrated in the ADV economies, while supply is mainly concentrated in the REST and BRIC countries, especially the Middle East and Russia (Figure 1.53).

The large discovery of shale gas will not only serve to expand the role of natural gas in our world energy mix but, perhaps even more importantly, will serve to increase supply where the demand exists. According to a 2011 study from the US Energy Information Administration, EIA on 32 countries and selected basins, the total reserves of technically recoverable shale gas in these countries amount to 6600 tcf (trillion cubic feet), while there is still potentially up to three to four times more reserves of gas not yet technically available. These huge available resources are equal to the existing proved reserves of natural gas and will serve to cover the world demand for another 60 years, giving almost 130 years supply in total at

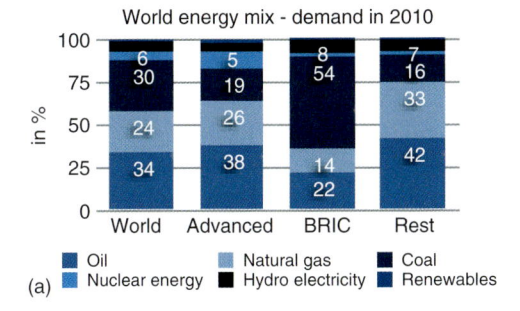

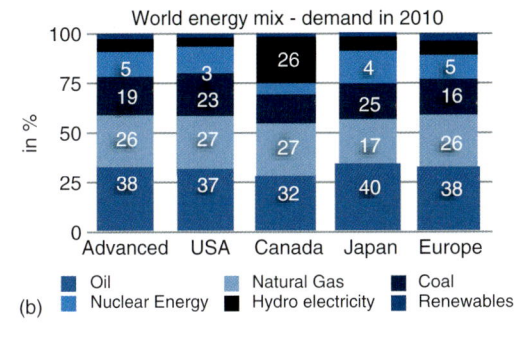

Figure 1.52 (a) World energy mix, demand in 2010. (b) ADV economies energy mix, demand in 2010. (Source: BP Statistical Energy Review 2011.)

86 | *1 Global Megatrends by 2050*

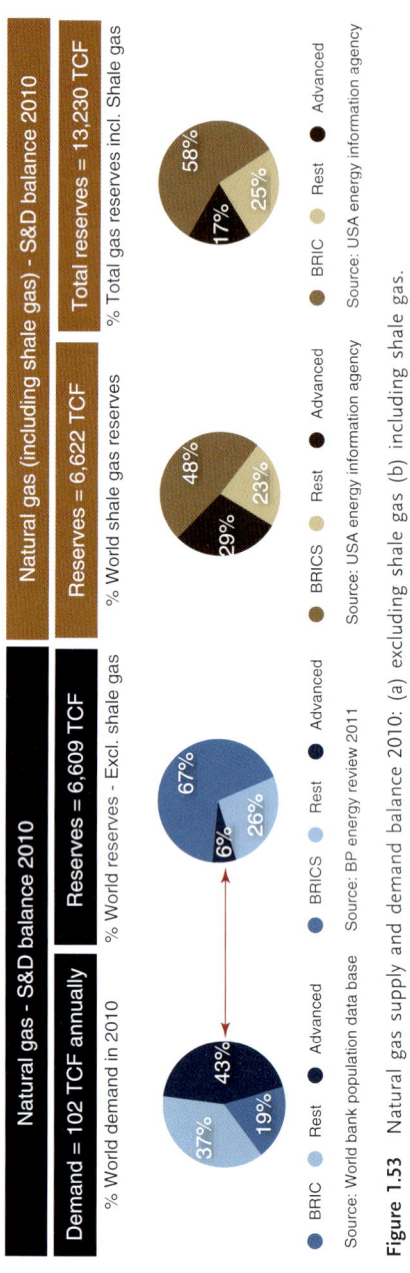

Figure 1.53 Natural gas supply and demand balance 2010: (a) excluding shale gas (b) including shale gas.

2010 world annual demand. Please see Table 1.14 listing all the major reserves of technically recoverable shale gas around the world.

Certainly the discovery of shale gas will change the outlook and future of natural gas globally. Until recently, the ADV economies accounted for a large part of the global demand, almost 43%, but only had 6% of the global reserves, the new discoveries of shale gas will serve to boost their reserves to almost 17%. Shale gas will serve to double the reserves of natural gas globally, but this is especially true in the ADV economies.

As we can observe in Figure 1.53 and Tables 1.14, 1.15 and 1.16, most of the technically recoverable shale gas appears to be in the REST and ADV economies. However, within the REST economies, new names will start appearing. Countries like Mexico, Australia, Argentina, Libya, Algeria, or the Sub-Sahara will start adding to the global supply. Within the BRIC, China with the largest expected reserves of shale gas globally, almost 1300 tcf, will be able to cover its demand for the next 400 years at its current demand rate.

Therefore, shale gas truly has the potential to change the supply and demand of natural gas globally, fostering demand and putting supply where it is most needed. Certainly the traditional gas producers will remain as large suppliers and the REST group will still keep 58% of the natural gas reserves, but in a world where demand for energy will increase in the next decades, especially in the BRIC economies, shale gas appears as a huge blessing for the energy industry.

Traditionally, natural gas has been the second most important source of energy after crude oil in the ADV economies, especially in North America and Europe, as coal was gradually replaced by oil and gas. The oil shocks of the 1970s and 1980s encouraged the US government to accelerate the introduction of natural gas into the energy mix. Indeed, during these decades natural gas demand grew very fast in the USA, increasing supply and ultimately increasing prices too, as we will see later. During the last decade prices of natural gas in the USA were becoming uncompetitive, until the arrival in 2007 of shale gas. In Europe, however, the reasons for the large use of natural gas are slightly different.

The traditional lack of local energy supply in Europe, and the efficiency and cleaner energy of natural gas fostered its introduction. Now the fact that Europe imports most of its energy, and especially natural gas, from abroad and especially from Russia, has made the discovery of shale gas reserves in Europe a very interesting milestone for the European energy industry.

In the USA, gas represents 27% of the energy supply, being the second largest energy supply after crude oil, with shale gas being around 14% of the 2010 USA natural gas consumption. Despite the development and introduction of shale gas in the USA being very recent, by 2010 it already accounted for almost 15% of the natural gas supply, highlighting its brilliant future in the years to come, and especially for its use in the chemical industry, although we will talk about this extensively when looking into the feedstocks of the chemical industry.

Unconventional gas that includes shale gas has been known and used in the USA for a very long time; however, it was not until 2007–2008 with the arrival of new drilling technologies that it became economically feasible. Horizontal drilling

Table 1.14 Estimated shale gas technically recoverable resources for select basins in 32 countries, compared to existing reported reserves, production and consumption during 2009.

	2009 Natural Gas Market[(1)] (trillion cubic feet, dry basis)			Proved Natural Gas Reserves (trillion cubic feet)	Technically Recoverable Shale Gas Resources (trillion cubic feet)
	Production	Consumption	Imports (Exports) (%)		
Europe					
France	0.03	1.73	98	0.2	180
Germany	0.51	3.27	84	6.2	8
Netherlands	2.79	1.72	(62)	49.0	17
Norway	3.65	0.16	(2156)	72.0	83
U.K.	2.09	3.11	33	9.0	20
Denmark	0.30	0.16	(91)	2.1	23
Sweden	–	0.04	100		41
Poland	0.21	0.58	64	5.8	187
Turkey	0.03	1.24	98	0.2	15
Ukraine	0.72	1.56	54	39.0	42
Lithuania	–	0.10	100		4
Others[c]	0.48	0.95	50	2.71	19
North America					
United States[d]	20.6	22.8	10	272.5	862
Canada	5.63	3.01	(87)	62.0	388
Mexico	1.77	2.15	18	12.0	681
Asia					
China	2.93	3.08	5	107.0	1275
India	1.43	1.87	24	37.9	63
Pakistan	1.36	1.36	–	29.7	51
Australia	1.67	1.09	(52)	110.0	396
Africa					
South Africa	0.07	0.19	63	–	485
Libya	0.56	0.21	(165)	54.7	290
Tunisia	0.13	0.17	26	2.3	18
Algeria	2.88	1.02	(183)	159.0	231
Morocco	0.00	0.02	90	0.1	11
Western Sahara	–	–		–	7
Mauritania	–	–		1.0	0
South America					
Venezuela	0.65	0.71	9	178.9	11
Colombia	0.37	0.31	(21)	4.0	19
Argentina	1.46	1.52	4	13.4	774
Brazil	0.36	0.66	45	12.9	226
Chile	0.05	0.10	52	3.5	64
Uruguay	–	0.00	100		21
Paraguay	–	–			62
Bolivia	0.45	0.10	(346)	26.5	48
Total of above areas	53.1	55.0	(3)	1274	6622
Total world	106.5	106.7	0	6609	

Source: [a]Dry production and consumption: EIA, International Energy Statistics, as of March 8, 2011.
[b]Proved gas reserves: *Oil and Gas Journal*, Dec., 6, 2010, P. 46–49.
[c]Romania, Hungary, Bulgaria.
[d]U.S. data are from various EIA sources. The proved natural gas reserves number in this table is from the U.S. Crude Oil, Natural Gas, and Natural Gas Liquids Reserves, 2009 report, whereas the 245 trillion cubic feet estimate used in the Annual Energy Outlook 2011 report and cited on the previous page is from the previous year estimate.

Table 1.15 World natural gas reserves 2010.

NATURAL GAS (incl. Shale gas)	Annual Demand (Trillion Cubic Feet)	Proved Reserves (Trillion Cubic Feet)	Proved Reserves Number of Year Covered Demand	Technically Recoverable Shale Gas	Shale Gas Number of Year Covered Demand	Combined Reserves Number of Year Covered Demand
USA	22.8	272	12	862	38	50
EUROPE	17.1	189	11	639	37	48
CANADA	5.63	62	11	388	69	80
BRASIL	0.66	12.9	20	266	403	423
INDIA	1.87	37.9	20	63	34	54
CHINA	2.93	107	37	1275	435	472
WORLD	106.7	6609	62			124

Source: BP Statistical Energy Review 2011 and USA EIA for the Shale gas data.

and hydraulic fracturing has enabled the extraction of large quantities of shale gas in the USA. Both techniques seem to be very well known and used in the industry, however, it is only very recently that they have started to be used together, and this was where the whole technical innovation came and, more importantly, what allowed producers to get access to the large quantities of shale gas. Horizontal drilling involves drilling vertically into the source, as for conventional gas, then drilling horizontally at right angles, giving access to new sources of unconventional gas (see Figure 1.54).

Hydraulic fracturing is a technique that involves the injection of water, sand, and chemicals into the underground formation to fracture the source rock, leading to release of the gas into the pipe that will extract the available gas. Figure 1.54 illustrates the geology and extraction of shale gas. However, the injection of chemicals underground and the release of several gases after the fracturing of the shale has led to demonstrations over ecological concerns. While in the USA these concerns seem to have been overcome, in Europe, and especially in France, there is still a long way to go.

At this stage we will not discuss the potential concerns and perils of shale gas, as we expect that new technical solutions will overcome any of these concerns, especially in the light of the vast resources available and the need for these clean energy resources.

In a global scenario of solid economic growth and stronger energy demand for the next decades, the clear and mounting concerns posed by climate change and energy supply will provide a tremendous support for shale gas. Global support for shale gas is expected to increase during the next decades, as the technology is improved and social awareness increases.

There are several reasons why shale gas has become important and will become even more important in the next decades for both the world and also for the chemical industry. These include large available reserves, global availability and competitive price and economics.

Large Available Reserves – 60 Years of Extra Global Demand.

According to the latest data from the 2011 BP Statistical Report, there has never been a shortage of gas nor is there expected to be any time soon, at least on a global basis. Indeed, at the 2010 consumption level of 106 tcf annually, and

Table 1.16 Risked gas in-place and technically recoverable shale gas resources: 32 countries.

Continent	Region	Country	Risked Gas In-Place (Tcf)	Technically Recoverable Resource (Tcf)
North America	I. Canada		1490	388
	II. Mexico		2366	681
	Total		3856	1069
South America	III. Northern South America	Colombia	78	19
		Venezuela	42	11
		Subtotal	*120*	*30*
	IV. Southern South America	Argentina	2732	774
		Bolivia	192	48
		Brazil	906	226
		Chile	287	64
		Paraguay	249	62
		Uruguay	83	21
		Subtotal	*4449*	*1195*
	Total		4569	1225
Europe	VI. Eastern Europe	Poland	792	187
		Lithuania	17	4
		Kaliningrad	76	19
		Ukraine	197	42
			1082	252
	VII. Western Europe	France	720	180
		Germany	33	8
		Netherlands	66	17
		Sweden	164	41
		Norway	333	83
		Denmark	92	23
		U.K.	97	20
		Subtotal	*1505*	*372*
	Total		2587	624
Africa	VIII. Central North Africa	Algeria	812	230
		Libya	1147	290
		Tunisia	61	18
		Morroco*	108	18
		Subtotal	*2128*	*557*
	X. South Africa		1834	485
	Total		3962	1042
Asia	XI. China		5101	1275
	XII. India/Pakistan	India	290	63
		Pakistan	206	51
	XIII. Turkey		64	15
	Total		5661	1404
Australia	XIV. Australia		1381	396
	Grand Total		22016	5760

*Includes Western Sahara & Mauritania

Advanced Resources International, Inc.

1.4 Energy Megatrends | 91

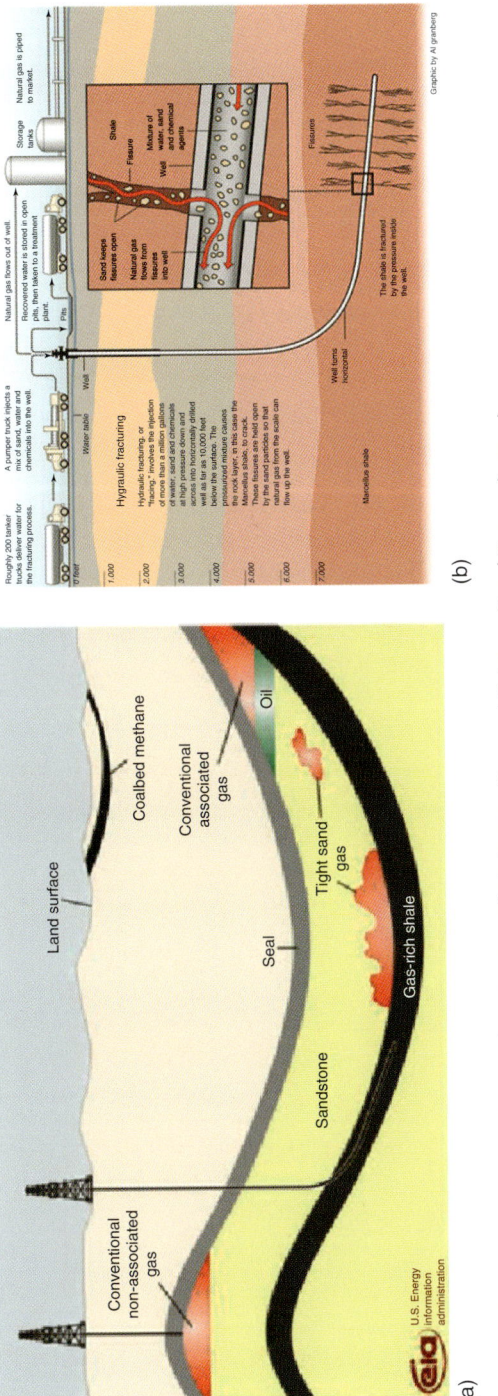

Figure 1.54 (a) Geology of shale gas. (b) Extraction of shale gas. (Source: US International Energy Agency.)

with the existing proved reserves of 6609 tcf, the world will have sufficient reserves for the next 60 years. However, two major additional factors should also be considered (Table 1.17).

First, the fact that the largest natural gas consumers, like the USA, Europe, or Canada, had less than 15 years of proved reserves of natural gas on their own soil, and in fact Europe has been importing most of its gas demand for decades already. Indeed, before shale gas, the major reserves of natural gas were mainly located in the BRIC (Russia) or REST (mainly the Middle East) economies. In other words, gas reserves were not located in the largest consumer countries. The recent discovery of shale gas not only served to double the global reserves of natural gas, but also spread the geographical location of natural gas to many new countries (Figure 1.55).

Considering the fact that the USA consumed 24.1 tcf in 2009, this untapped new reserves of gas could serve to supply the USA for another 36 years of consumption, increasing the current USA reserves fourfold.

Secondly, and perhaps even more important, is that, especially in the BRIC and emerging economies, the demand for energy is expected to keep growing over the next decades. Thus shale gas has become a huge window of hope and fresh energy, not only for the ADV economies that were seeing fast reduction in their local reserves, but also because in a more global economy, the BRICs and the world in general could have extra reserves to cope with the upcoming increases in demand.

Table 1.17 US ethylene expansions based on shale gas as of November 2012.

COMPANY	Project	Location	Tones	Timing
Westlake Chemical	Expansion	Lake Charles, Louisiana	108863	H2 2012
Dow Chemical	Restart	St Charles, Louisiana	390000	2012
INEOS	De bottleneck	Cholocate Bayou, Texas	115000	2012
Westlake Chemical	Expansion	Lake Charles, Louisiana	115000	2014
LyondellBasell	Expansion	Laporte, Texas	386000	2014
Aither Chemical	New cracker	US Northeast	300000	2016
Shell	New cracker	US Northeast	World-Scale	2016–2017
Chevron Phillips	New cracker	Cedar Bayou, Texas	1500000	2017
Dow Chemical	New cracker	US Gulf Coast	World-Scale	2017
Sasol	New cracker	Lake Charles, Louisiana	1 to 14 Million	N/A
LyondellBasell	Expansion	Channelview, Texas	N/A	N/A
SABIC	New cracker	US	World-Scale	N/A
Formosa Plastics	New cracker	Point Confortm Texas	N/A	N/A
Braskem	New cracker	US	N/A	N/A
Occidental Chemical	New cracker	Ingleside, Texas	N/A	N/A
TOTAL			7 to 85 Million Tones	N/A

Source: ICIS News.

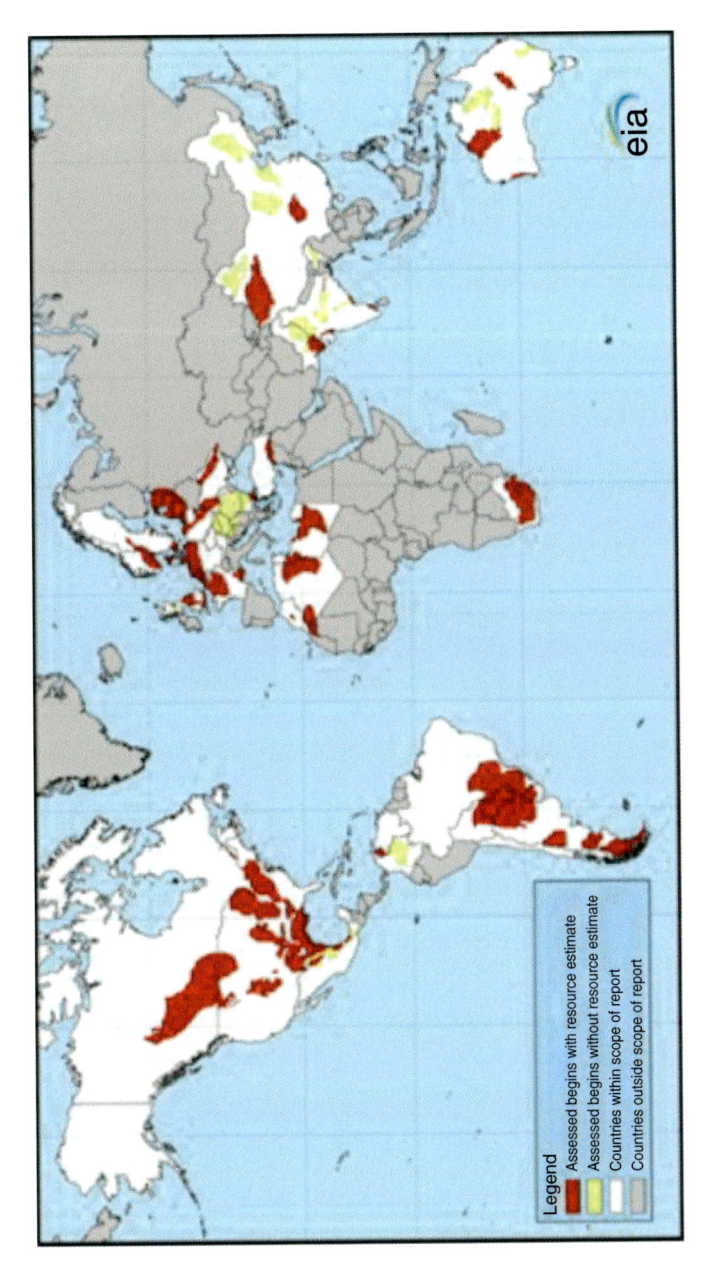

Figure 1.55 Map showing locations of shale gas reserves. (Source: US International Energy Agency.)

Shale gas has the potential to double or triple the current reserves of natural gas during the next decades. For some countries the increase will be dramatic, especially in the BRIC and ADV economies, see Table 1.15. The USA as the largest natural gas consumer, will see its gas reserves triple, allowing an extra 40 years of demand at the current rate. In China, the country with the highest potential for shale gas globally, the increase would be even more overwhelming, as China has the potential to cover its current demand for more than 400 years at the 2010 rate.

India will see its total reserves triple and Brazil will have a huge boost of their reserves. At the current demand levels, India will be able to cover its demand for 53 years and Brazil for almost 400 years. Based on these large expectations we will see a boom in the commercialization of shale gas around the world. We have seen it already in the USA and we can expect something similar in other countries.

Global Availability

The fact that shale gas is widely available and can be found in many new countries, is another reason why we believe shale gas will develop fast around the world. As you can see in the map, courtesy of the USA IEA, there are reserves of shale gas in all five continents. Figure 1.55 relates to Table 1.16, both parts of the same 2011 US Energy Information Agency study on shale gas in selected countries. Based on this study, and only for the 32 selected countries, the EIA estimated almost 6600 tcf of shale gas was available, this relates to the countries marked in red on the map. However, the study has not yet considered all the areas in yellow or gray. It may still take years for the full potential of shale gas to be known and for it to be fully extracted but there is the potential for another 22 000 tcf. The fact that shale gas appears to be available almost globally will not only provide a lot of energy security to new countries, but will also accelerate the technological progress toward extracting the gas in safely and efficiently.

Competitive Price and Economics

Finally, and as important as availability, is its relative cost position versus incumbent energies. As we learnt from previous energy transitions, to be finally introduced energies need to be widely available, competitive, and sustainable for long periods of time. Shale gas appears to be in an excellent situation, as not only is it widely available, but it comes with huge reserves and its price is cheaper than current natural gas sources around the world, with the only exception being the Middle East. Please see Figure 1.56 for comparative natural gas prices around the world.

During the 1990s, the US government promoted the use of gas as a clean and cheap energy, after the oil shocks of the 1970s and 1980s. These large incentives made the USA the largest consumer of gas in the world, accounting for almost 21% of world demand in 2010. This served to stimulate many industries based on gas, including the petrochemical industry; however, the incentives also served to accelerate demand and ultimately prices. Indeed,

*Source: BP Statistical Energy Review 2011

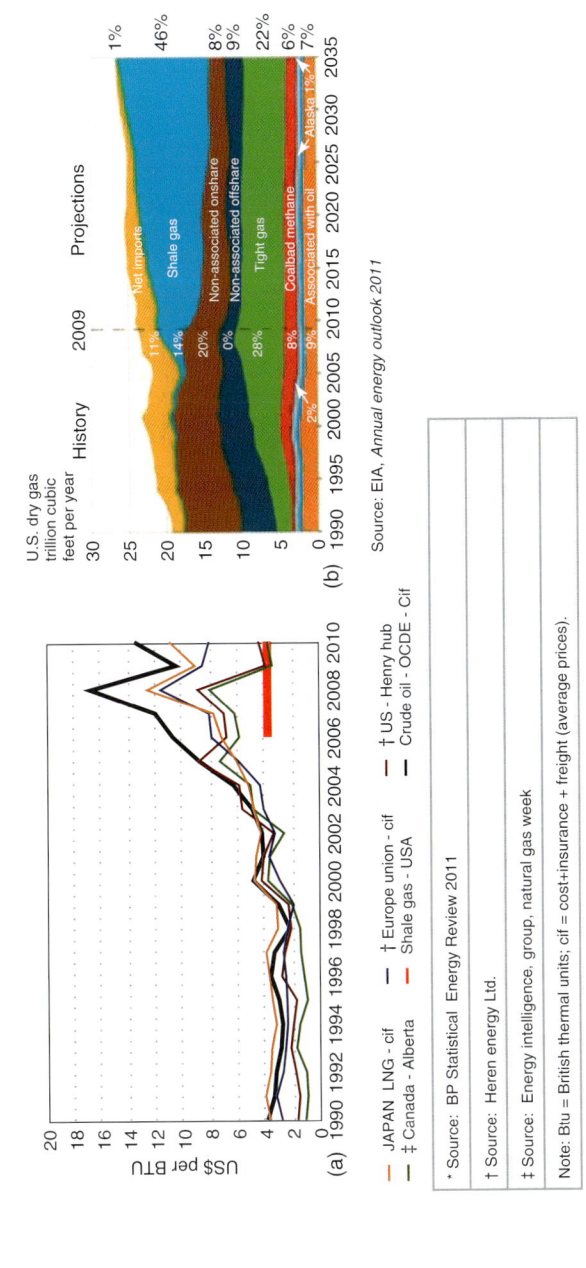

† Source: Heren energy Ltd.

‡ Source: Energy intelligence, group, natural gas week

Note: Btu = British thermal units; cif = cost+insurance + freight (average prices).

Figure 1.56 (a) Selected natural gas prices versus oil parity. (b) US natural gas supply, 1990–2035.

USA natural gas prices almost quadrupled from 1.64 to 8.85 dollar per BTU from 1990 to 2008.

The discovery and use of shale gas in the USA has served not only to alleviate the pressure on prices for the largest gas consumer in the world, but also to give additional life to a very large gas-based industry. Since shale gas was discovered and started to be used, natural gas prices in the USA have declined by 50% to $4.39 per BTU in 2010, while serving to grow a new energy and petrochemical industry around this development.

In Figure 1.56, we can see that, with the exception of the Middle East, USA shale gas would be the cheapest globally, even below the current USA Henry Hub prices.

Conclusions As a result of the large "shale gas" reserves and competitive cost position in the USA, many industries, including the chemical industry, have started to blossom again. Indeed, several large chemical producers, including Westlake Chemical, the Dow Chemical Company, Shell Chemicals, SABIC, and Lyondell Basell, among others, have announced plans for expansion and new crackers in the USA, something completely unthinkable just a few years ago. On the back of shale gas, up to 8 million ethylene tonnes of additional capacity has been announced in the USA for the next decade, almost 30% of its current capacity (Table 1.17).

Since the "shale gas" reserves can be found globally, not only in the USA, we will follow with interest their impact not only on the natural gas market but also on the chemical industry. Meanwhile, the first consequences of shale gas are quite noticeable in the USA.

- **Prices** – Since 2008, US natural gas prices declined from $8.85 to 4.35 per BTU between 2008 and 2010, due partly to the economic slowdown but mainly to the introduction of shale gas. In future lower prices could be expected as supply keep improving.
- **Supply** – according to the EIA, see Figure 1.56, despite its recent introduction, in 2010 shale gas already accounted for 14% of the US supply. EIA estimate that by 2035, 46% of the US gas supply will be based on shale gas.
- **Investment** – several industries have been investing heavily in shale gas, first, to extract it and commercialize it, and secondly to start making use of it. We have already mentioned some notable examples in the chemical industry, but these can also be extended to other industries.

With respect to the chemical industry the positive consequences, as we will review in more detail in the following chapters, are huge. The fast introduction of shale gas into the US energy mix has already had a clear impact, not only on gas prices and supply, but also on the preferred feedstock of the chemical industry. A cheaper and more available natural gas has already altered the feedstock mix of the USA chemical industry in its favor.

According to the chemical consulting firm (CMAI), in 2005 ethane accounted for almost 45% of the ethylene feedstock, and naphtha for almost 30%. The introduction of shale gas during 2007 and 2008 has served to rapidly accelerate the transition toward still more ethane versus naphtha. Since 2007–2008 the use of naptha has declined rapidly, and in 2011 ethane represented almost 60% of the ethylene feedstock in the USA.

Obviously the capacity for the US chemical industry to have access to an abundant and globally competitive feedstock is a significant improvement for the long term competitiveness of the US chemical industry, bringing a tremendous opportunity to create wealth and jobs. The rising prices of natural gas during the last decade in the USA, led to the belief that the US basic petrochemical industry would never be able to be competitive again, especially in comparison to cheap ethane crackers in the Middle East. Shale gas has also become a game changer for the chemical industry, and although the USA will still not be more competitive than cheaper crackers in the Middle East, certainly, shale gas will improve the overall competitiveness of the US chemical industry.

According to the American Chemical Council (ACC), shale gas has indeed the potential to create significant wealth and jobs for the USA and the chemical industry. In its new report "Shale Gas and New Petrochemical Investment: Benefit for the Economy, Jobs, and U.S. Manufacturing," the ACC uncovered a tremendous opportunity for the USA (Table 1.18). ACC estimates that ethane supply could still increase by another 25%, supporting the US petrochemical industry and creating the opportunities listed in the table.

However, we should also be aware that shale gas discoveries, extraction, and use are in their very early stages, so in order to confirm the full potential of shale gas many years of operation are needed. The potential is certainly very high but, as with any other new technology and energy, time and experience will be required in order to assess the real and full potential.

A matter of concern is that it has been observed that sometimes shale gas wells tend to peak production very early and then decline very fast, implying that in order to get a stable supply, many wells may have to be drilled in quick succession.

Table 1.18

- 17 000 new knowledge intensive, high-paying jobs in the U.S. chemical industry
- 395 000 additional jobs outside the chemical industry
  - 165 000 jobs in industries related to increase U.S. chemical production
  - 230 000 jobs from new capital investment on the chemical industry
- 4.4 billion more in federal, state and local tax revenue, annuallyFe(II)–EDTA
- 32.8 billion increase in U.S. chemical production
- 16.2 billion in capital investment by the chemical industry
- 132.4 billion in U.S. economic output
  - 83.4 billion related to increased chemical production
  - 49 billion related to capital investment by the U.S. chemical industry

Source: Shale Gas and New Petrochemicals investment: Benefits for the Economy, jobs and US Manufacturing – American Chemistry Council – Economics & Statistics – March 2011.

Additionally, and in 2011, some shale gas exploration in the UK, near Blackpool in northwest England, is believed to have triggered some small earthquakes. According to the company leading this exploration, Cuadrilla Resources, apparently tremors were triggered by pumping a vast amount of water at high pressure 3 km underground through the drill holes. After the last tremor in May 2011, Britain suspended fracking and commissioned a study to evaluate the situation and potential shale gas activities.

Indeed, for Europe the developments of shale gas will come later than in the USA. In Europe some initial concerns from an ecological and property rights point of view, in combination with some geology risks and potential higher cost, may delay the whole development of shale gas. Shale gas in Europe tends to be found deeper, reserves tend to be thinner and the clay content tends to be higher than in the USA (Table 1.19).

However, Europe, as the world's second largest consumer of natural gas and the largest net importer of gas globally, might not be able to afford to neglect the positive side of shale gas.

Shale gas is certainly one of the most positive developments in the energy sector in the USA and most probably globally, however, we are in the very early stages of this technology and we should be prudent in our forecast and expectations.

The need for energy in general, and clean energy specifically, will be a tremendous incentive to obtain the full potential for shale gas. However, our understanding of the reserves and extraction technology has to be improved. Technology has to be solid and risks and concerns addressed and mitigated. We believe the industry will be able to solve most of these initial issues, as incentives and interest will be vast and clear.

We will discuss in more detail the impact of shale gas on the chemical industry, but we can be sure that shale gas specifically, and natural gas in general, are poised to have a large role in the world energy mix and in the chemical industry during the decades to come.

Table 1.19 How Europe's shale gas compares to the USA.

Country	Place	Thickness (ft)	Depth (ft)
US	Barnett (core)	100–600	6500–9000
US	WWoodford	120–345	6000–13000
US	Fayetteville	20–2000	1000–7000
US	Haynesville	200–300	10000–13500
US	Marcellus	50–2000	4000–8500
Germany	NW German Posidonia	50–200	6500
Netherlands	West Netherlands Epen	50–82	4900–21325
Poland	Polish Baltic Depression	>328	8200
Poland	Lublin Trough	325–6500	7545

Source: F/Geny, Can Unconventional Gas be a Game-Changer for European Gas markets? OIES, Dec 2010

1.5
Climate Change

"Climate changes appear to be the first global problem for mankind, putting at risk human progress, while having the potential to hinder economic growth"

Even if I knew that tomorrow the world would go to pieces, I would still plant my apple tree.

Martin Luther King

Despite the calamities brought about by two World Wars and the emergence of global terrorism, the last century brought an unprecedented period of peace and wealth. Indeed, the last century witnessed the longest and greatest period of economic growth in human history; on the back of economic, political, and social integration. Integration that accelerated global trade and served to coin the term *"globalization."* The last century also saw the emergence of fascinating technological developments in all scientific areas, from medical to aerospace, transportation, energy, communications, including the internet, as well as the massive use of fossil fuel as a major source of energy.

During the last century the world GDP multiplied more than 10 times, while the population trebled. Similar growth was observed during the second half of the century when the world population grew by 150% and the World's GDP by almost 500%. However, all this prosperity did not come alone but with it came climate change, to be the first truly global and perhaps most complicated challenge for our humanity and our earth in the decades to come. We will try to explain the basis of climate change, its background, origins, potential consequences, and solutions.

The enormous economic growth experienced by our world, especially since the end of the Second World War brought with it a tremendous increase in energy demand. Looking only at the most recent period between 1965 and 2010, we can observe in Figure 1.57, some of the major consequences of this massive economic growth. The world economy increased more than fivefold during the last 45 years, improving living standards and quality of life around the world, however, and unfortunately, that massive economic growth also brought some more dangerous consequences, like an unprecedented and record high level of energy demand and the highest CO_2 concentration in the earth's history.

According to data from the latest BP Statistical Review, world energy demand grew by almost 219% during this period, reaching a staggering figure of 12 002 Mt. At the same time world CO_2 Emissions also grew, by 182%, reaching the record high figure of 33 Gt of CO_2 emissions globally, resulting in the highest level of CO_2 concentration in our atmosphere, 390 ppm, according to the Mauna Loa Observatory (Figure 1.58).

The increase in world energy demand resulted in an extraordinary increase in fossil fuel demand, indeed crude oil, carbon, and natural gas accounted, and still

1 Global Megatrends by 2050

World overview: 1965 – 2010		
Population growth = + 150%	**Climate change**	**Economic growth = + 450%**
World energy demand in MM tones	World CO$_2$ emissions in ginatons	World CO$_2$ concentration (PPM)
1965: 3,766 — 219% — 2010: 12,002	1965: 11,74 — 182% — 2010: 33,16	1965: 320 — 22% — 2010: 390

World energy demand (Million tons of equivalent): 1965–2010, rising from ~3750 to ~15000.
Source: BP energy statistical review 2011

World CO$_2$ emissions (Gigatons of CO$_2$): 1965–2010, rising from ~10 to ~40.
Source: BP energy statistical review 2011

Annual avg. CO$_2$ at Mauna loa observabory - (PPM) (Avg CO$_2$ in PPM): 1965–2005, rising from ~300 to ~400.
Source: Scripps insitution of oceanography annual mean CO$_2$ concentration at the Mauna loa observatory in (PPM)

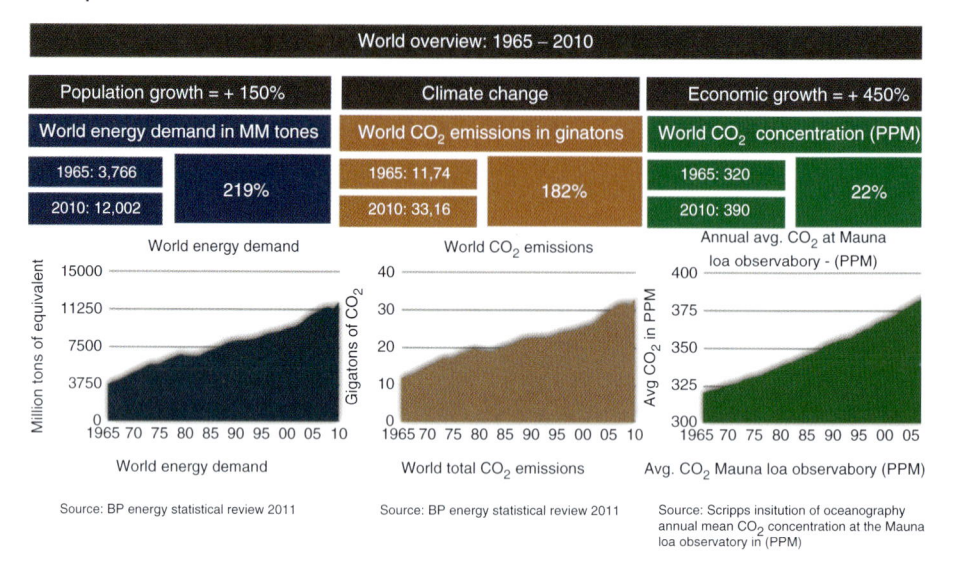

Figure 1.57 World overview 1965–2010.

account, for almost 90% of the world energy demand during the last century, see Figure 1.49.

This massive use of fossil energies resulted in a huge increase in CO$_2$ emissions. During the last 420 000 years, the levels of CO$_2$ in our atmosphere oscillated between 150 and 300 ppm. However, since the industrial revolution, and especially during the last century with the world's massive economic growth, the levels of CO$_2$ grew to well above 300 ppm, reaching 390 ppm in 2010.

As Figure 1.58 illustrates, the problem associated with the variations in the levels of CO$_2$, is the subsequent variations in temperature. According to the Mauna Loa Observatory, our earth has seen not only an unprecedented increase in the levels of CO$_2$ emissions and concentration in our atmosphere, but also a temperature change of more than 1.2 °C around the world.

Historically, levels of CO$_2$ greater than 300 ppm have been associated with severe changes in temperature, including some of the ice age cycles. Looking at the global temperature evolution, we can see how the global temperature of the earth has been increasing very fast since 1860, by almost 1 °C globally.

We can observe fluctuations in most of the parts of the world, varying from plus to minus 2 °C, depending on the area, but especially noticeable in the northern hemisphere. According to the experts there is a correlation between the increase of CO$_2$ emissions, the levels of CO$_2$ concentration in our atmosphere, and the increase in temperature and temperature variation in our planet. They call this phenomenon global warming or climate change, and although CO$_2$ is not the only gas influencing climate change it is certainly the most important one with the largest contribution.

When looking at to the major drivers of climate change, the Intergovernmental Panel on Climate Change (IPCC) during its Fourth Assessment Report (AR4)

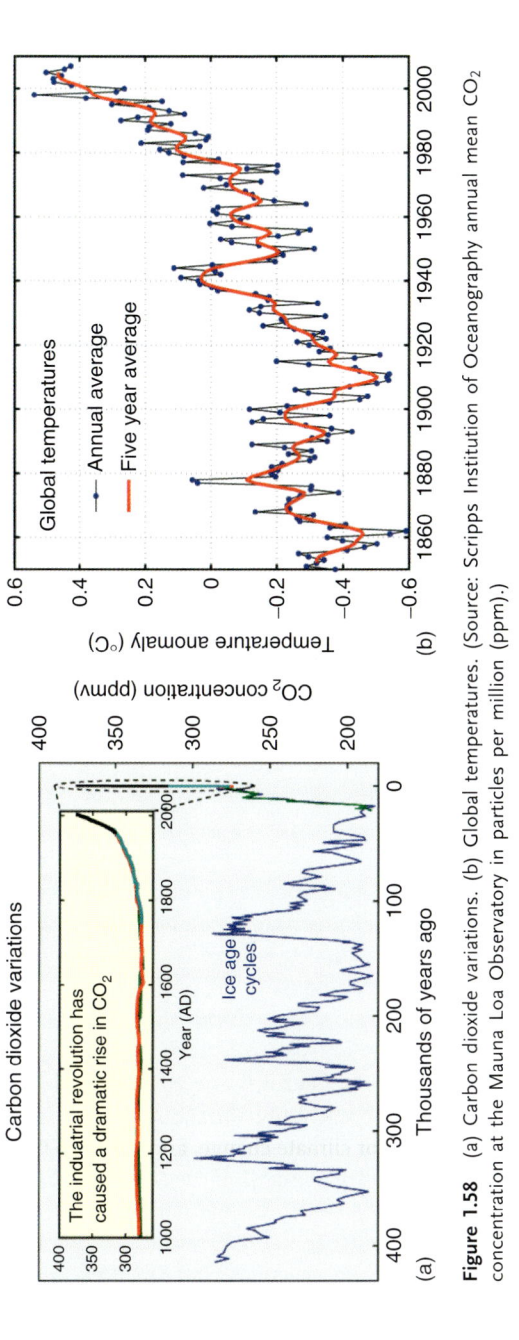

Figure 1.58 (a) Carbon dioxide variations. (b) Global temperatures. (Source: Scripps Institution of Oceanography annual mean CO_2 concentration at the Mauna Loa Observatory in particles per million (ppm).)

established the sectors with the highest contribution to climate change. Perhaps it is worth noting, that so far we have been talking about a correlation between CO_2 emissions and CO_2 concentration level and variations in the temperature or climate change. However, in reality, CO_2 is only one of three major gases impacting climate change. These are normally called *greenhouse gases*, although most people mean only CO_2 when alluding to them, as CO_2 accounts for almost 72% of the total greenhouse gases. However, we should be aware that methane (18%) and nitrous oxide (9%) are also greenhouse gases and have also been growing during the last century, see Figure 1.59.

However, these two gases have normally been disregarded, not only because there is less of them but also because almost 50% of their emissions are related to agricultural, animal, and human emissions and, therefore, are much more difficult to eradicate. Methane is released into the atmosphere from both natural and human sources. Natural sources include animals, volcanic eruptions, earthquakes, some land plants, as well as human and animal waste. Livestock worldwide have been estimated to contribute more that 100 million tonnes of methane gas to the atmosphere each year. Industrially, methane can also come from the burning of different fossil fuels. When looking at the three gases together, experts talk about CO_2-e or CO_2-equivalents. As an example, if in 2010 the world CO_2 concentration was 390 ppm then the CO_2-e was around 490 ppm.

In this book when we refer to climate change or greenhouse gases we will be referring to just CO_2 emissions, however, we should be aware of the whole concept behind greenhouse gases and the existence of CO_2 and CO_2-e.

Looking now at the world total greenhouse emissions, power stations with 21%, industrial processes with 16%, and transportation fuels with 14% appear to be the largest contributors and the key sectors for reduction. As CO_2 is the major constituent, experts and governments have been focusing on reducing the CO_2 emissions (i.e., the Kyoto Protocol) (Figure 1.59).

Understanding the relationship between global warming, greenhouse gases, CO_2 emissions and its major sector contributors is critical in order to understand where the social, political, and economic pressure, as well as potential solutions will focus in the near future. The fact that sectors with the highest CO_2 emissions are at the core and base of the world economic growth and the world economy, has been one of the key reasons why governments around the world, as for instance USA or China, have been reluctant to compromise on CO_2 targets for reductions. Indeed USA, the largest CO_2 contributor globally, has not yet signed the Kyoto Protocol (an international agreement aimed at the reduction of CO_2 emissions) and Canada, one of the largest CO_2 contributors on a per capita basis just withdrew from the Kyoto Protocol in December 2011. The BRIC economies will face similar struggles, trying to trade off economic development with CO_2 emissions.

At this point you might still be wondering: (i) why we should worry about greenhouse gas emissions and potential increases in temperature, (ii) even if indeed we should worry, which countries, industries, or companies should do something about it, and (iii) what is the outlook for greenhouse emissions and climate change and what are its potential consequences. We will devote the

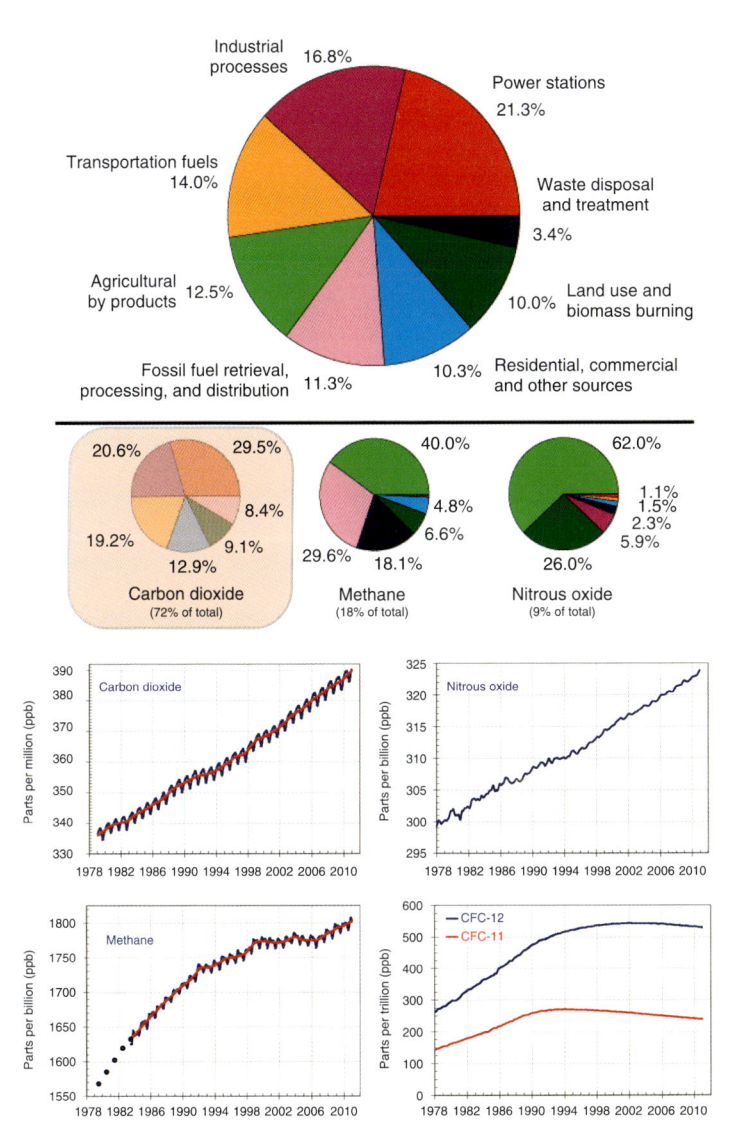

Figure 1.59 Annual greenhouse gas emissions by sector. (Source: Intergovernmental Panel on Climate Change (IPCC).)

following pages to trying to answer some of these questions, while highlighting complexities and difficulties behind some of the potential solutions.

In the next chapters we will also discuss the potential impact for the chemical industry from several angles, as an energy consumer, energy transformer, and as a provider of products and technical solutions to other industries.

For the first question, there are several interesting sources, however, none of them present very positive scenarios. To understand the potential relationship between emissions and climate change, and the potential consequences of these for our world, let us share some simple documents that clearly explain these relationships and the potential impact. According to the IPCC, if the world remains at 350–400 ppm of CO_2 or its equivalent 445–490 CO_2-e, the earth's temperature will increase by 2.0–2.4 °C, see Table 1.20. In that sense we should remember that in 2010 the level of CO_2 in our atmosphere was already 390 ppm, the highest level since the industrial revolution.

The IPCC also established different scenarios for higher levels of emissions, scenarios that contemplated temperature increases up to 4 °C, when the level of CO_2e reaches 590–710 ppm or 485–570 ppm of CO_2.

Once we have established the potential relationship between emissions and climate change, the next logical questions are: what are the potential consequences of rising temperatures and to what extent could they happen?

In 2006, the economist Nicholas Stern, chair of the Grantham Research Institute on Climate Change and the Environment at the London School of Economics, released a 700 page report, the so-called "Stern Review on the Economics of Climate Change", where he presented clear evidence of climate change, its potential consequences, and the cost to society to mitigate these. Certainly for anyone interested in further understanding the basis of climate change, this report in combination with the reviews from the IPCC are a must to read.

On page 5 of the Executive Summary of the Stern review, Stern presented an interesting chart, see Figure 1.60, where he summarized the potential consequences for different aspects of human life, including food, water, ecosystems, or extreme weather events, of changes of the earth's temperature relative to a pre-industrial era.

Minor variations of temperature due to climate change, could trigger already dramatic changes in in crop yields, water availability, or more extreme weather events. Higher temperature variations, and above 4 degrees Celsius could put at risk complete cities, or the complete front cost lines around the world.

In this table Stern certainly explains the potential correlation among emissions, variations in temperature and climate change. However, at this stage we should also be aware that in reality the scientific community cannot really anticipate all

Table 1.20 Climate change scenarios.

Temperature increase	All GHGs	CO_2	CO_2 emissions 2050 (% of 2000 emissions)
(°C)	(ppm CO_2 eq.)	(ppm CO_2)	(%)
2.0–2.4	445–490	350–400	−85 to −50
2.4–2.8	490–535	400–440	−60 to −30
2.8–3.2	535–590	440–485	−30 to +5
3.2–4.0	590–710	485–570	+10 to +60

Source: International Panel for Climate Change – Fourth Assessment Report (AR4) – Climate Change 2007: Synthesis Report http://www.ipcc.ch/publications_and_data/publications_ipcc_fourth_assessment_report_synthesis_report.htm

Figure 2 Stabilisation levels and probability ranges for temperature increases

The figure below illustrates the types of impacts that could be experienced as the world comes into equilibrium with more greenhouse gases. The top panel shows the range of temperatures projected at stabilisation levels between 400 ppm and 750 ppm CO_2e at equilibrium. The solid horizontal lines indicate the 5 - 95% range based on climate sensitivity estimates from the IPCC 2001[2] and a recent Hadley Centre ensemble study[3]. The vertical line indicates the mean of the 50th percentile point. The dashed lines show the 5 - 95% range based on eleven recent studies[4]. The bottom panel illustrates the range of impacts expected at different levels of warming. The relationship between global average temperature changes and regional climate changes is very uncertain, especially with regard to changes in precipitation (see Box 4.2). This figure shows potential changes based on current scientific literature.

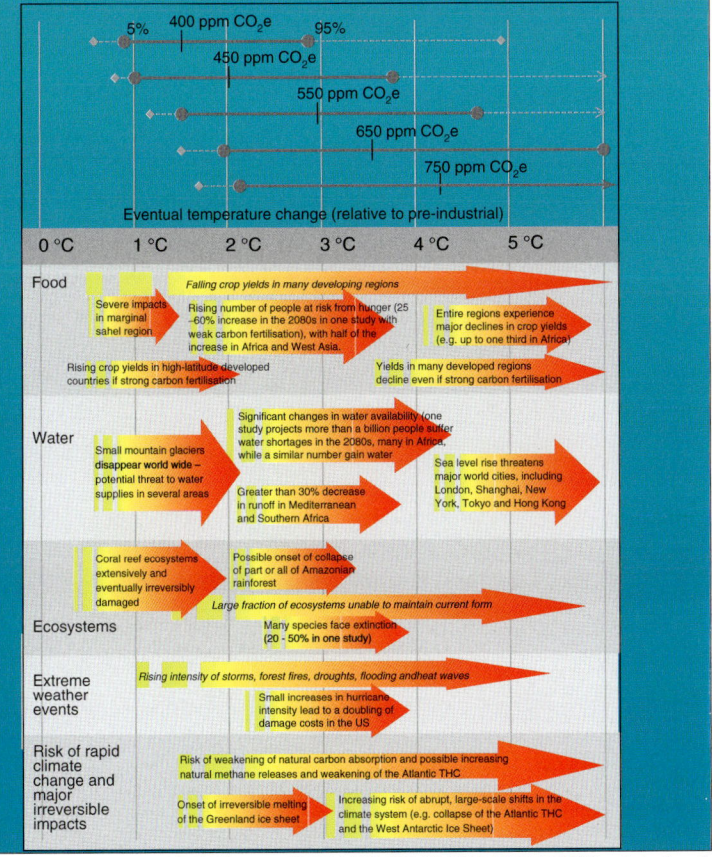

http://siteresources.worldbank.org/INTINDONESIA/Resources/226271-1170911056314/3428109-1174614780539/SternReviewEng.pdf

---

[2]Wigley, T.M.L. and S.C.B. Raper (2001): 'Interpretation of high projections for global-mean warming', Science 293: 451-454 based on Intergovernmental Panel on Climate Change (2001): 'Climate change 2001: the scientific basis. Contribution of Working Group I to the Third Assessment Report of the Intergovernmental Panel on Climate Change' [Houghton JT, Ding Y, Griggs DJ, et al. (eds.)], Cambridge: Cambridge University Press.

[3]Murphy, J.M., D.M.H. Sexton D.N. Barnett et al. (2004): 'Quantification of modelling uncertainties in a large ensemble of climate change simulations', Nature 430: 768 - 772.

[4]Meinshausen, M. (2006): 'What does a 2°C target mean for greenhouse gas concentrations? A brief analysis based on multi-gas emission pathways and several climate sensitivity uncertainty estimates', Avoiding dangerous climate change, in H.J. Schellnhuber et al. (eds.), Cambridge: Cambridge University Press, pp. 265 - 280.

Figure 1.60 The Stern Review on the Economics of Climate Change.

consequences of these changes, as this is the first time that humans and the earth experience all these changes in full awareness, and all the potential ramifications and iterations of these changes are simply unknown. The V Assessment report on Climate Change from the IPCC is due in 2013, and there we should be able to see the latest updates and progress in this area.

So if the correlation exists, the evidences are so clear, and the potential perils so large then what has been done and, more importantly, what is the future outlook for the world CO_2e emissions. Certainly this is a very complicated topic and before entering into the potential solutions and outlook, we should review the origins of the problem.

As previously stated, the enormous and unprecedented economic growth experienced by our world during the last century triggered a huge consumption of energy, mainly fossil fuels. Consumption that triggered a large amount of greenhouse gas emissions and ultimately climate change. However, when we start reviewing these statements from a country perspective, the first thing to be noticed is that not all parts of the world benefitted equally from this unprecedented economic growth, and neither the energy demand nor emissions were equal.

From 1965 to 2010, the world released almost 1 billion tonnes of CO_2 emissions, while generating almost $1300 trillion between 1970 and 2010, measured on a 2005 constant basis. However, while the ADV economies, with just 14% of the world population, enjoyed 73% of the world cumulative GDP and generated 52% of the world CO_2 emission, the BRIC economies, with 44% of the world population, enjoyed just 9% of the world cumulative GDP, while releasing 22% of the world CO_2 emissions. See Figures 1.61 and 1.62 on the world cumulative economics of climate change.

Behind this massive disparity, there is a significant difference in wealth, energy consumption, and CO_2 emissions on a per capita basis. Disparity that lies at some of the core issues on facing climate change and has been the subject of large debates and controversy. The historical correlation among energy demand, CO_2 emissions, and economic wealth and prosperity, has made the reduction of CO_2 emissions a topic of severe confrontation, especially in the short term, when reduction of emissions was directly related to prosperity.

Perhaps it is also worth noting that, despite the fact that the BRIC economies consumed almost half of the energy consumed by the REST of the world during the last 45 years, 14% for the BRIC vs. 26% for the REST, the fact that the BRIC, and specifically China, still relies on a large consumption of coal, explained why the level for the CO_2 emissions for the BRIC and the REST are very similar, 22 and 26%, respectively. This small illustration highlights again the complexity behind climate change, and its potential solutions (Figure 1.62)

From a country or national point of view, the complexity is indeed quite large. The fact that among the 10 largest CO_2 releasers, the first 7 are either ADV or BRIC countries (excluding Brazil), also adds tremendous complexity to any potential solution. Although it is true that the BRIC are among the largest producers of CO_2 in the world with 37% of the emissions in 2010 – especially since China surpassed US as the largest contributor in 2010 – versus 38% for the ADV economies, the

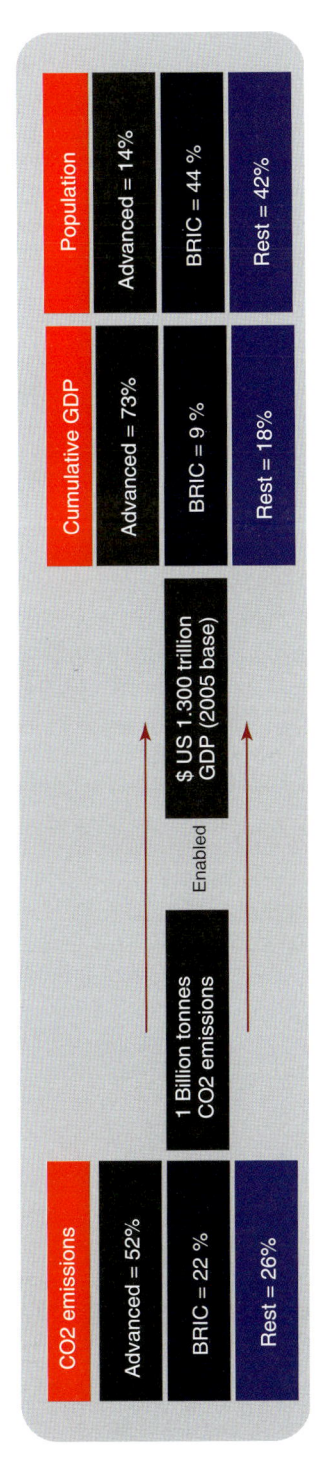

Figure 1.61 The cumulative economics of CO_2 emissions, 1965/70 to 2010. Source: British Petroleum – World Energy review 2011.

108 | *1 Global Megatrends by 2050*

| Population = 6,8 billion | Energy demand = 353 million MT | CO₂ emissions = 1 billion | Total GDP = $ 1,28 billion (2005) |

% Population in 2010: BRIC 14%, Rest 44%, Advanced 42%
Source: World Bank Population data base

Cumulative demand 1965-2010: BRIC 26%, Rest 14%, Advanced 60%
Source: BP Energy Review 2011

Cumulative emissions 1965-2010: BRIC 26%, Rest 22%, Advanced 52%
Source: BP Energy Review 2011

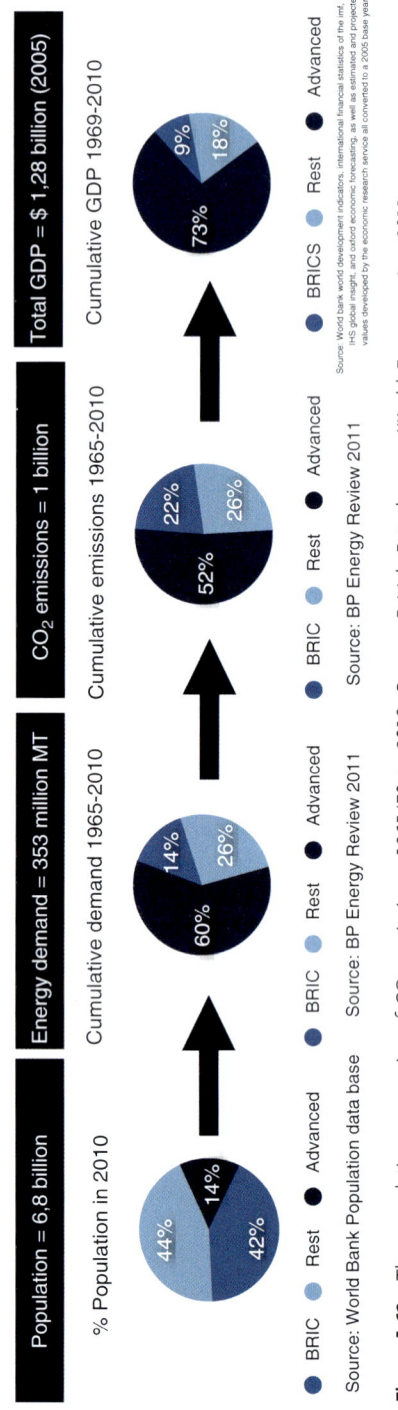

Cumulative GDP 1969-2010: BRICS 18%, Rest 9%, Advanced 73%
Source: World bank world development indicators, international financial statistics of the imf, IHS global insight, and oxford economic forecasting, as well as estimated and projected values developed by the economic research service all converted to a 2005 base year.

Figure 1.62 The cumulative economics of CO₂ emissions, 1965/70 to 2010. Source: British Petroleum – World Energy review 2011.

reality is that on a per capita basis the BRICs are still 4–10 times below the ADV economies.

The average annual per capita CO_2 release in the ADV economies in 2010 was 12 MT, while in the BRIC and the REST it was 4.2 and 2.8 Mt, respectively. Despite clear reductions in the ADV economies during the last decades, the reality is that the USA and Canada still remain above 17 Mt of CO_2 per capita annually, while China remains at 6.2 Mt, Brazil at 2.4 Mt, and India at 1.5 Mt. The large disparity of CO_2 emissions on a per capita basis, presents significant challenges on how climate change should be addressed and by whom (Figure 1.63).

When international organizations, countries, and societies started to look into climate change and wonder about the countries and the industries where reductions could be accomplished, the first realization was that reducing CO_2 emissions could hinder economic growth and prosperity, as some of the major related industries were at the core of our economies and society. The second unpleasant realization was that when looking into the countries with the largest emissions, the differences on a per capita basis were very large. China, the current largest emitter of CO_2 globally, utilizes three times less than the second (USA) or the third (Europe) emitters. Reducing its emissions further could be devastating for its economy, especially when its GDP per capita still remains 10 times lower than those of the USA or Europe. The third and perhaps most controversial realization was that some of the countries, and especially those with the largest CO_2 per capita, were simply not ready to face major reductions of CO_2 emissions, and that the USA, and Canada since 2011, as the largest CO_2 emitters on a per capita basis, decided not to be part of the Kyoto protocol. Interestingly, the USA and Canada's per capita emissions are currently more than four times the world per capita emissions, 4.8 Mt in 2010. However the major challenge is still to come. As the world realizes that we cannot afford not to fight against climate change and we will need to reduce our world CO_2 emissions considerably, the realization of how big the cut needs to be might be something simply unbearable. According to the IEA latest studies, our world average per capita CO_2 emissions will need to be reduced from the current 4.8 Mt annually to almost 1.5 Mt annually.

Therefore, the clear need to avoid climate change and reduce world CO_2 emissions, not only poses significant technological challenges, but also creates tremendous concerns about the ability of the world economy to keep growing at the forecast levels, see Section 1.2.

In June 2008, McKinsey, the consultancy firm, released a paper "The productivity challenge: curbing climate change and sustaining economic growth" that described some of the potential consequences for the world economy when fighting climate change and reducing emissions to sustainable levels. In Chapter 2 we will explore this and other similar scenarios in detail, but for the time being let us share McKinsey's insight to give you a quick flavor of the challenge ahead of us.

They stated some of the basic trade-offs the world will need to face when deciding to fight climate change. According to McKinsey, if the world reduces its annual CO_2e emissions to a world sustainable level of 20 Gt of CO_2e annually, a sustainable level previously defined in "The Stern Review on the Economics of

110 | *1 Global Megatrends by 2050*

World CO$_2$ emissions in 2010 (*Thousand tonnes*)

Rank CO$_2$ emissions	Total CO$_2$ in mill MT
1 - China	8.3 million
2 - USA	6.1 million
3 - Europe	4.1 million
4 - Russia	1.7 million
5 - India	1.7 million
6 - Japan	1.3 million
7 - Canada	0.6 million
16 - Brazil	0.5 million
World 2010	33 million

Rank CO$_2$ emissions	Total CO$_2$ in mill MT
Advanced	12,201 million
BRIC	12,004 million
Rest	8,753 million
World 2010	33 million
Area	World %
Advanced	38%
BRIC	37%
Rest	26%
World 2010	33 million

(a)

Per capita annual CO$_2$ emissions in 2010 (*Metric tonnes*)

Rank CO$_2$ emissions	Annual per capita CO$_2$ in MT
1 - USA	19.9 MT
2 - Canada	17.8 MT
3 - Russia	12.0 MT
4 - Japan	10.3 MT
5 - EU 27	8.3 MT
6 - China	6.2 MT
7 - Brazil	2.4 MT
8 - India	1.5 MT
World CO$_2$	4.8 MT

Rank CO$_2$ emissions	Annual per capita CO$_2$ in MT
Advanced	12.2 MT
BRIC	3.3 MT
Rest	3.0 MT
World 2010	4.4 MT
Scenarios	2050
500 PPM	2.18 MT
450 PPM	1.53 MT
World 2050 - BAU	12.8 MT

(b)

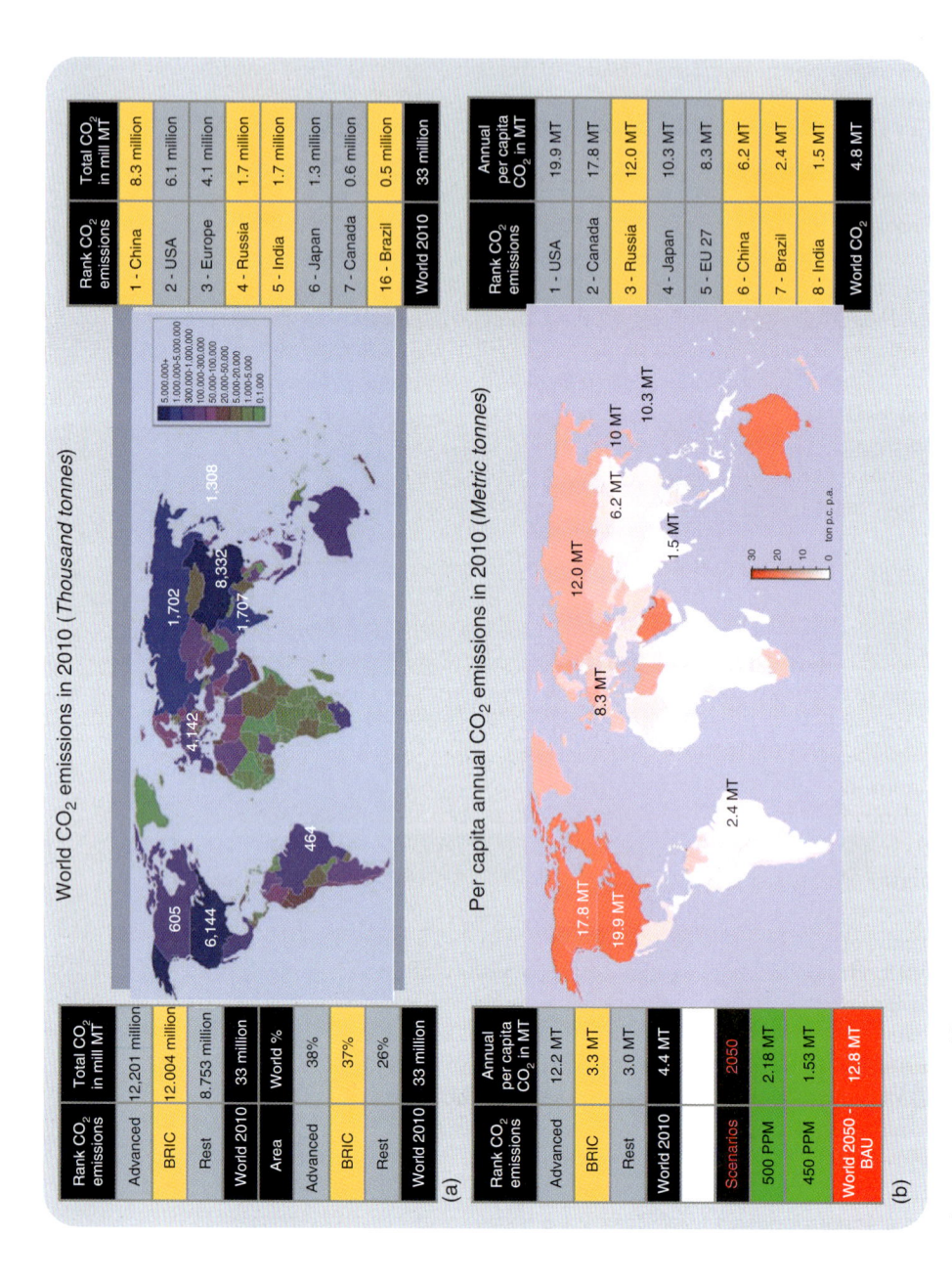

Figure 1.63 (a) World CO$_2$ emissions in 2010. (b) Per capita annual CO$_2$ emissions in 2010 in Metric Tons (MT). (Source: BD Statistical Energy Review 2011, and own calculations.)

1.5 Climate Change

Climate Change" then an annual reduction of CO_2e emissions of 2.4% annually until 2050 will be needed. In other words reducing CO_2e emissions to the suggested 20 Gt per year by 2050, while maintaining the forecast World Economic Growth, will require an increase in carbon productivity of 10 times versus current standards (see Figure 1.64).

Reducing emissions and maintaining growth implies carbon productivity must increase by ten times

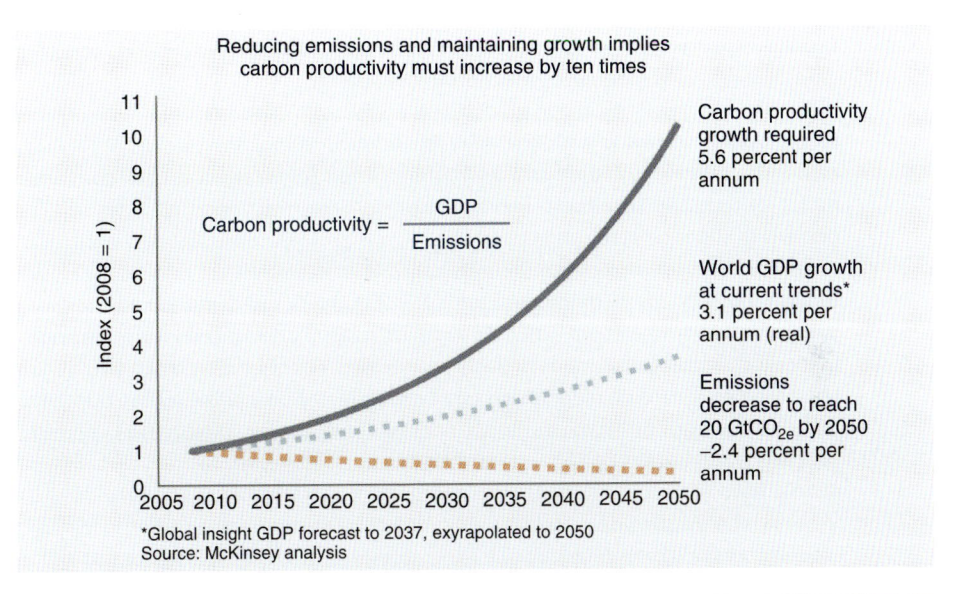

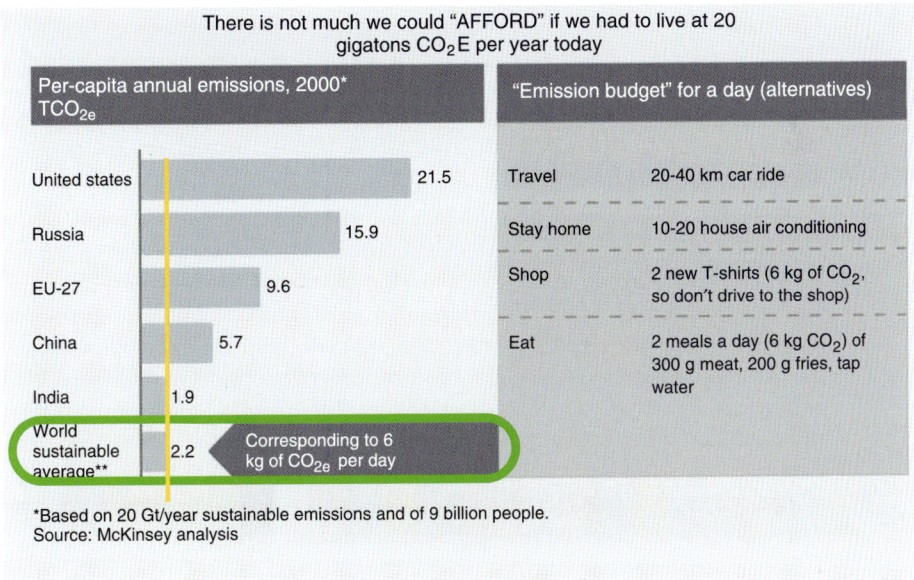

$$\text{Carbon productivity} = \frac{GDP}{\text{Emissions}}$$

- Carbon productivity growth required 5.6 percent per annum
- World GDP growth at current trends* 3.1 percent per annum (real)
- Emissions decrease to reach 20 $GtCO_{2e}$ by 2050 −2.4 percent per annum

*Global insight GDP forecast to 2037, exyrapolated to 2050
Source: McKinsey analysis

There is not much we could "AFFORD" if we had to live at 20 gigatons CO_2E per year today

Per-capita annual emissions, 2000* TCO_{2e}	
United states	21.5
Russia	15.9
EU-27	9.6
China	5.7
India	1.9
World sustainable average**	2.2 — Corresponding to 6 kg of CO_{2e} per day

"Emission budget" for a day (alternatives)	
Travel	20-40 km car ride
Stay home	10-20 house air conditioning
Shop	2 new T-shirts (6 kg of CO_2, so don't drive to the shop)
Eat	2 meals a day (6 kg CO_2) of 300 g meat, 200 g fries, tap water

*Based on 20 Gt/year sustainable emissions and of 9 billion people.
Source: McKinsey analysis

Figure 1.64 The carbon productivity challenge. (Source: McKinsey Institute.)

Considering that in 2010 the world reached 33 Gt of CO_2 emissions and the world per capita average was 4.5 Mt, if the world is to remain at 20 Gt by 2050, with an expected world population of 9 billion, the world per capita CO_2 should be reduced to a mere 2.2 Mt, or 50% of what we have today.

Considering that in 2010 the USA and Canada current CO_2 emissions per capita were around 19 Mt annually, and in Russia, Europe, and Japan around 10 Mt, you can imagine with what sort of challenge we will be confronted. Even China per capita CO_2 emissions in 2010 were above the suggested target at 4.8 Mt and only Brazil and India were below the 2.2 Mt per capita.

A tenfold increase in carbon productivity is certainly a challenge with formidable consequences, especially for the ADV economies. McKinsey in the same study shared how our daily life could be if constrained to live with a CO_2 per capita of 2.2 Mt annually, and these scenarios did not seem very hopeful, especially for citizens of the most modern economies.

With 2.2 Mt of CO_2 per capita, or 6 kg of CO_2 a day, a citizen could choose daily between traveling 20–40 km by car, having 10–20 h of air conditioning, or eat two meals a day of 300 g meat with tap water. As you can imagine, living at these levels seems almost unthinkable for anyone in the developed world, and perhaps this is the reason why climate change appears to be a tremendous challenge for humanity if the world does not manage to get this 10 times carbon productivity increase, because going back on our habits is simply almost impossible, but facing climate change seems as undesirable as facing a life with just 6 kg of CO_2 per capita.

More recently the IEA, established this sustainable target at *14 Gt of World CO_2 emissions by 2050, implying annual CO_2 per capita emissions of 1.53 Mt, or just 4000 g of CO_2 per capita per day. This scenario is still more aggressive that that forecast by McKinsey some years ago.*

Putting our world on a CO_2 "super diet" of 4000 g of CO_2 per day will be a gigantic effort, especially when considering our current levels and our projections for 2050 and the "BAU" scenario. In 2010, the world consumed 13,283 g of CO_2 per capita per day but according to our BAU scenario that number could increase to 28,300 g by 2050.

Living with 4000 g of CO_2 per capita per day will be a significant challenge, and more details will be provided in Chapter 5 where we will review the future of the Chemical Industry by 2050.

Indeed, the challenge is so large and overwhelming that it is even difficult to explain it and put it into the right perspective. It is extremely interesting to notice that the last time our world was living in an economy with 14 Gt of CO_2 emissions was 40 years ago, in 1970. However, at that time things were very different with a much smaller population than expected for 2050. In 1970, the world population was 3.7 billion instead of the projected 9.1 billion. The world GDP in 1970 was $15 trillion (2005 basis) instead of the projected 280 trillion (2009 basis). The crude oil demand in 1970 was 45.4 million barrels a day, versus the projected 85 million barrels a day by 2050 on the IEA sustainable scenario (14 Gt of 450 ppm).

As you can observe having the same CO_2 emissions in 2050 as in 1970, when the population, economy, and crude oil demand were very much smaller, will be significantly challenging. As the world realizes the potential consequences of climate change, its relationship with greenhouse gases, and the fact that despite the efforts to reduce them they still continue increasing, more drastic measures and actions will be required.

From an industry point of view, and as we have already seen in some industries such as the automotive or tire industries, more and stringent regulation should be expected. More comprehensive and accurate life cycle analysis will be required, and as the world becomes wealthier and more knowledgeable about the consequences of climate change, consumers and citizens will become more conscious of climate change and its consequences. Citizens will demand more from governments and industry.

Higher awareness of climate change, greenhouse gases emissions, and higher regulation, will have a clear and strong impact on the chemical industry and its companies, especially for sectors where the capacity to attain the required levels of greenhouse gases might be difficult or not feasible.

The capacity, for any given company, industry, or sector, to reduce emissions and to lead in this area, will become a critical competitive advantage globally, but especially in ADV economies. *The capacity to gain that competitive advantage, will put the chemical industry in an extraordinary position to create a triple win situation*: (i) by increasing the value for citizens and society, by reducing greenhouse gases; (ii) by increasing the value for its customers, enabling them to get higher market share and higher margins, and (iii) for the chemical industry itself, which will be in a unique position to expand margins across the whole value chain as well as for itself.

This statement will become especially true if unfortunately, in the end, the world is not able to achieve the required levels of greenhouse gas reductions and has to suffer the consequences of climate change. Climate change will be one of the most critical megatrends for our world and for the chemical industry, probably the single largest opportunity for the industry to unleash its full potential, and certainly the one that will allow the industry to maximize the value created by technology and innovation.

Finally, we would like to conclude this segment with the following business case in the tire industry, to illustrate the difficulties governments and industries can have when they start facing tenfold carbon productivity increases. Even when governments are able to set the right set of regulations, all parties are committed, and the incentives and penalties are clear, the reality is that accomplishing severe cuts in greenhouse emissions in short periods of time is certainly very challenging. The benefits for consumers, citizens, and companies might be very positive, however, the difficulties will remain very high as we will see.

1.5.1
Business Case – EU Tire Labeling – CO_2 Emissions Reduction in the Tire and Automotive Industry

The purpose of this case is to illustrate the complexities that governments, industries, and the chemical industry can face when trying to reduce CO_2 emissions, even when all parties are committed and have clear targets.

For this case we have selected the reduction of CO_2 emissions in the transportation industry in Europe and, to be more specific, in the tire industry. During this case we will extract several interesting patterns and lessons that will serve us to illustrate not only the real complexities to reduce emissions but also how industries can benefit form it.

The tremendous complexity behind climate change and reducing emissions will require a more collaboration and innovative solutions across industries in the value chain. No longer will any single company in the value chain be able to attain the best performance. Climate change and the need to attain 10 times higher carbon productivities, will not only change our individual industries and their products, but also how they work with each other. In some cases potential solutions go well beyond incremental changes, and for exponential or radical changes collaboration among industries and players will become a must. The companies able to cope with this tremendous challenge, collaborate, and deliver and outperform on the new emissions requirements will be the winners of the future. Companies not able to comply with the minimum levels of emissions will simply go out of business. So, more than ever, the balance between rewards and penalties is crucial. Companies over delivering will gain rapid market share, increasing revenues and margins, margins that will enable to them to keep investing in better technology as the challenge and target will keep increasing over time. Companies not performing according to the minimum established by the regulations will not be able to sell, and those still in compliance but not at the highest levels will be disregarded by most selective, educated, and wealthy consumers.

Therefore, climate change and the need to address the increased carbon productivity will change completely the business game within industries and among suppliers, customers, and the whole value chain. Excellence and collaboration will score over quantity and isolation. After this small preamble, let us now start framing the case.

As previously stated, the transportation industry is one of the largest emitters of CO_2 on a global basis and also one of the sectors where emissions will keep growing. As the world becomes wealthier not only will people tend to drive further but also societies will tend to have more cars. In 2010 ADV economies had around 700–900 cars for 1000 inhabitants, China and India, with much larger populations were still well below 100 cars for 1000 inhabitants. On this basis the transportation sector, already one of the largest releasers of CO_2 globally, will, as the world economy keeps growing, especially in the BRIC and emerging economies, keep growing and adding additional CO_2 emissions.

As the economy becomes not only wealthier but also more integrated, the transportation of goods and people will accelerate the emissions around the world. If we extrapolate some of the projections from the previous megatrends, and specifically the one that stated that by 2050 50% of the population will be middle class with an average GDP per capita of more than $30 000, you can imagine that the projected needs in terms of transportation and mobility will be simply enormous. That is why, the transportation, and more specifically the automotive and tire industry, has been and will be the target of much regulation and attention.

According to a study from the Directorate-General for Energy of the EU, in 2007 24.2% of greenhouse emissions were generated by the transport sector, just behind energy production with 30% of the total. Within the transport area, road transportation was the largest segment, accounting for the vast majority of the transport emissions with 71% of the total, followed by maritime with 13.6% and air transportation with 10.7%. Thus road transportation with 17% of the total greenhouse emissions in Europe was the second largest emitter (Figure 1.65).

In order to understand the complexity of cutting emissions in this transport sector, and very similar to what we saw when reviewing the major contributors to greenhouse emissions, we should consider both absolute values as well as efficiency ratios. In this case air plane transportation has less total emissions than road and maritime transportation. However, when we look at the efficiency ratios of the different modes measured by freight (CO_2 created per tonne and per kilometer), and by person (CO_2 created per person per kilometer) we can see that air transportation is the least CO_2 efficient transportation mode for both freight and person per kilometer.

According to a recent study from the IEA, where they looked at the freight and passenger CO_2-eq efficiency in 2007 of the different transportation models globally, see Figure 1.66, shipping appears to be the most efficient method of transporting freight, followed by rail, road, and air. Looking at the passenger efficiency, again rail appeared the most CO_2-eq efficient, followed by buses, two-wheeled vehicles, passenger vehicles, and air planes.

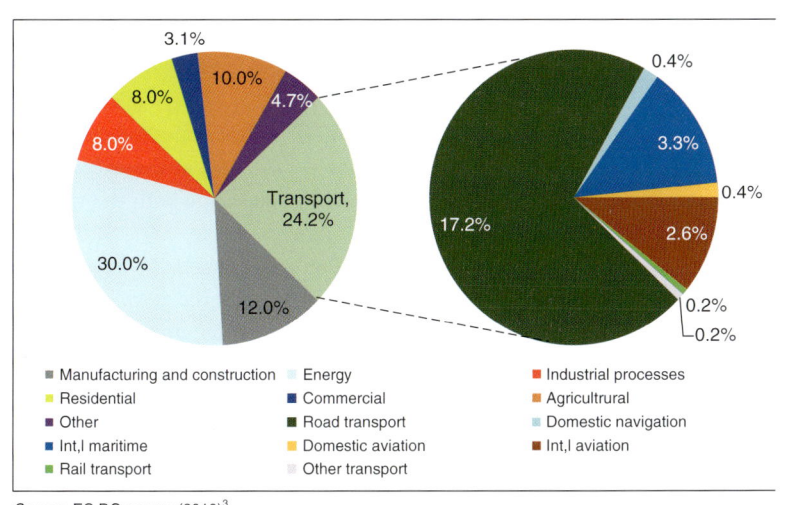

Source: EC DG energy (2010)[3]
Notes: International aviation and maritime shipping only include emissions from bunker fuels

Figure 1.65 EU27 greenhouse gases by sector and mode in 2007. (Source: Based on historic data from DG Energy (2010) EU energy and transport in figures Statistical Pocketbook 2010 Luxembourg, Publications Office of the European Union 2010. Publication and data available at http://ec.europa.eu/energy/publications/statistics/statistics_en.htm.)

116 | *1 Global Megatrends by 2050*

Global - green house-gas efficiency of different modes, freight and passenger 2007

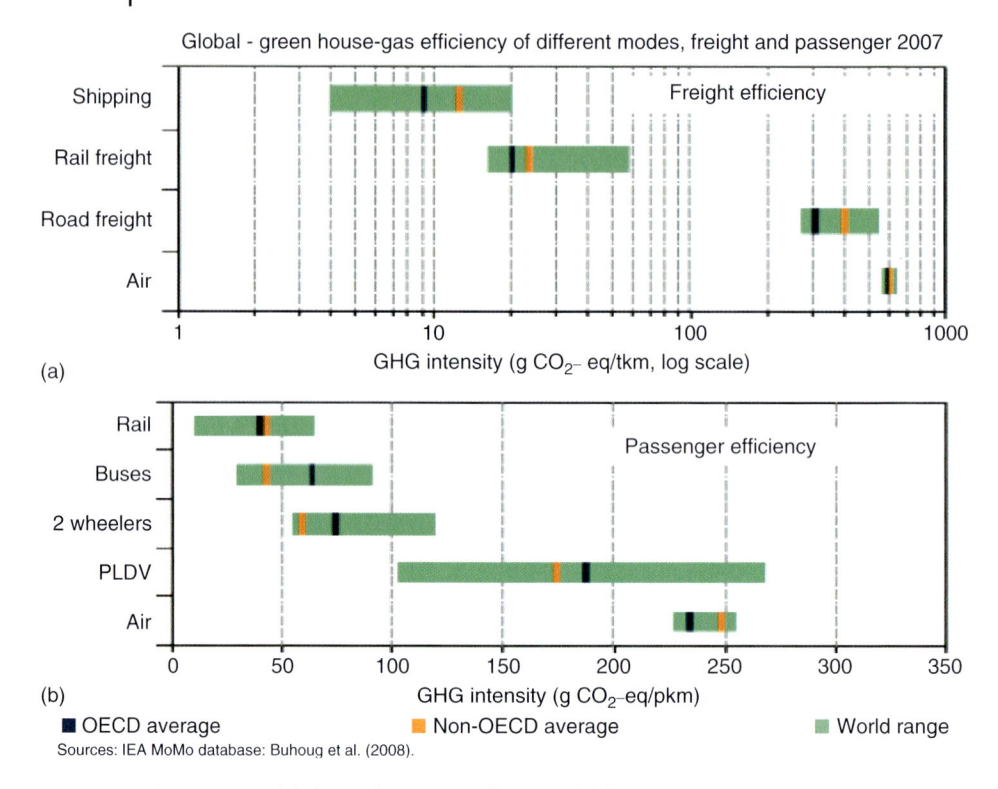

(a) Freight efficiency — Shipping, Rail freight, Road freight, Air; GHG intensity (g CO$_2$-eq/tkm, log scale)

(b) Passenger efficiency — Rail, Buses, 2 wheelers, PLDV, Air; GHG intensity (g CO$_2$-eq/pkm)

■ OECD average   ■ Non-OECD average   ■ World range
Sources: IEA MoMo database: Buhoug et al. (2008).

Figure 1.66 Global greenhouse gas efficiency of different modes of transport: (a) freight, (b) passenger. (Source: IEA MoMo database.)

In both cases, the IEA provided two averages, one for members of the OCDE and one for non-members, as well as a world range. The first thing to notice, and we will see this also in other parts of the business case, is that the average for OCDE members tends to be better than that for non-members, with the notable exception of rail and air transportation where the ratios were the same. From that we can validate the fact felt intuitively that most developed economies already have more CO$_2$-eq efficient performance. That validation gives us some simple room for improvement in the future, as in most of the cases the potential improvement can be accomplished by investing in better infrastructure or better equipment, in other words the technology might be available already. That case seems extremely clear when looking at the efficiency of passenger light-duty vehicles (PLDV) across the world. We can observe a huge range globally, from the most efficient vehicles at 100 g of CO$_2$-eq per passenger per kilometer to the worst performer at more than 270 g of CO$_2$-eq per passenger per kilometer (Figure 1.67).

As previously discussed this huge range makes us believe that a 10 times higher carbon productivity might be very difficult to accomplish, not only for the potential cost attached to it but, more importantly, because the technologies might not yet be available. However, looking at the large range of emissions efficiency of cars,

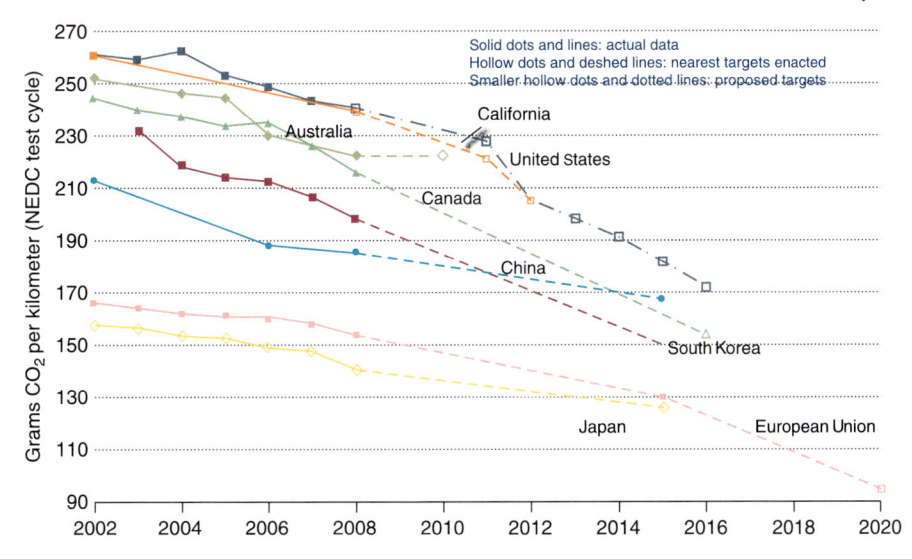

Figure 1.67 Passenger vehicle greenhouse gas emissions, fleet average performance and standards by region. Source: International Council on Clean Transportation *http://www.theicct.org/*

we could argue in some cases that large improvements might already be available. In other words, from the 250 g, or in the best case the 100 g per passenger per kilometer of CO_2-eq, to the 25 g or 10 g per passenger and kilometer (10 times carbon productivity), might certainly be very challenging, expensive and even technologically impossible (Figure 1.67). However, if we could start by moving the worst performers to the level of the current best performers we might have already taken a big step in the right direction, and achieved perhaps even five times greater carbon productivity for the worst performers.

Under the scenario where there is a clear need to reduce emissions globally and where transportation accounts for a large part of the emissions in Europe and globally, and the efficiency range across regions was vast and the global expectations were of higher emissions rather than lower – (as more and more people were expected to use cars and globalization would foster more and longer movements); Governments around the world embarked on multiple regulations with one aim: to cut emissions from the transportation industry.

For this case study we will just look into the European Regulation pertaining to the tire industry, the so-called European Tire Labeling. However, and just for illustration purposes, we would like also to share some of the regulations and CO_2 emissions targets imposed on the automotive industry across the world. Figure 1.67 shows for some selected countries the existing legislation as a continuous line, and some of the expected legislation of passenger car emissions reduction as a dotted line. As you can see most of the most ADV economies have been imposing and succeeding inn implementing very stringent targets on CO_2 emissions during the last decade, targeting even further reductions by 2020.

Figure 1.68 Some major tire labelling initiatives. Source: Japanese Tire Labeling – Japanese Automotive & Tire & Manufacturing Association *http://www.jatma.or.jp/english/labeling/outline.html*. European tire Labeling – Regulation (EC) No 1222/2009 of the European Parliament and of the Council of 25 November 2009 on the labelling of tyres with respect to fuel efficiency and other essential parameters *http://europa.eu/legislation_summaries/energy/energy_efficiency/en0005_en.htm*. USA Tire Labeling Proposal – US National Highway Traffic Safety Administration *http://www.nhtsa.gov/DOT/NHTSA/.../Label_Examples.pdf*. *http://www.nhtsa.gov/Laws+&+Regulations/Tires*

1.5 Climate Change

Perhaps one of the most significant cases is that of the EU, not only because Europe started from one of the lowest levels of CO_2 emissions per passenger car globally with 170 g of CO_2 per kilometer, but also because after a first legislation aiming at reduction to 130 g per kilometer in 2015, Europe has been considering an extension of the existing legislation, where the target would be close to 100 g km^{-1} (Figure 1.67). Additionally, the European Legislation calls for at least 5 g of CO_2 reductions from the tire. With that potential target by 2020, Europe will have reduced its CO_2 emissions per passenger car from 170 to 100 g in just 20 years.

Despite the clear commitment of the EU to reduce CO_2 emissions in the passenger car area, and the capacity of the industry to deliver on these aggressive targets several questions still remain. Would these reductions be enough in the light of climate change? Will other countries follow, as after all climate change is a global problem? What are the potential cost or benefits for the European automotive industry versus its competitors, and also for the European citizens?

We will answer most of these questions when we review the potential consequences of the European Tire Labeling, meanwhile, and just to put things in perspective on this remarkable progress from the EU and European Automotive Industry, the reality is that if we will need to live with 1.5 Mt of CO_2 annually, or in other words with 4000 grams of CO_2 per capita daily, driving the car 40 km per day will simple use up our daily CO_2 allocation. Therefore, certainly the progress and commitment has been remarkable but could that be enough? or more aggressive and global regulations will be needed?

Along these lines, and as has the automotive industry, the European Commission has been studying for years how to reduce the CO_2 emissions and fuel consumption from the tires of the car. Tires not only account for 20–30% of the total fuel consumption but it is also estimated that fuel consumption can be reduced by up to 10% for cars equipped with low rolling resistance tires in place of the worst performing ones. That can be translated into a reduction of 5 g km^{-1} of CO_2 and a reduction of up to 3–4% in fuel consumption annually.

With all these elements in mind the EU established the European Tire Labeling, a compulsory legislation for all tires sold in Europe as of October 2012, where several minimum standard criteria had to be met simultaneously and in different stages. Please see Figure 1.68 with some of the major tire labeling initiatives around the World, including the European one and some new ones under discussion.

Under the European Tire Labeling, all Passenger Cars (C1), Light Duty Vehicles (C2), and Heavy Duty Vehicles (C3), sold in Europe via the replacement or OEM market, as of October 2012, have to be in compliance with certain minimum standards for each of the selected criteria. The selected criteria include "rolling" resistance, fuel efficiency or CO_2 reductions, "wet grip" or braking distance on wet roads, and "noise," to provide a quieter ride and reduce noise pollution (see Figure 1.69). These regulations will be implemented in two stages, the first as of October 2012 and the second as of October 2016. For each of the stages, new minimum and more stringent standards will be implemented for the selected criteria. Non-compliance with the regulation will mean the tire producer cannot

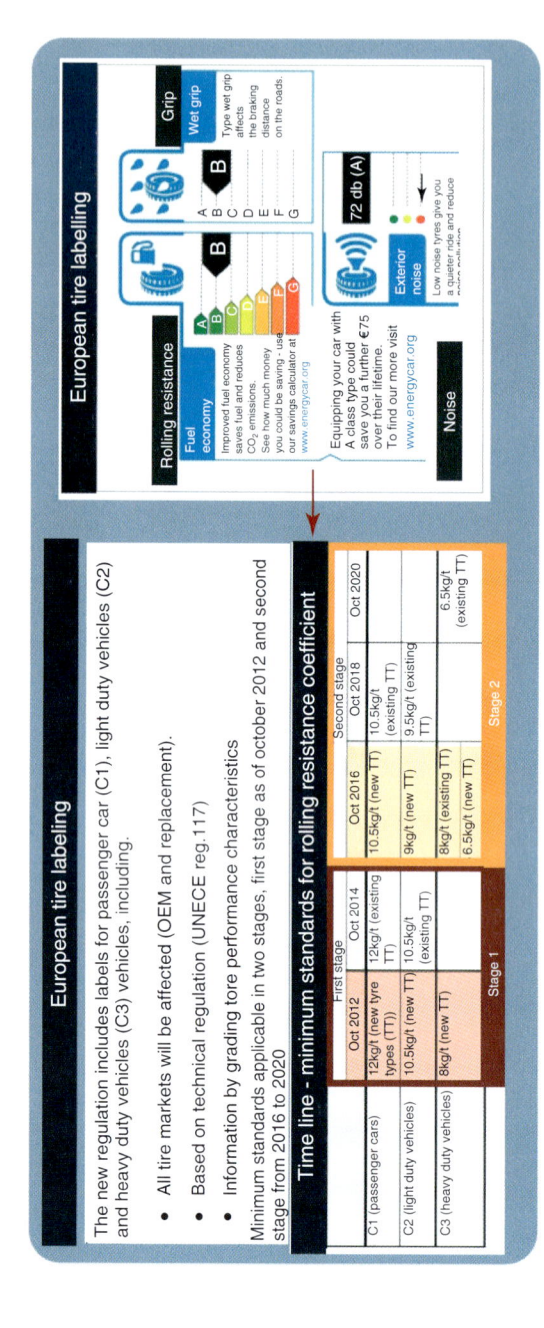

Figure 1.69 European tire labeling. (Source: European Commission – Extract from the European Tire Labeling.)

1.5 Climate Change | 121

sell the tire in Europe, and, in the case of violation of the regulation, severe penalties can be imposed on the different involved parties.

In order to illustrate how the regulation will work and the potential impact on the different stakeholders let us review in more detail the first criterion, rolling resistance. Under this criterion, for the first stage and before October 2012, all passenger cars (C1) (new type) sold in Europe have to be below a minimum standard of 12 kg per tire. If a tire is above this standard then it could not be sold in Europe. If the tire performance is exactly 12 kg per tire, then the tire will have a "G" grade, or minimum grade on the rolling resistance scale and will be able to be sold in Europe, assuming all other criteria are met.

The regulation not only defines a minimum standard for each criterion, but also establishes a seven steps ranking for each of the criteria below the minimum standard going from a "G" level to an "A." So, for instance, for rolling resistance, tires could be rated from G with 12 kg per tire to A, the best rolling resistance, with 6 kg per tire (see Figure 1.70).

In order now to understand how stringent and tough was this regulation for the tire suppliers, in 2008, the European Policy Evaluation Consortium (EPEC) released a document called "Impact Assessment Study on Possible Energy Labeling Tires" where they validated the performance of the European Tires versus the first stage of the European Tire Labeling. On this first appraisal, and for rolling resistance, the EPEC found that applying the standards at Stage 1, 40% of the summer tires in Europe and almost 65% of the winter tires were non-compliant. Additionally, the best summer tire only managed to be rated D, with none sold in Europe rated C to A.

Applying the minimum criteria at Stage 2, where rolling resistance should be 10.5 kg or less, 90% of the tires were non-compliant.

With 40% of the summer tires out of compliance by 2012, 90% by 2016, and not even one single tire rated above D, you can imagine how challenging are the

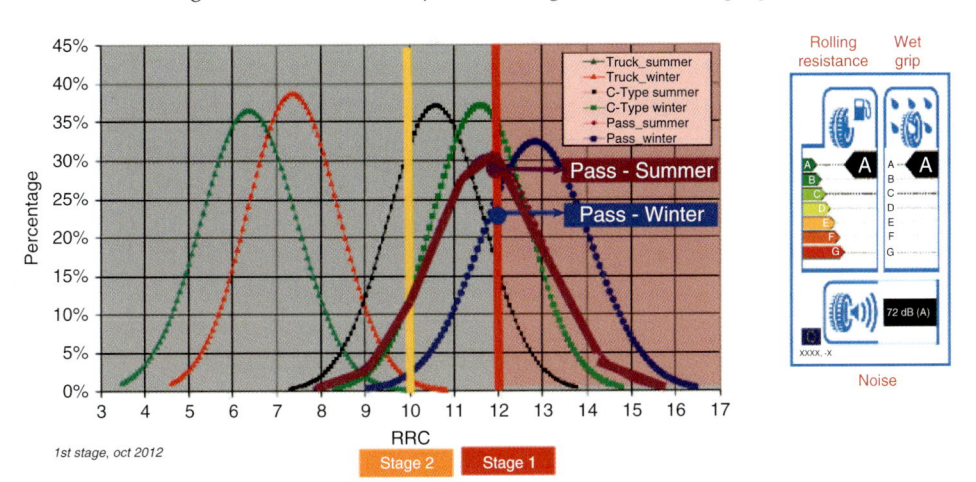

Figure 1.70 Rolling resistance state of the art for truck, C-type and passenger. (Source: EU Policy Evaluation Consortium, 2008.)

standards established by the European Tire Labeling Union (Figure 1.70). It is perhaps worth mentioning that, according to the EPEC study, not even 1% of the Chinese tires imported into Europe were in compliance with the European Tire Labelling at that time.

On looking at the combined performance of the European tires versus two criteria, rolling resistance and wet grip, it was found that 38% of the passenger tires were out of compliance, 57% were on the next level, 5% were in the top quartile, and none were at the A/B level (Figure 1.71). Considering the fact the European Passenger and Light Truck tire market accounted for almost 32% of the global market in 2008, with almost 350 million tires sold annually, according to the Michelin Fact-book data for 2008, the fact that 38% of the passenger tires were out of compliance with Stage 1 of the regulation posed a severe challenge for the tire industry. Considering that 70% of the 350 million tires are passenger tires and the rest light truck tires, and of these, 38% were out of compliance versus the tire labeling, you can imagine the dimension of the challenge. Basically, the performance of almost 90 million passenger car tires would have to be upgraded in order to be saleable again in Europe.

As tire producers became aware of the legislation – in some cases they were also leading parts of the discussions – they started to look into the different ways to enhance the tire performance in order to be in compliance with the upcoming legislation. However, as many people in the tire and rubber industries know, improving one of the parameters alone might be quite simple, the problem comes when it is necessary to improve several parameters at the same time. In other words, reducing the CO_2 emission and fuel consumption due to the tire might be quite simple to achieve by having the tire more inflated or using a plainer tread on the tire. However, by doing this to improve the rolling resistance, the wet grip (braking distance/safety of the tyre) will be significantly reduced, as will the life of the tire by abrasion. This is what is known in the tire and rubber industry as the *magic triangle*. A triangle that implies a trade-off between three major properties in the performance of the tire: rolling resistance, wet grip, and abrasion. In normal conditions increasing one of the parameters will imply a reduction in the others, unless new materials or technologies serve to enhance the three parameters at the same time. Thus the EU, aware of this trade-off, established a regulation where at least two of the parameters have to be improved simultaneously, especially those related to CO_2 emissions (rolling resistance) and safety (wet grip).

When tire producers started to address the challenges posed for the upcoming regulation, they first looked again into the fundamentals of tire performance, meaning design, construction, materials, and so on. Looking specifically at how to improve rolling resistance, the theory recommended to address the following aspects and in this order of relevance: 50% tread, 20% belt construction, 10% carcass, 10% sidewalls, 5% bead, and 5% inner-liner (Figure 1.72).

In a typical case of an 80:20 rule, or how to obtain 80% of the results with 20% of the effort, most tire producers started to look into the design and composition of the tread as the fastest way to accomplish quick improvements in rolling resistance. Improvements in the tread design could be done with the tire producers

1.5 Climate Change

		Rolling resistance rating (RRC in kg/t)						WGI only
		A 6 to 7	B 7 to 8	C 8 to 9	D 9 to 10	E 10 to 11	F 11 to 12	G Above 12%	
Wet Grip Rating (WGI)	A Above 1.45	0%	0%	0.4%	0.3%	0.8%	1.3%	1.2%	4.0%
	B 1.45–1.30	0%	0.3%	2.4%	1.9%	5.4%	8.9%	8.1%	27.0%
	C 1.30–1.15	0%	0.6%	5.3%	4.1%	11.7%	19.3%	17.6%	58.6%
	D below 1.15	0%	0.1%	0.9%	0.7%	2.1%	3.5%	3.2%	10.5%
RRC only		0.0%	1.0%	9.0%	7.0%	20.0%	33.0%	30.1%	

WGI rating	RRC rating	
A	A/B	0%
B	C/D	5.0%
C	E/F	57.0%
D	G	38.0%

Figure 1.71 EPEC combined tire performance versus rolling resistance and wet grip. Source: Impact Assessment Study on Possible Energy Labeling of Tires Annexes to the Impact Assessment To the European Commission Directorate-General Transport and Energy Specific Contract: DG TREN No TREN/D3/375-2006 Under Framework Contract No. DG BUDG No BUDG06/PO/01/LOT no. 2 ABAC 101922 – EPEC 31st July 2008 www.epec.info

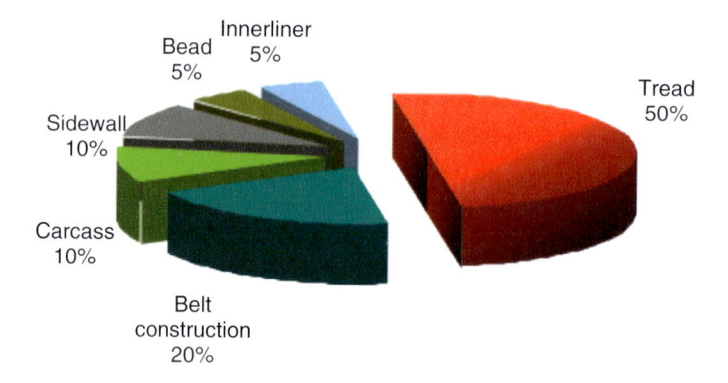

Figure 1.72 Mastering rolling resistance into a tire. Source: Own elaboration.

own technology, however, tread improvements based on enhanced compounded materials, like synthetic rubbers, fillers, or oils had to be done in collaboration with others. Similarly for the belt construction, tire producers should work intensively with wire or metal producers.

As tire and material producers started to work together on how to improve the tread and rolling resistance of the tire without compromising, but even improving, its wet grip or noise, new materials like silica or rubbers like SSBR (solution styrene butadiene rubber) or Nd-PBR (polybutadiene rubber) started to be massively introduced. However, the realization that these levels were simply the initial ones, and more stringent ones would follow in the next years and decades, led to the acknowledgment that larger improvements will only come with much higher technical collaboration, collaboration that might need to go beyond the typical supplier–consumer relationship, even across the whole value chain. Therefore, industries and companies will need radical transformations to comply with carbon productivity, and not only incremental improvements.

In our case, the rolling resistance of the tire will need to go by regulation from above 12 kg for a non-qualified tire in 2008 to 6 kg per tire to get the highest ranking in October 2012. However, if we need to achieve a 10 times greater carbon productivity by 2050, tire rolling resistance might need to go to 1.2 kg per tire by 2050. This is certainly a large over simplification, used just for illustration purposes and without real scientific evidence, but just to emphasize how much collaboration and technical collaboration might still be needed. However, under a potential scenario like that, you might argue whether indeed we will need incremental or radical changes and even if one of the members of the value chain will be able to achieve such improvement. You can also question whether there will be available technologies to enable such kind of improvements, or simply if tires and cars might need to be completely replaced by other transportation alternatives.

As tire companies will manage to achieve and beat the minimum standards – we expect most of them to be in full compliance – several lessons can be extracted from this business case:

1) **Regulators and governments** have a tremendous role to play in setting the right targets for society and companies. Governments have the opportunity to take care of the interest of citizens and society by leading industries to solve the right problems, giving them the proper background, incentives, and support.
2) **Citizens** will enjoy some of the best tires in the world, with the best fuel efficiency and lowest CO_2 emissions. They might need to spend a bit more on their tires, but they will achieve even greater savings from lower fuel consumption and reduced CO_2 emissions.
3) **Tire producers**, after a very challenging time will manage through technology and innovation to comply with the regulation or they may disappear. For those that manage to be competitive and comply with the regulation, their prospects will be extremely positive as they will manage not only to grow in the European market but also globally. According to previous experience in other industries where labeling systems were introduced; see Figure 1.73 with experiences for cooling systems, ovens and washing machines; players able to reach the highest rates on the labeling were rewarded with higher market share and profits through the years.
4) **Material and chemical producers** have been forced to upgrade their technologies, and way of working. Thanks to that, not only can they now supply some of the most competitive products to the tire industry around the world, but also they are among the most competitive companies in the world. This will give them the opportunity to raise their market share and profits, expanding their technological advantage into other regions and industries.
5) **The tire market** will benefit from a higher degree of transparency. In the past customers were buying based on the brand, associating brands with higher quality and prices, despite this not always being necessarily true. Now thanks to the European tire labeling, they will be able to decide based on objective performance criteria and price. Performance will determine the potential success of a tire producer. This may allow less well-known tire producers to quickly gain a market share, if their products present a higher performance than those from the branded producers. After all, markets will reward the right attributes, like performance or price, rather than brand or product availability or positioning.
6) **The European tire industry market**, thanks to the regulation, will become one of the most competitive markets in the world. All companies on the value chain will be in an extremely competitive position globally, furthermore, they will generate external synergies to many other adjacent industries, like the automotive industry, or many others.
7) **Society** will benefit in multiple ways. Citizens equipped with better tires will not only reduce their emissions but also their energy demand. Less CO_2 emissions can also have significant impact on the overall health of society, but perhaps even more important these days, competitive industries will be able to grow globally and create jobs in Europe and globally. Regulations that stimulate technology, research, and innovation always benefit societies, creating positive synergies in universities and research centers. Finally, Europe will have the

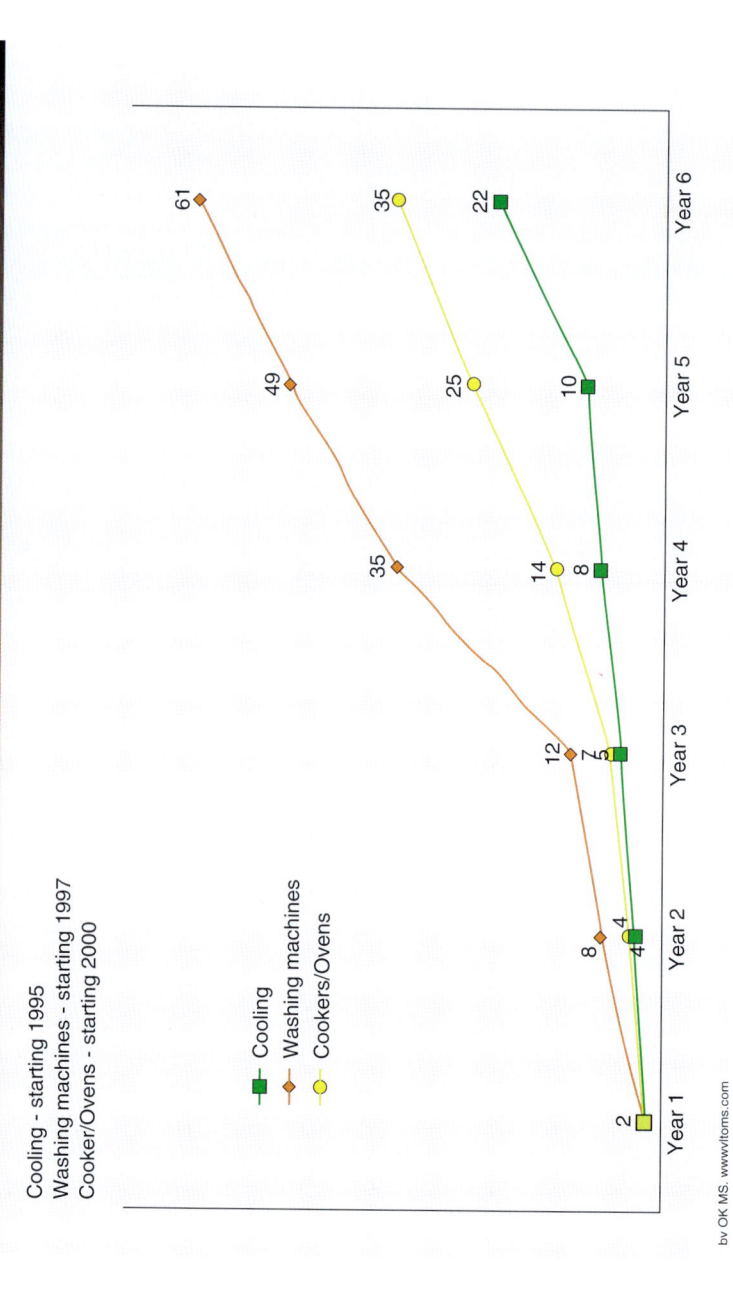

Figure 1.73 Percent market share for A and B performers. (Source: European Commission – Impact Assessment of European Labeling into several European industries – Stockle, Gfk (2006).)

most efficient and "clean" tire market in the world, the best tire products and lowest CO_2 emissions.

8) **A triple Win–Win** – The European Tire Labeling appears to be a triple win, for citizens, consumers, and companies. Citizens will enjoy less CO_2 emissions in the atmosphere. Customers will enjoy better tires with better fuel economy, making real savings on fuel consumption while reducing CO_2 emissions. Successful companies across the value chain, from tire producers, to rubber, fillers, or other related producers, will see their market share, prices, and profits increase. Europe will be rewarded for setting the right regulation and take the lead on this effort globally. After all, the European Tire Labeling might not only contribute to cutting CO_2 emissions in Europe, but might also serve to enable to keep in Europe the most technological and competitive tire industry in the world.

After reviewing all the positive lessons from the Tire Labeling in Europe, you might argue that, in principle, it seems an excellent example of how to address climate change and reduce CO_2 emissions while upgrade an industry. Indeed, it has been a very positive experience but there have also been some negative aspects and criticisms. First, the enormous cost and pressure imposed on the whole industry created significant criticism, especially from the large tire producers. Secondly, many under-performing companies might be forced to disappear, as they could not comply with the legislation. Thirdly, and perhaps the major criticism of the European labeling, was the acknowledgment from non-European tire companies that the upcoming labeling might prevent them selling in the European Market. In 2009, the USA started to impose economic tariffs for a numbers of years for all tires imported from China. China used to export 30% of their local production to the USA. Obviously these tariffs created a significant amount of discomfort and severe complaints from the Chinese Government. Meanwhile Europe started to define the European Tire Labeling for implementation in 2012. As in 2008 90% of the Chinese tires were estimated to be not suitable for into Europe. Non-European tire producers perceived this as a protectionism measure, while Europe might argue that it was simply addressing the upcoming CO_2 concerns and climate change. In any case, and as consequence, the EU with its labeling raised the performance standards across the whole industry, and after that several tire labeling initiatives similar to the European one will be implemented across the world. Japan has started already, while the USA, Korea, China, and Brazil will gradually start during the next decade.

Despite the clear success of the European Tire Labeling, perhaps the key lesson from this business case is not the enormous value and positive impact the right legislation can create for citizens, customers, and companies, but how even very successful regulations, with all stakeholders committed and delivering on stringent targets might not be enough to reduce CO_2 emissions in the short term.

According to a recent study from the EU on the outlook for CO_2 emissions across the different sectors in Europe, despite the fact that Europe will be able to reduce its total CO_2 emissions from the 1990 levels by 65–95%, depending on the different scenarios, the contribution from the transportation segment will continue

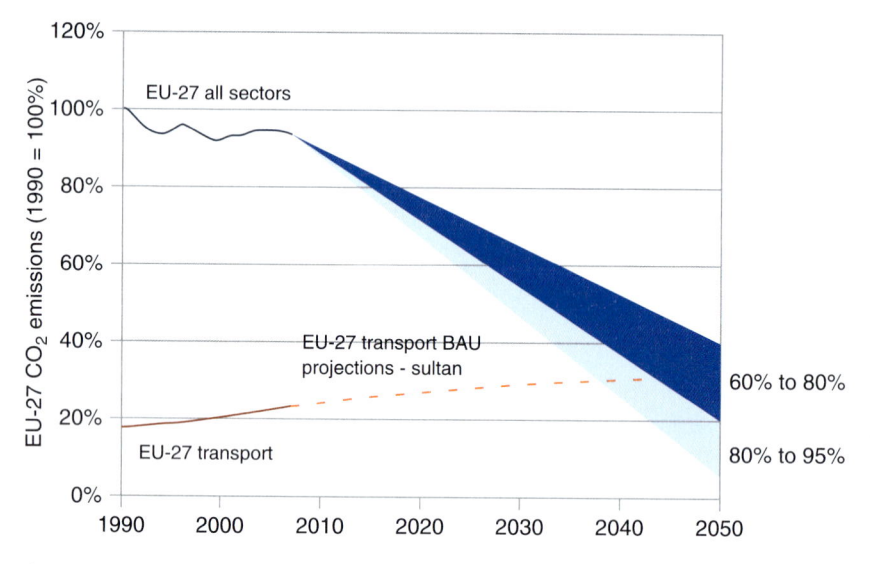

Figure 1.74 Outlook 2050 – EU27 CO_2 emissions. (Source: EC DG Energy (2010) and SULTAN Illustrative Scenarios Tool.)

increasing until 2050 (Figure 1.74). Indeed the total CO_2 emissions from the EU transportation segment are expected to increase from approximately 1.6 Mt CO_2e in 2010 to more than 2 Mt CO_2e in 2050, implying that further regulations and more aggressive targets will be needed.

When looking at the major CO_2 contributors within the European transportation sector by 2050, see Figure 1.75, the car industry, despite all the legislation and aggressive targets for the reduction of emissions on both the tire and automotive industries will remain stable in its overall contribution. So there will be no real reduction during the next decades. Additionally, the increase in emissions from the maritime sector and heavy trucks will keep emissions from the transportation sector increasing.

At this time there are very valuable lessons from this business case, lessons the chemical industry can learn from and apply into other products, industries and regions:

1) Climate change presents the challenging need to reduce emissions in sectors that are still growing and are critical for economic growth. On this basis, optimizing while growing present a tremendous complexity
2) Good regulation, like the European Tire Labelling, can have the potential to create value across the whole value chain creating a triple win for citizens, customers, and companies but might not be enough.
3) Good regulation, stakeholder commitment, the right targets, and outstanding technology might not be enough in the short term, and more regulation and reductions might be needed. More radical technologies and complete game-changing solutions will be required.
4) Considering the fact that maritime shipments and truck transportation are at the core of the supply chain of the chemical industry and these segments will

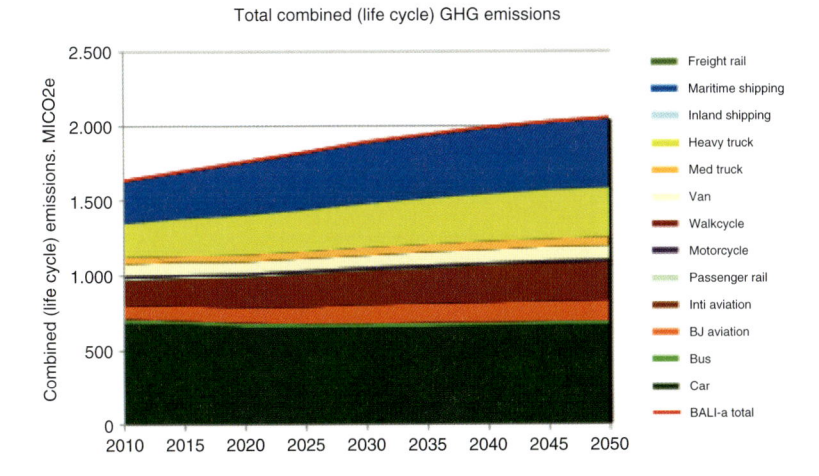

Total combined (life cycle) GHG emissions

Figure 1.75 Outlook 2050, transportation segment, total combined (life cycle) greenhouse gas emissions. (Source: SULTAN Illustrative Scenarios Tool, developed for the EU Transport GHG: Routes to 2050 project.)

Notes: International aviation and maritime shipping include estimates for the full emissions resulting from journeys to EU countries, rather than current international reporting which only include emissions from bunker fuels supplied at a country level (which are lower)

be the focus of tremendous attention; we might expect considerable impact for the chemical industry too

5) In a growing world and global environment the 80/20 rule is not enough, reductions might need to come in all segments, all geographies and at the same time.
6) Finally, collaboration and coordination across all the industry players and the value chain, including governments, will be needed, as incremental steps will not be sufficient.

For the convenience of the reader, additional data on World CO_2 emissions from 1965 to 2010 are shown in Figure 1.76.

1.6
Wild Cards

"Because "ALL" seems possible these days."

Only two things are infinite, the universe and human stupidity, and I'm not sure about the former.

Albert Einstein

Reflecting on the major megatrends shaping the World by 2050 it is easy to realize that those presented in this book might represent some of the most significant and predictable ones, but certainly not all of them.

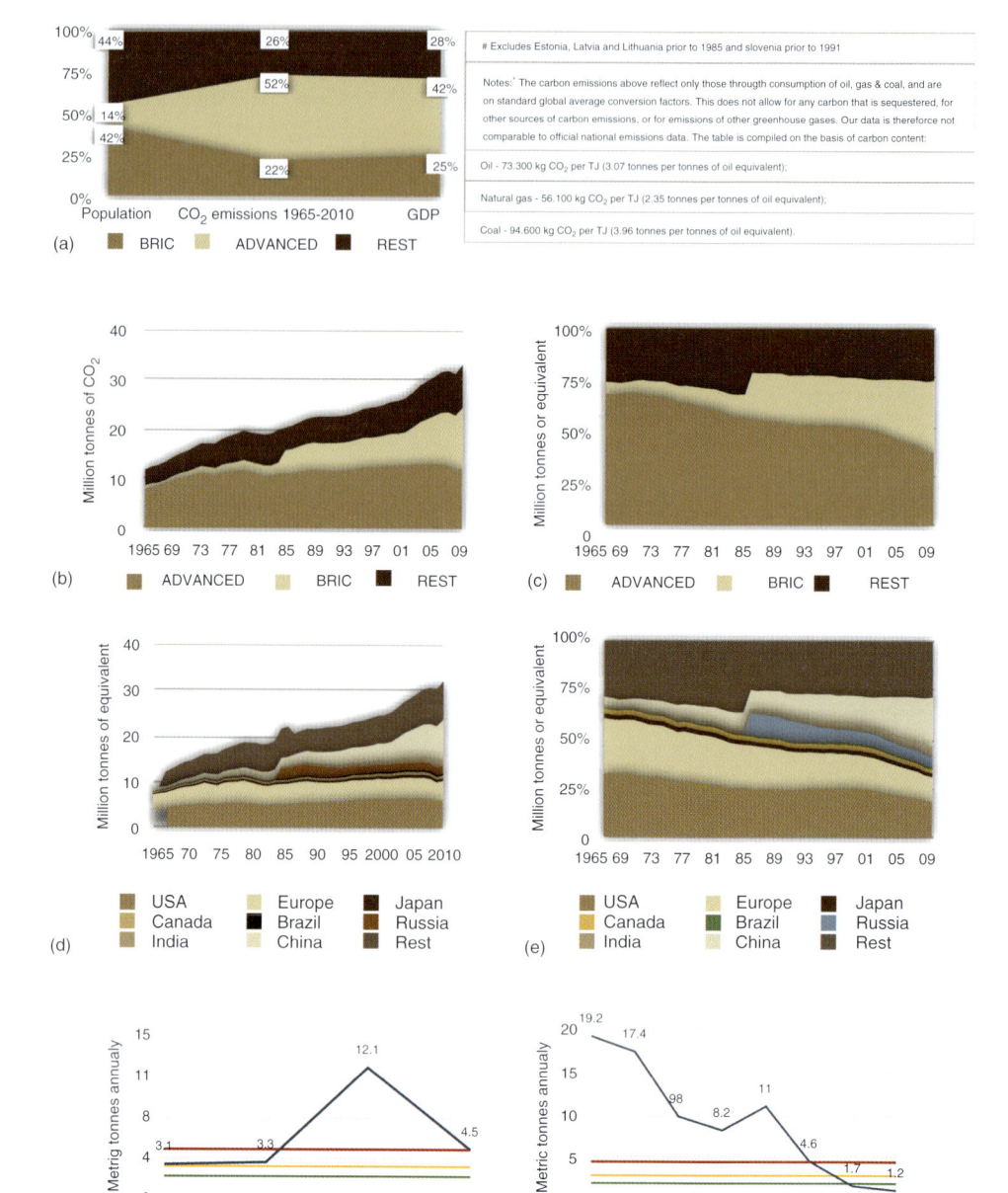

130 | *1 Global Megatrends by 2050*

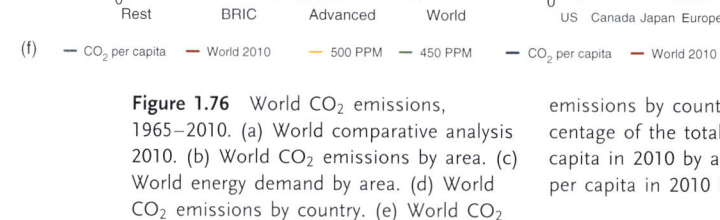

Figure 1.76 World CO_2 emissions, 1965–2010. (a) World comparative analysis 2010. (b) World CO_2 emissions by area. (c) World energy demand by area. (d) World CO_2 emissions by country. (e) World CO_2 emissions by country expressed as a percentage of the total. (f) CO_2 emissions per capita in 2010 by area. (g) CO_2 emissions per capita in 2010 by country.

Additionally, and with the current speed of innovation, we all agree there will be many other disruptive trends that could emerge that we did not capture here, but that could still have a very large influence on our world. As a simple reflection on the difficulties in forecasting the future, let us think about some of these potential and radical wild cards, cards that could significantly change our world by 2050:

1.6.1
Political

- **One global government**. As the world ten to become more global and national identities tend to disrepair, countries around the world start to align into major groups, first economical groups, later one politically. As times pass, governments and society realize that more coordinated actions among government and groups will require one global government for the best sake of the mankind and our earth, our culture in general tends to become more global in nature, national identities tend to disappear. Indeed, countries around the world start initially to realign into major groups like America, Europe, Asia, and so on, and just start to realize we need just one global government for the whole world.
- **The European Union and the Euro disappear**. The issues around the European Sovereign Debt trigger several countries to leave the Euro and eventually Europe. The tension, originated by the European crisis, starts to foster the renewal of historical confrontations and rivalries, creating the possibility for a new European war and the complete destruction of Europe.
- **The third World War occurs on the back of large imbalances across the world**, imbalances that are accelerated by the need to fight climate change. However, this time the war will be a nuclear one and the world will disappear.
- **Fundamentalism and terrorism extends, reducing liberties globally**. As the world becomes involved in another period of structural transformation and fast changes, dictatorial regimes will try to increase fundamental positions to defend themselves in the long term.
- **Africa** after centuries of being forgotten will become the fastest growing economy in the world.

1.6.2
Social

- **Humans become immortal**. As computational capacity keeps increasing exponentially, all sciences achieve exponential developments too. Improvements in medicine, robotics, and the computational capacity to analyze, forecast, and prevent illness and disasters will make humans immortal. This could not only change forever the way we live and work, but also all upcoming projections
- **Humans acquire the technology to live on other planets**. Fostered by the need to avoid climate change, the world starts focusing on ways to live outside planet earth.
- **Religions are eradicated from society**. As technology overtakes all areas of life and, thanks to technological progress, mankind can create life and extend its own

life until, becoming immortal, man starts to abandon religions. That certainly creates severe crises of identity, not only for mankind in general but also for many countries and societies around the world.
- **People can create families while being single**. Thanks to technological, social, and mental progress, societies will have the potential to start reproducing themselves in a completely different manner. That could certainly alter all population and economics patterns projected for the future.
- **People live in big communities without governments**. As people become wiser and healthier, they question the roles of governments and whether actually they would be better off governing themselves in good faith and in a more efficient way.

1.6.3
Technological

- **The Earth is destroyed by a meteor impact or a nuclear war**
- **Climate changes** erase several critical species from Earth and ultimately human life disappears from Earth. Global pandemics spread around the world and our world is confronted with new illnesses.
- **Medical advances extend human life forever**
- **New radical sources of energy appear**, after crude oil is completely depleted.

1.6.4
Transportation

- Regular air planes can start flying into space
- Regular cars disappear to be replaced by small planes
- Rockets can fly at the speed of light, and humans start conquering other planets
- Humans can teletransport themselves around the world
- To avoid climate change, road transportation is replaced by magnetic transportation.

1.7
Accelerators – Information Technology and "Singularity"

> *"The speed of change will continue being exponential. The merge of humans"*

> *It's not a faith in technology. It's faith in people*
> Steve Jobs

In this last section on the megatrends, the accelerators, we would like to capture some of the latest and most radical thoughts on innovation and technology. Ideas that could certainly alter the speed of change and completely change most of the trends previously discussed, especially when looking at the world by 2050.

1.7 Accelerators – Information Technology and "Singularity"

This book will use none of these accelerators on any of the megatrends and scenarios for the chemical industry in 2050, however, as we cover a very long period of time, we deemed it necessary to acknowledge their existence, and the fact that indeed they could accelerate and alter significantly some of the megatrends and, by default, our scenarios.

Therefore we will present some of the most pervasive ones, leaving you the option to believe in them or not and to use them or not in your own projections for the chemical industry.

During the last century humanity has seen an unprecedented amount of change, unique in human history. As we saw during the review of the major megatrends, during the last century the world witnessed a massive increase in wealth, population, education, health, and wealth levels, leading to massive changes in trade with globalization, transportation with millions of cars, planes, and so on, and in the way people communicate and live, with the arrival of the radio, television, internet, and, most recently, social media. Additionally and during the last 30–40 years *information technology, computer science, and the internet* have served as key enablers for all other changes, accelerating them tremendously.

Although many people will agree with the idea that information technology has served to accelerate and enable change in all areas of our society, it seems difficult to find real ways to probe or justify that, perhaps because it is so obvious to us. Thus we have tried to find two simple examples that perhaps can help us to visualize what information technology has done and could do for society in the years to come.

According to Mr. Michio Kaku, a prestigious professor at Princeton University and author of the bestseller *The Physics of the Future*, "In 2011 a smart mobile phone has the same computational power as the whole of NASA in 1969, when NASA landed two men on the moon".

Many people have already noticed this massive progression in technology, especially in the ADV economies. When we look at the first mobile phones introduced just 40 years ago, versus the current and most advanced ones, we can notice the massive changes and progression. Perhaps what we did not know is what was behind this fast progression, how computers, mobile ones, have not only become faster, better, and smaller, but even cheaper.

In 1965 the Intel co-founder, Gordon E. Moore published the famous paper "Cramming more components onto integrated circuits," where he noted that the number of components in integrated circuits had doubled every year from the invention of the integrated circuit in 1958 until 1965 and predicted that the trend would continue "for at least 10 years", while prices would continue falling from 1962 to 1970 (Figure 1.77).

Lately, and based on this statement, some professors coined the now famous Moore's law that describes a long term trend in the history of computing hardware, where the number of transistors that can be placed inexpensively on an integrated circuit doubles every two years. This trend has continued for more than half a century and although experts expect a decline, the number of transistors is still doubling every three years. A graphical illustration of Moore's law is shown in Figure 1.78).

134 | *1 Global Megatrends by 2050*

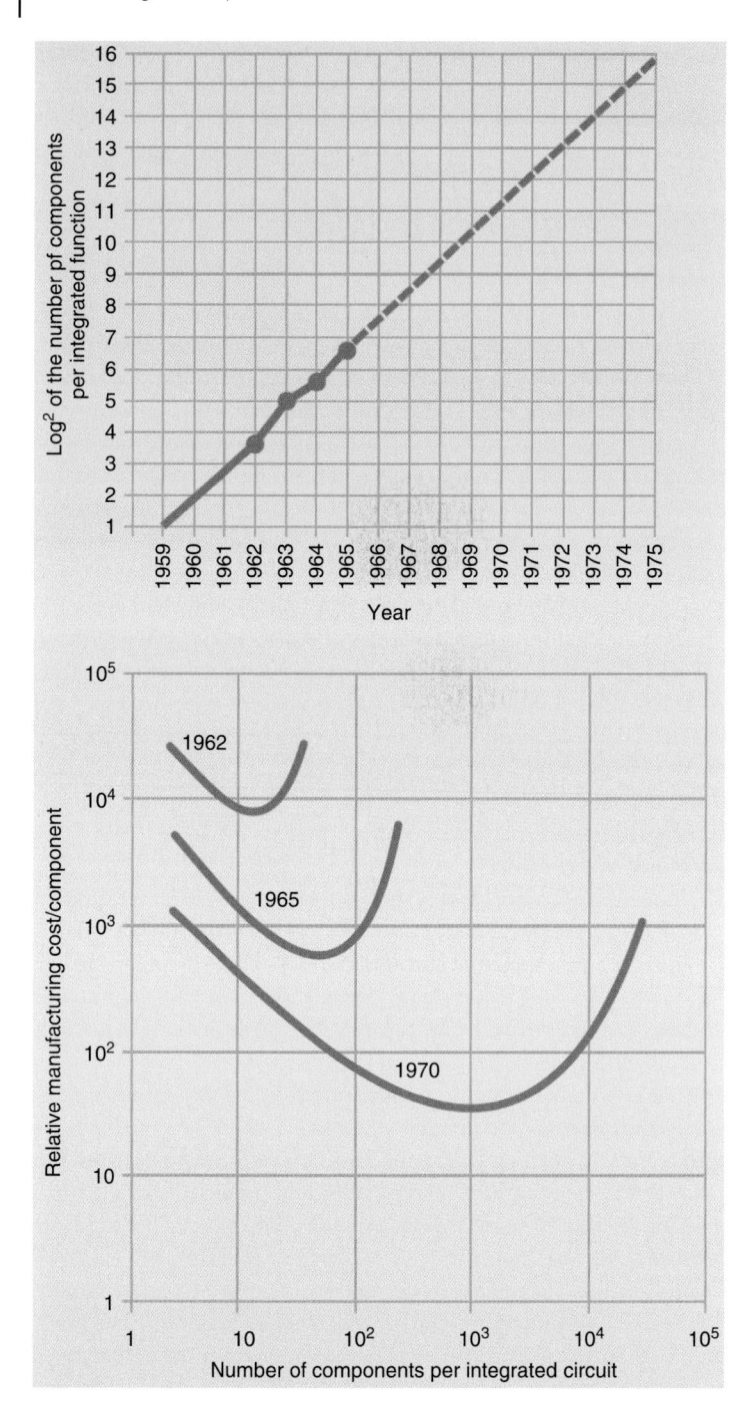

Figure 1.77 Moore's projections. (Moore, G.E. 1965, Cramming more components onto integrated circuits, Electronics, 38(8).)

This exponential growth in terms of transistor capacity has enabled tremendous progress in computational capacity, capacity that when linked to the internet has accelerated the path of innovation and change in our society.

Building on Moore's Law and the fact that computational capacity keeps growing very fast, Vernor Vinge, a famous science fiction writer, coined the term "singularity" and argues that artificial intelligence, human biological enhancement or brain/computer interfaces could be possible causes of the singularity.

As human minds have a computational or technological limitation, while computing power has an exponential growth, at a certain time, not far away, computational capacity in computers could be larger than in humans. Indeed by 2023 computing power in computers will be larger than in a human brain, and by 2045 the computational power of one computer could be as large as for the whole world population. That point has been defined as singularity.

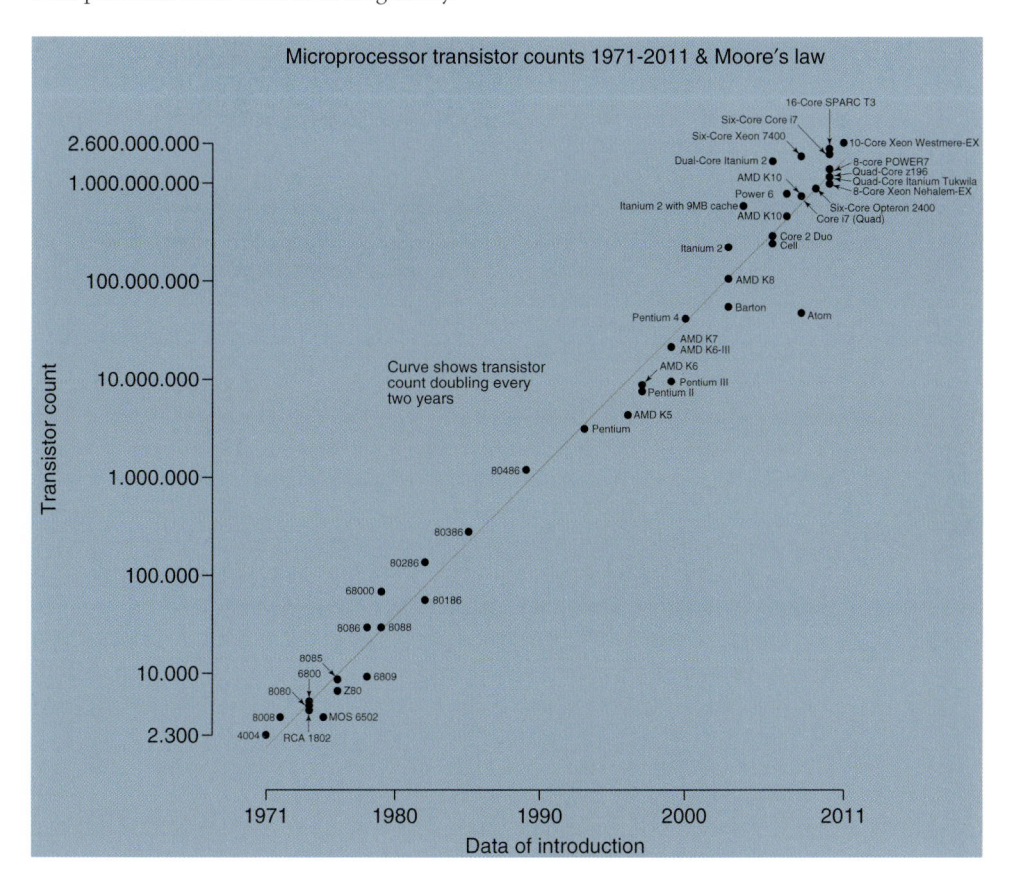

Figure 1.78 Moore's law. Source: Publication: "Cramming more components onto integrated circuits" By Gordon E. Moore – Director, Research and Development Laboratories, Fairchild Semiconductor division of Fairchild Camera and Instrument Corp. Published in Electronics, Volume 38, Number 8, April 19, 1965. http://www.intel.com/pressroom/kits/events/moores_law_40th/

Technological singularity refers to the hypothetical future emergence of greater-than-human intelligence through technological means. Since the capabilities of such an intelligence would be difficult for an unaided human mind to comprehend, the occurrence of a technological singularity is seen as an intellectual event horizon, beyond which the future becomes difficult to understand or predict. Nevertheless, proponents of the singularity typically anticipate such an event to precede an "intelligence explosion," wherein super intelligences design successive generations of increasingly powerful minds. That growth is illustrated in Figure 1.79.

Despite the logical controversy and suspicions that the concept of "singularity" can create, the fact that computer power will continue increasing in the next decades, and computer and human brain power could work even closer together or even merge has created very interesting projections in multiple fields. Projections that if they were to become true, could have a huge impact on some of the previously mentioned trends, and even if they become only partially true could still have a large impact on our megatrends and potential scenarios for the chemical industry by 2050.

Let us quickly review the potential impact of "singularity" on some fields and then discuss briefly what could be the impact on our megatrends and potential scenarios for the chemical industry.

"Singularity" could impact multiple fields in our society, from computer science and social networks to artificial intelligence and robotics, to nanotechnology, biotechnology, medicine, forecasting analysis, or human life expectancy. Let us cover below some of them, while presenting what potential impacts they could have on our society.

- **Artificial Intelligence and Robotics**. "Singularity" could accelerate the introduction of intelligent machines, eventually even robots, that could sense, learn, act, and move independently. Machines and robots that could work with artificial intelligence, able to learn and communicate. Robots that could be ubiquitous in our daily life, being used at home, in medicine, in transportation, education, and so on.
- **Computer Science and Social Networks**: Exponential growth in computing is a fundamental enabler of the changing technologies. Singularity could impact the way the internet searches, presents, and stores information. New versions of the internet could carry out an intelligent search not only by words but by the content and context of your search. Imagine a new internet able not only to search across the 24 billion web pages available, but to give you a solution to your search by building knowledge across multiple web pages at once. Some of the new versions include semantic web, ubiquitous networking and sensing, cloud computing
- **Medicine and Neuroscience** singularity could support several areas in this field. Imagine a world where doctors could have timely access to information on the status of your body, detecting illness and issues in real time, and solving them with customized treatments and regenerative medicine, where doctors could replicate or regenerate our organs or muscles, and robots could insert these seamlessly into our bodies.

In a world where "singularity" is reached and so much change and progress attained we could see many potential consequences for humans and society.

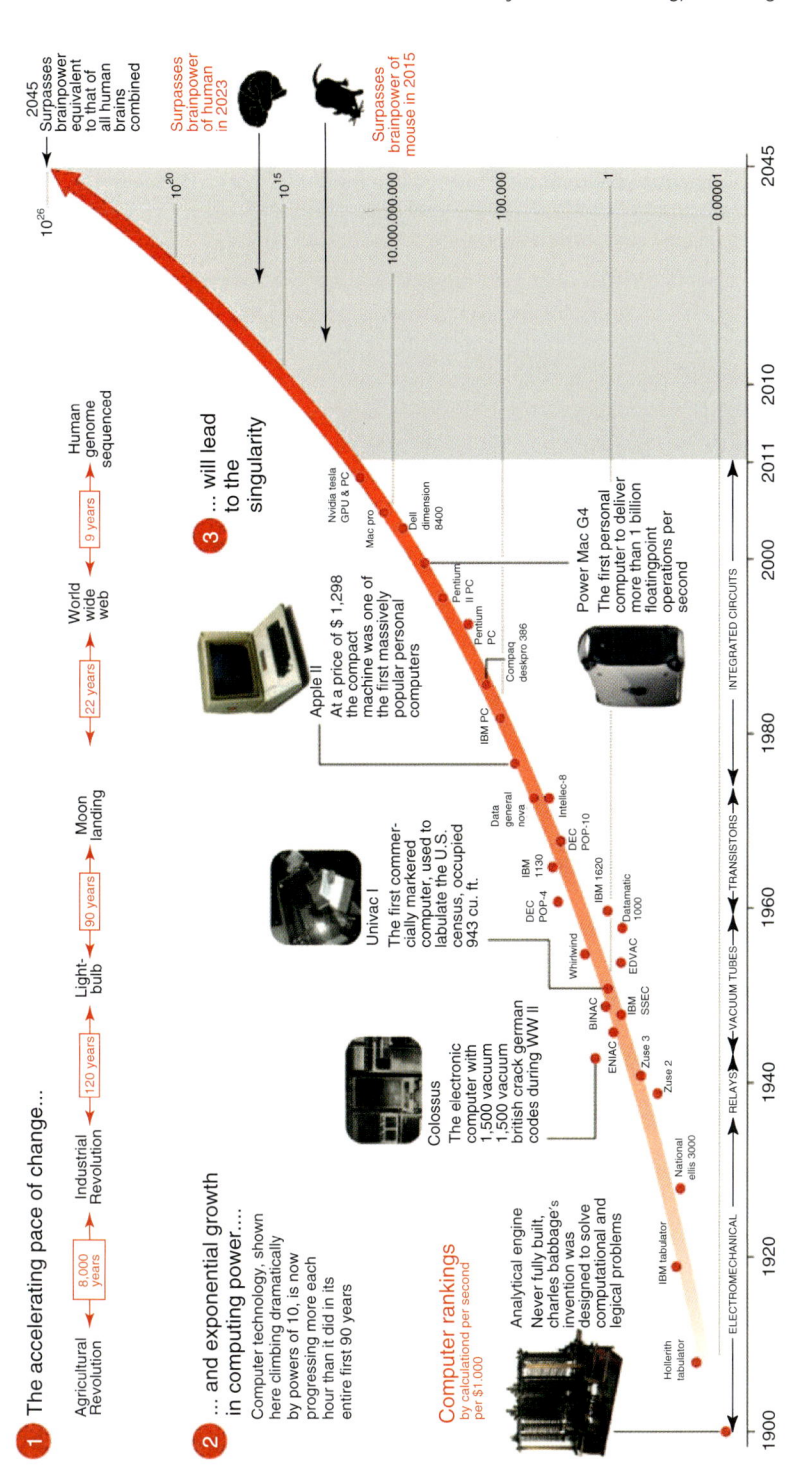

1.7 Accelerators – Information Technology and "Singularity" | 137

Figure 1.79 Singularity projections. (Source: Time magazine – February 21st, 2011 – Vol. 177. No 7.)

Perhaps one of the most notorious and pervasive is the potential for humans to live much longer than today. In the social megatrends we share the view that the world average life expectancy would go up to 75 years by 2050. However "singularity," thanks to much better and efficient medicine could extend life expectancy much longer than that, at least for some large parts of society.

Sonia Arrison, a futurist, and relevant member of the singularity way of thinking, recently presented a book called *100 Years Plus* where she reviewed the chances for human population to have a life expectancy of more than 100 years and up to 150 years by 2050. She argues that the increased levels of wealth, and consequently health, in combination with the potential progress "singularity" could bring to medicine could expand significantly the human life span. That could certainly bring multiple advantages but also multiple issues and changes, such as the way we live, relate to each other, consume, and perceive life in general. That situation could require significant adjustment in the way we live, work, study, marry, or have children and family.

For obvious reasons we did not consider this scenario in our megatrends, but as you can imagine the impact of such a situation would not be minor, including a world of more than 9 billion people. Under this scenario, a large number of people in the ADV economies and in the rest of the world would live much longer than estimated in our current scenarios, a fact that could have significant impacts on society, the economy, and all the projections we will make in the next chapters.

Singularity and a much higher level of computer power could help to solve most of the upcoming challenges. Supercomputers could help us to run much faster and more accurate projections in all fields, from economics, financial, medical, or social. They could search for more complex solutions, when we look for better or more efficient sources of energy, ways to reduce CO_2, or how to solve the major illnesses of our society like cancer, Human immunodeficiency virus infection/acquired immunodeficiency syndrome (HIV/AIDS), or large pandemics.

In general, a world that reaches singularity could be a much "better" world but is a world that is much more difficult to forecast and predict. Singularity could impact most of our existing megatrends, with much older and smarter people. A world that will become more populated and wealthier than projected today and perhaps with different problems. Climate change, CO_2, energy, or some critical health issues perhaps will be or could be solved much faster than today, as supercomputers could enable better technical analysis and developments to solve some of our much larger issues.

For obvious reasons we will not consider singularity and some of these extreme scenarios yet, although we clearly recognize the already large influence computer power has had during the last 40 years in enabling and accelerating human progress, and its real capacity to do it again during the next 40 years, accelerating even further and faster all fields of human progress.

For those readers interested in the subject we recommend the following reading.
What is the singularity?
http:/ginst.org/overview/whatisthesingularity/

Appendix: Climate Change

United Nations – Framework Convention on Climate Change

Climate Change – Background
http://unfccc.int/essential_background/the_science/items/6064.php

The year 2007 marked a new and disturbing global realization. The world learnt that climate change was human-made, definitely happening, and that the collective global effort so far to keep greenhouse gases to a "safe" level was grossly insufficient.

The IPCC had released its Fourth Assessment Report (AR4), in the wake of an unusual number of severe weather-related disasters, and at the head of an almost unbroken series of the hottest years on record. Any child under the age of 10 in 2007 was part of this worrying global trend: almost every year he or she had lived on Earth had been among the hottest in living record.

These are some basic well-established links:

- the concentration of greenhouse gases in the earth's atmosphere is directly linked to the average global temperature on earth;
- the concentration has been rising steadily, and mean global temperatures along with it, since the time of the industrial revolution; and
- the most abundant greenhouse gas, carbon dioxide, is the product of burning fossil fuels.

Greenhouse gases occur naturally and are essential to the survival of humans and millions of other living things, through keeping some of the sun's warmth from reflecting back into space and making earth livable. But it is a matter of scale. A century and a half of industrialization, including clear-felling forests and certain farming methods, has driven up quantities of greenhouse gases into the atmosphere. As populations, economies, and standards of living grow, so does the cumulative level of greenhouse gas emissions.

AR4 took stock of where we are and what we now know. Thanks to the IPCC, here's a quick snapshot of what we know:

The average temperature of the earth's surface has risen by 0.74 °C since the late 1800s. It is expected to rise another 1.8–4 °C by the year 2100 if no action is taken. That is a fast and intense change in geological time. Even if it "only" gets another 1.8 °C hotter, it would be a larger increase in temperature than in any century in the last 10 000 years.

- About 20–30% of plant and animal species are likely at higher risk of extinction if the global average temperature goes up by more than 1.5–2.5 °C.
- Nine of the last ten years were the hottest years on record, according to the US National Oceanic and Atmospheric Administration (NOAA). 2005 and 2010 tied for first place. In second place was 1998.
- The average sea level rose by 10–20 cm over the 20th century. An additional increase of 18–59 cm is expected by the year 2100. Higher temperatures cause the ocean volume to expand. Melting glaciers and ice caps add more water. And

as the bright white of ice and snow give way to dark sea green, less and less rays from the sun are reflected back into space, intensifying the heating.

What we know we don't know, and what we don't know we don't know.

But, these days, it is what we don't know that is the most worrying – because you cannot properly prepare for what you cannot foresee. Knock-on effects of even small changes in many natural systems show just how delicate a balance nature strikes.

Scientists talk about "tipping points," where a gradual change suddenly moves into a self-fueling spiral. How much methane is trapped in the melting permafrost and in sea-beds in a warming ocean, and, if some or all of that methane is released, what effect will it have on the global temperature and climate? If the ice cover in the poles keeps shrinking so that there is less bright white surface and more dark liquid sea surface, how much more heat from the sun will the dark surface trap, and how much less can the ice packs reflect back into space? Sea mass expands when warm – how much will this add to the rise in sea level?

Each of these is among the simplest examples of potential vicious cycles identified by scientists.

There is also another unknown. At some point, bright children ask questions about electricity, light and heat, and, inevitably, "where does oil come from?". The simple answer is that, hundreds of thousands of years ago, before humans, the animals and plants that died accumulated on the bottom of water bodies, mixing with sand and mud. Sediment kept piling over the top of that, and the heat and pressure eventually transformed it into oil, petroleum, or natural gas. These are trapped in porous layers in the earth, prevented from escaping by a non-porous layer of rock.

That's the leading scientific theory, in any case, and no one has the definite answer on whether the world's oil reserves will, eventually, run out.

Right now, coal, oil, and natural gas power the economies of the world; almost all modern human endeavor produces carbon dioxide. This makes climate change extremely complex, tied up with other difficult issues, such as poverty, economic development, and population growth. Clearly, dealing with climate change is not easy. It is not about to get any easier but ignoring it would be worse.

2
The World by 2050

> *The ultimate measure of a man is not where he stands in moment of comfort and convenience, but where he stands at times of challenge and controversy.*
>
> Mr. Martin Luther King, Jr.

2.1
"A Much Larger, Wealthier, Healthier, and Sustainable World"

By 2050 our world will be a "much larger, wealthier, healthier, and sustainable world", and certainly a better place to live. A world with more than 9 billion people and an estimated Gross Domestic Product (GDP) of US $280 trillion (2009 dollars) will host the wealthiest society in human history.

That society will not only enjoy the highest GDP per capita in human history, with US$ 30 675 (2009 dollars) but, more importantly, the highest life expectancy in our history. The average world life expectancy will be raised from the current 67 years to 75 years, while a child born in 2050 will have a life expectancy well beyond 100 years. The average life expectancy in the ADV economies will be above 80 years. Progress in wealth, education, and medicine, will support the creation of not only the largest but also the wealthiest society in human history.

Millions of people will be pulled out of poverty, and the world will have a huge super middle class with more than 50% of the population with an average GDP per capita of US$ 30 675 on a 2009 basis, similar to the current average in Europe today. The world urbanization ratio will keep increasing and millions of people will move into cities. The world will have the largest amount of megacities, those with more than 10 million people, in human history, including several huge megacities with more than 30 million people. Tokyo in Japan is expected to become the largest city in the world with more than 36 million people. Can you imagine how many countries will be smaller than just one city? Can you imagine how it would be to live in such a megacity? Urbanization's challenges and opportunities will be formidable and the synergies paramount.

The enormous progress in computational programs and information technology will not only accelerate all aspects of our life, from medicine, to transportation, education, business, telecommunications, energy, and/or chemicals production

The Future of the Chemical Industry by 2050, First Edition. Rafael Cayuela Valencia.
© 2013 Wiley-VCH Verlag GmbH & Co. KGaA. Published 2013 by Wiley-VCH Verlag GmbH & Co. KGaA.

but also foster the convergence of technologies, industries, and societies. Sciences, devices, and industries will start converging as never seen before. Computers and the internet will connect all aspects of our life, and our capacity to measure, share, analyze, and understand the status of our world and ourselves will not only be instantaneous but also much more accurate than today.

This extraordinary growth will result in multiple changes in all aspects of human life, from political to economic and social. Politically, the expectation that the BRIC economies will be among the top 10 largest economies in the world, especially after China and India will become the two largest economies, will serve to reshuffle the current international order. Additionally, the expectation that new emerging countries like Mexico, Turkey, Indonesia, or Nigeria will be among the top 10 largest economies will accelerate that trend. Having said that, the ADV economies, the USA, Europe, Japan, and Canada, will reach levels of wealth never seen before. For instance, the USA with a larger population than today, will have one of the highest GDP per capita in the world with a staggering average GDP of more than US$ 97 000 (2009 dollars), doubling its current GDP in 2010. At these levels, you can start questioning how super wealthy could that society and their citizens be? What could they afford? How would they like to live and work?

Meanwhile countries like China, India, or Brazil will also see their GDP per capita multiplying by up to nine times. So if you are now in one of these countries, try to imagine how life could be with nine times more income than today; and if you are in India try to imagine yourself with seventeen times more money in your pocket. How much wealthier would we feel? Additionally, and even more importantly, millions of people will be pulled out of extreme poverty around the world. Indeed the GDP per capita for the groups of countries defined under the REST, basically all other countries with the exception of the ADV and BRIC economies, will see their GDP per capita tripled from US$ 3926 in 2010 to US$ 16 687 by 2050 (2009 dollar). In parallel, and under the ever growing shadow of globalization, huge corporate mega-economies will continue to grow. If in 2008, 50% of the world largest economies were companies, by 2050 that percentage might be even bigger, but even more important companies will become much bigger than countries. If in a world economy of US$ 63 trillion (2011 dollar), the two largest companies by market capitalization (Apple and Exxon) each accounted for more than US$ 400 billion in 2011; in a world of US$ 280 trillion (2009 dollar) could you imagine how large multinationals could become?

Under this new scenario, International Organizations will need to adjust to the new international order, so we could expect higher weight from the BRICs and emerging economies in these organizations and in our general daily lives. New organizations will also need to be created. Organizations where not only the BRICs will need to have a bigger role, but also large multinationals and large corporate mega-economies, as large economies will need to be represented.

Socially, the convergence of globalization and social media in a much wiser, wealthier, and healthier population will provoke some of the most pervasive changes of all – the way our world is governed. Social media and the internet will enhance real-time communication among the different stakeholders, fostering a real-time democracy. So far, citizens have been exercising their power over governments and companies through periodical elections or referendums in democracies, and through consumption in the case of companies. Social media and the internet have opened a new direct and instantaneous channel of communication with governments and companies. Social media might have the power to increase the numbers of democracies around the world, as well as improving the quality of these. Democracies might go to new models that better reflect citizens' real-time opinions, while government's and company's behavior will be scrutinized and validated in real time. In this society, reputation and corporate social responsibility will become paramount, and much more critical than today. Power will be given back to the citizens and real-time democracies with much people involvement and decision making should be expected.

Ultimately, all this will imply a tremendous and unprecedented growth for all industries in all geographies, but more especially in those related to the BRIC and the REST economies. Besides the BRICs, large countries like Indonesia, Mexico, Turkey, Nigeria, Korea, or Vietnam, just to mention some, and the whole African continent will be exposed to a massive growth during the next 100 years. Most of the industries will see spectacular growth, especially the most commodity oriented ones, however, none will be subject to a higher challenge than the energy industry. The convergence of two conflicting trends will put the energy industry in a unique position. The formidable growth our world is expected to have during the next 40 years will trigger an unprecedented growth in energy demand. If, by 2010, the world energy demand amounted to 12 002 million tones of oil equivalent (MTOE); by 2050 that demand, according to our business as usual (BAU) scenario could theoretically more than triple. However, the clear need to reduce greenhouse emissions by half from today's levels, in order to avoid climate change, will imply a drastic reduction of emissions and, therefore, energy demand, especially from fossil fuel energies (currently 80% of our world energy).

A world poised to host the wealthiest society in human history, will have a huge obligation to preserve all the positive aspects derived from the expected economic growth, while facing the enormous challenge to avoid climate change. In other words, to grow while reducing energy demand and greenhouse emission. Carbon productivity of 10 times will not only require the commitment of all involved parties, including governments, industries, and citizens, but also much greater awareness of the impact of our actions and the consequences of these. In a sustainable world, with just 4000 g of CO_2 per capita per day (equivalent to driving a car with 130 g CO_2 emissions per km for 30 km); citizens will need to learn about their carbon footprint and their alternatives. Industries and regulators can provide the tools and incentives to reduce greenhouse gas emissions, however, this will only start occurring when all citizens are aware of the trade-offs, the real impact of their actions. At the same time the energy industry will need to deploy gigantic amounts of resources on alternative energies, while the chemical industry will find itself in a challenging position too; facing massive demand and spectacular growth, developing new products to curb emissions, while cutting its own emissions and energy consumption.

I believe in a sustainable world, because there is no other way to achieve all the expected positive aspects of the upcoming economic growth, and there is no other way to preserve our quality of life. As challenging as it is, I am confident the world will cut emissions and avoid climate change. There is no alternative!

Under that urgent need, we believe all stakeholders and our world will deliver. Therefore, I believe the next 40 years will present one of the most challenging and complex experience of human history, with our world facing a truly global issue that puts at risk the sustainability of our planet. The rewards for our society are so large, the risk so unbearable, and the stake so high that our society will not have any other option but to make it happen, and I believe we will make it happen. Success in this challenge will enable the largest, wealthiest, and healthiest society in human history, but even more importantly, "a sustainable world." This is something this generation can simply not deny to the upcoming ones; a challenge where we all need to fight together.

2.1.1
Methodology

During the next pages we would like to guide you through two different scenarios until we will arrive at our vision for the World by 2050 – "A larger, wealthier, healthier, and sustainable world."

The first scenario will be the business as usual one (BAU), and will imply a non-sustainable model of growth. With this scenario we will project how the world could look by 2050 under the assumptions that the world will grow in accordance with the available projected economic and population forecast, and without changes in terms of energy demand and greenhouse emissions. In some cases we will introduce some modifications or adjustments to the scenario, in order to have a more realistic or conservative scenario. These modified scenarios will be called modified or adjusted BAU scenarios.

The second scenario will be based on a "sustainable model of growth, and reductions of energy demand and greenhouse emissions will be applied." This scenario will be based on the latest energy projections from the International Energy Agency (IEA) and will be complemented with minor calculations and analysis from our own model.

In Chapter 1 we presented third party projections for the world population, the economy, energy demand, life expectancy, and climate change, however, we never put them all together, nor did we validate the potential interaction among them or the likelihood they all could still occur at the same time. In this chapter we will review how all these trends could operate together and we might challenge some of the existing projections and assumptions. Additionally, for some of these topics we will share our own projections.

Ultimately, the whole purpose of this chapter is to present an overview of how the world could and should look by 2050, on what basis and with which assumptions. Thus, when we start looking at the future of the chemical industry and the potential impact of these megatrends on it we will be able to make the right projections for the industry. On the one hand we will try to estimate how this new world could have an impact on the chemical industry, and on the other hand we could estimate how the chemical industry, as an enabler of other industries and provider of solutions, could influence and improve the world by 2050.

Therefore, without spending too much time on the preamble and introduction let us start the process of defining some of the key parameters that define our world as we know it today, and start confronting these with some of the projections that have been made for 2050. For this purpose we have defined our world around three major pillars or inputs and one final output.

For the first pillar, the social one, we have selected two simple indicators that summarize the basic state of our world – the world population and the world GDP. For the second indicator we have selected one critical indicator that tends to reflect the stage of development and industrialization of a society. For that purpose we have selected energy demand as the second indicator. Within this pillar we will also cover the demand for crude oil and natural gas, as the largest sources of energy in our society today. As a third pillar we have selected one simple indicator of the potential sustainability of that society. For that purpose we have chosen greenhouse emissions or climate change. For that indicator we will track the world total CO_2 emissions and the levels of CO_2 concentration in the atmosphere. Finally, as major outputs or result of all these inputs, we have chosen two indicators that could potentially reflect the quality of life of a society. For that purpose we have considered two indicators, life expectancy and GDP per capita.

These pillars and indicators are certainly not the only ones, and perhaps not even the best ones, but considering the purpose of our analysis and the expected megatrends we deemed them as the most appropriate and simple ones. Now, on this framework we will start reviewing how the world looked in 2010, and how it could look in 2050, depending on our two base scenarios.

2.2
Status of the World – 2010

In 2010 the world, despite the severe impact of the world economic recession that began in 2008, hosted the largest and wealthiest society in human history (see Figure 2.1). The world, with a total population of 6.8 billion people and GDP of US$ 63 trillion ($75 trillion in 2009 dollar) not only managed to triple its population during the last century but also managed to increase its GDP per capita and life expectancy to levels never seen before.

By 2010 the world reached and impressive average GDP per capita of US$ 9210 and an average life expectancy of 65 years; showing a tremendous improvement versus last century. However still there were significant differences in GDP per capita and life expectancy across the world, and major inequalities remained a critical area of concern and improvement for society.

The massive growth during the last century, especially during the second half, did not come alone. The world energy demand reached 12 002 million tones of oil equivalent (MTOE), including 87 million barrels of crude oil demand daily and

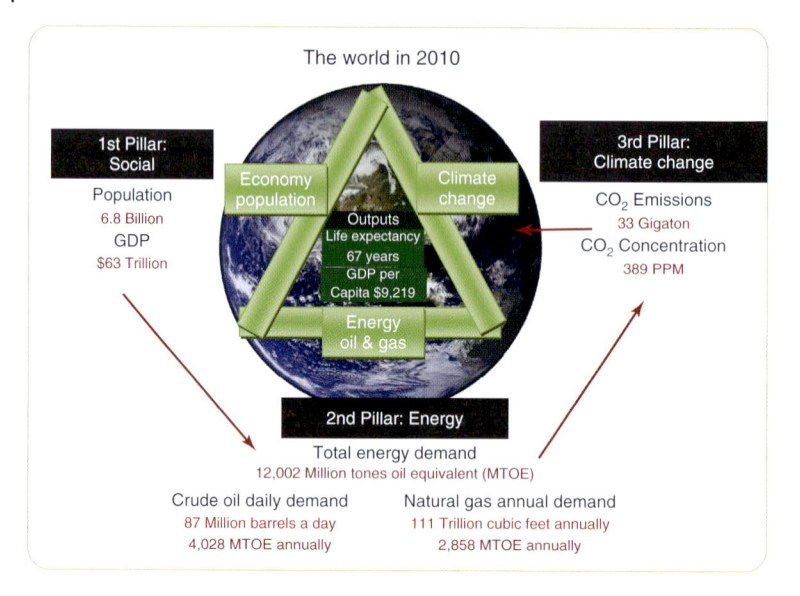

Figure 2.1 Status of the world in 2010.

111 trillion cubic feet (tcf) of natural gas annually. Both record high figures on historical basis. Unfortunately, as a negative consequence of all that development, greenhouse emissions and climate change also appeared.

By 2010 the world total CO_2 reached 33 Gt annually; and the concentration of CO_2 in the atmosphere reached 389 ppm at the Mauna Loa Observatory in Hawaii, both figures being the highest levels in the Earth's history.

At this time and despite the still large differences between countries, we could certainly state that the world as a whole reached highest level of wealth and quality of life in human history, after an unprecedented period of growth unique in humankind.

2.3
The World in 2050

2.3.1
BAU Scenario

Under the "BAU" scenario we will present a "larger, wealthier, healthier, but unsustainable world." See Figure 2.2. For this scenario we will consider the population forecast from the World Bank and the economic forecast from PriceWaterHouseCoopers, while considering no additional improvement in the observed ratios of efficiency in energy demand and carbon productivity. This is the BAU scenario, where the world population and economy will keep growing but energy efficiency and carbon productivity will remain as observed for the last 40 years.

Under these assumptions and for this scenario we could foresee a world with a very large population of 9.1 billion (+34%) and a GDP of $280 trillion (+345%).

2.3 The World in 2050

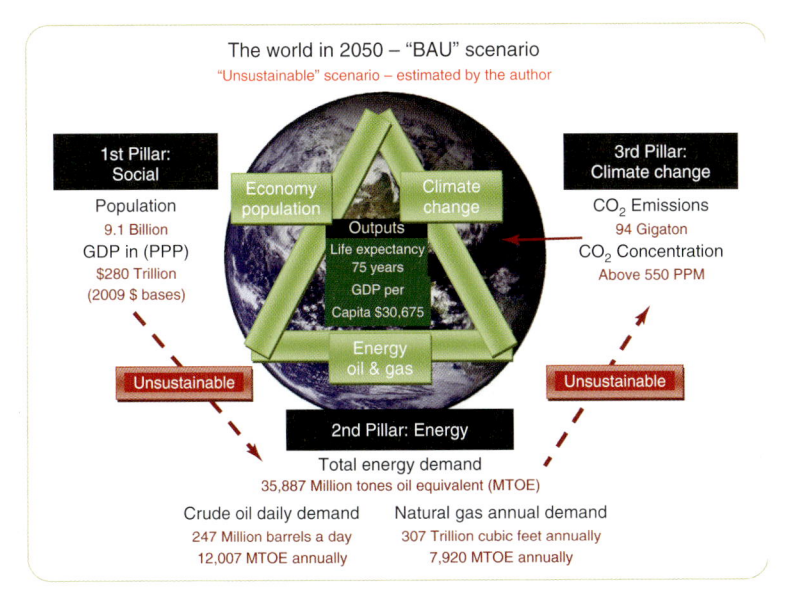

Figure 2.2 Status of the world in 2050 – BAU scenario. Source: Own elaboration, based on several sources including World Bank and United Nations Economic and Social projections. PWC economic projections for 2050. Energy and CO_2 projections estimate by the author under the BAU scenario.

Energy demand, assuming the observed energy efficiencies remain equal to those experienced during the period 1970–2010, will rise to 35 887 million metric tones of oil equivalent. That figure will include a demand of 247 million barrels a day of crude oil, or 12 007 million metric tones of oil equivalent (equivalent of 2010 total world energy demand); and a demand of 307 trillion cubic feet of natural gas or 7 920 million tons of oil equivalent. CO_2 emissions, assuming the same carbon productivity as during the last decades, could reach 94 Gigaton (Gt). Under this scenario the CO_2 concentration level could go well above 550 ppm, implying variations in temperature of 3 to 6 °C.

Life expectancy will continue to increase globally, with an expected world average of 75 years by 2050, 8 years higher than in 2010. Additionally, significant parts of our population with start having life expectancies well above 100 years, especially in the ADV economies. That will be a massive achievement in terms of quality of life, achievement that will be complemented with a huge increase in the average GDP per capita. The world average GDP will triple from the current $9219 in 2010 to the projected US$ 30 675 (2009 dollar). The distribution of wealth will still be unbalanced in favor of the ADV economies, with their average GDP per capita still double that in the BRIC economies; however, the percentage increase in wealth in the BRIC and REST economies will be much higher than in the ADV economies.

The ADV economies, with just 12% of the world population, will reach a staggering average GDP per capita of $78 000 (2009 dollar). The BRIC economies with 38% of the population will also enjoy massive improvements in GDP per capita, reaching an average of $36 000 (2009 dollars base). Finally, the REST, with 50% of the population will also reach an unprecedented average GPD per capita of almost $17 000 (2009 dollar). All these impressive improvements in wealth mean that millions of people around the world will be removed from poverty forever. Indeed, a world super middle class will be created and our society will enjoy the highest level of wealth in human history.

Under the BAU scenario the world energy demand is also expected to triple from 2010 levels, however, not all the areas will be subject to the same sort of increase (see Figures 2.3 and 2.4). Indeed, the BRIC (+340%) and REST (+324%) economies are expected to show the bulk of that growth in demand, while the ADV (1%) economies on the back of lower growth and higher energy efficiency will see their energy demand remain practically the same.

Energies with the highest expected demand increases are coal, crude oil, and natural gas. The strong growth expected in the BRIC economies, especially China, largely consumers of coal and crude oil justify these figures. For further details on the energy and crude oil projections under the BAU scenario please refer to the Figures 2.3, 2.5 and 2.6.

However, despite all the optimistic figures expected for economic, social, and quality of life aspects this BAU scenario will be completely unsustainable. A projected energy demand of 35 887 million tons of oil equivalent (MTOE) might be attainable but challenging to achieve, especially when considering a projected oil demand of 247 million barrels of crude oil a day. However, the projected amount of CO_2 emissions generated by all that demand will be unsustainable, leaving the world at the mercy of the worst scenarios for climate change.

From an energy perspective and scenario a 200% increase in energy demand might certainly be challenging and expensive but not impossible. An increase in coal demand of 245%, although undesirable from an emissions perspective might be possible. Hydroelectricity (207%) and natural gas (177%) under the "BAU" scenario are not only possible and desirable but actually more will be expected under the sustainable scenario. Similarly for renewable (113%) and nuclear (65%) energy, where the projected demand under the "sustainable" scenario might be larger. However, an increase in crude oil demand by 198% to almost 247 million barrels a day might not only be challenging for the amount of investment required, undesirable from an emission perspective but also might elevate oil prices to historically high and unsustainable levels (Figure 2.5).

Disregarding at this time the fruitless and endless discussion about the potential availability or not of such large amount of crude oil – after all the world crude oil reserves has remained flat at 40 years demand coverage during the last decades; despite massive increases in crude oil demand – is the world capacity and need of refining capacity what appears to us of even greater concern.

In 2010, the world refining capacity was 92 million barrels a day, while the world crude oil demand was 87 million barrels a day. Under the BAU assumption the world crude oil demand will grow to 247 million barrels a day, the refining capacity should grow annually at 35% every decade, from 2010 to 2050.

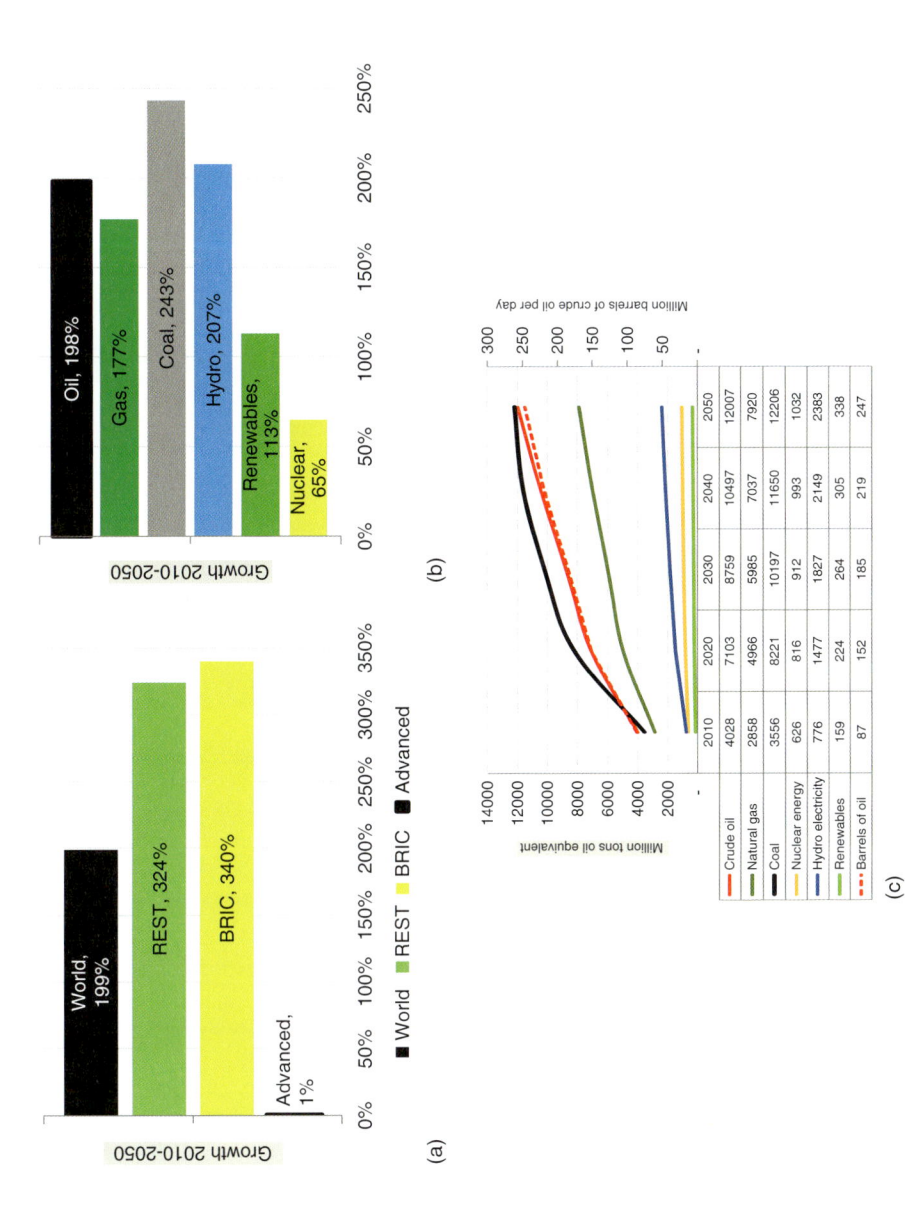

Figure 2.3 BAU scenario by 2050 – world energy demand. (a) by region, (b) by source of energy, (c) as MTOE plotted against year for various energy sources.

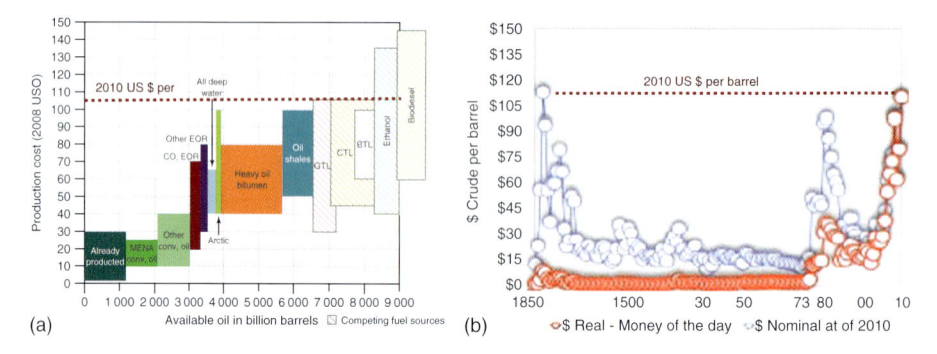

Figure 2.4 (a) Crude oil production cost curve. Source: IEA – International Energy Agency. (b) Crude oil historical prices (nominal vs. real). BP Statistical Energy Report 2011.

As you can imagine, increasing the refining capacity at this speed is not only tremendously expensive but perhaps even impossible. During the 1960s and the 1970s, the decades with the fastest recent growth in refining capacity, the world refining capacity grew by 46 and 55%, respectively. However, during the last three decades this figure has been gradually declining, with an average growth during the 1990s and 2000s of just 10%. Based on these recent experiences, we could conclude that a sustainable growth in refining capacities at 35% for the next four decades might not only be impossible and tremendously costly but also very unlikely.

Assuming now, and for illustration purposes, that the world will have enough crude oil reserves and manages to increase its refining capacity at the suggested rate of 35% in order to be able to extract the suggested 247 millions of barrel of oil per day; the fundamental question would be at what price will this crude oil be produced sold? and if that price would be still sustainable versus other alternatives?

According to a recent analysis from the IEA, that measured the potential cost to extract the existing crude oil reserves versus some potential alternatives, we can observe how higher crude oil prices are expected but also how some alternatives could be starting to become competitive a certain levels.

According to this analysis, see Figure 2.4(a), the first 1000 billion barrels of crude oil produced had a production cost of approximately US$ 30 a barrel. The following 2000 billion barrels will be available at a production cost of US$ 27 for the first 1000 billion barrels and US$ 40 for the following 1000 billion barrels. After that, the production cost for the following billion barrels will increase rapidly to US$ 70 and 80 a barrel, reaching the peak with "Arctic" crude oil at US$ 100 per barrel.

Considering that in 2011 the price of crude oil reached US$ 110 a barrel, and based on the US EIA (Energy Information Administration) estimated production cost for crude oil, we could infer that the current margin on crude oil would be around US$ 70 a barrel. Of course this is a very simple extrapolation and approximation, however, the conclusion we would like to arrive at is not how much margin crude oil is currently generating, but how much margin should we add to the production cost so we can estimate the potential price for the next billions of crude oil.

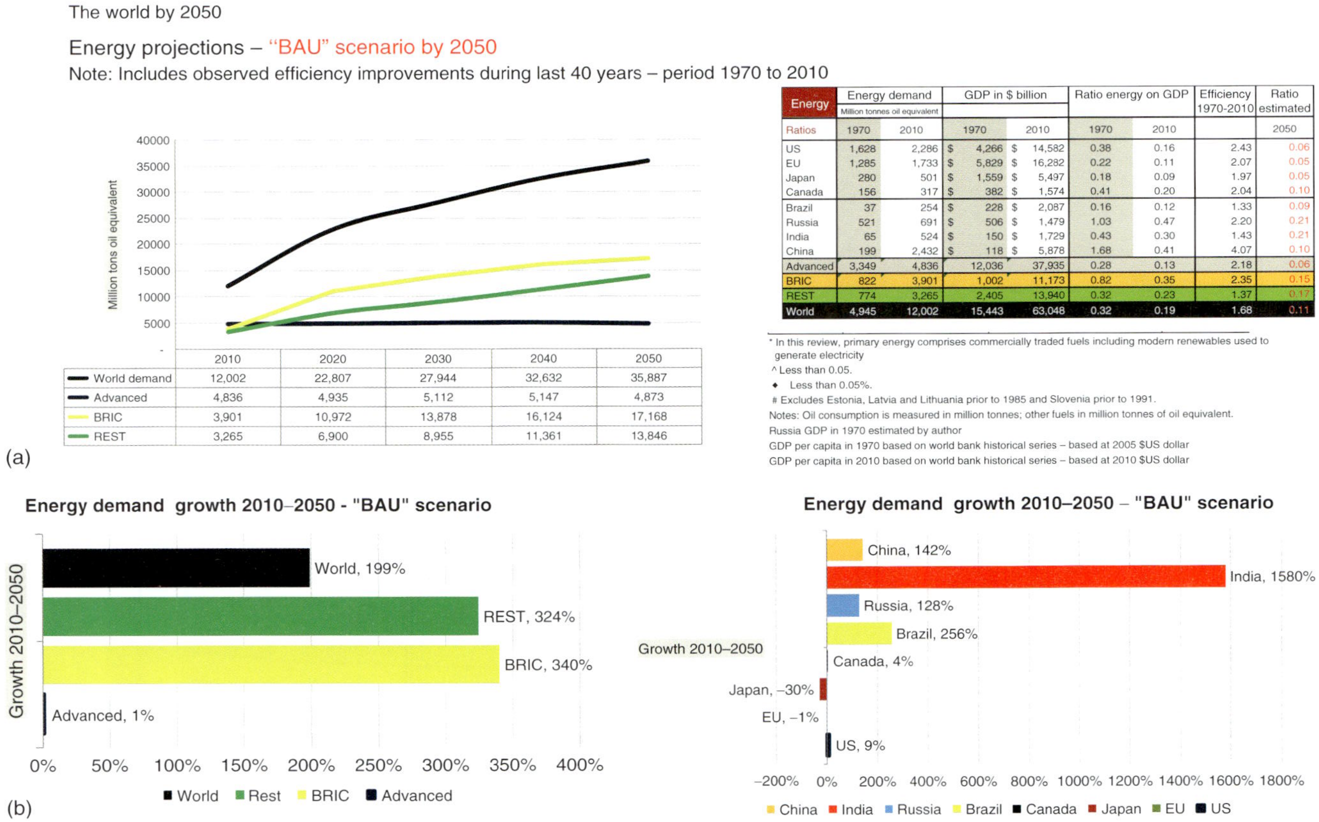

2.3 The World in 2050 | 151

The world by 2050
Energy projections – "BAU" scenario by 2050
Note: Includes observed efficiency improvements during last 40 years – period 1970 to 2010

Energy	Energy demand (Million tonnes oil equivalent)		GDP in $ billion		Ratio energy on GDP		Efficiency 1970-2010	Ratio 2050 estimated
Ratios	1970	2010	1970	2010	1970	2010		
US	1,628	2,286	$ 4,266	$ 14,582	0.38	0.16	2.43	0.06
EU	1,285	1,733	$ 5,829	$ 16,282	0.22	0.11	2.07	0.05
Japan	280	501	$ 1,559	$ 5,497	0.18	0.09	1.97	0.05
Canada	156	317	$ 382	$ 1,574	0.41	0.20	2.04	0.10
Brazil	37	254	$ 228	$ 2,087	0.16	0.12	1.33	0.09
Russia	521	691	$ 506	$ 1,479	1.03	0.47	2.20	0.21
India	65	524	$ 150	$ 1,729	0.43	0.30	1.43	0.21
China	199	2,432	$ 118	$ 5,878	1.68	0.41	4.07	0.10
Advanced	3,349	4,836	12,036	37,935	0.28	0.13	2.18	0.06
BRIC	822	3,901	1,002	11,173	0.82	0.35	2.35	0.15
REST	774	3,265	2,405	13,940	0.32	0.23	1.37	0.17
World	4,945	12,002	15,443	63,048	0.32	0.19	1.68	0.11

* In this review, primary energy comprises commercially traded fuels including modern renewables used to generate electricity
^ Less than 0.05.
♦ Less than 0.05%.
Excludes Estonia, Latvia and Lithuania prior to 1985 and Slovenia prior to 1991.
Notes: Oil consumption is measured in million tonnes; other fuels in million tonnes of oil equivalent.
Russia GDP in 1970 estimated by author
GDP per capita in 1970 based on world bank historical series – based at 2005 $US dollar
GDP per capita in 2010 based on world bank historical series – based at 2010 $US dollar

(a)

	2010	2020	2030	2040	2050
World demand	12,002	22,807	27,944	32,632	35,887
Advanced	4,836	4,935	5,112	5,147	4,873
BRIC	3,901	10,972	13,878	16,124	17,168
REST	3,265	6,900	8,955	11,361	13,846

(b) Energy demand growth 2010–2050 - "BAU" scenario
- World, 199%
- Rest, 324%
- BRIC, 340%
- Advanced, 1%

Energy demand growth 2010–2050 – "BAU" scenario
- China, 142%
- Russia, 128%
- Brazil, 256%
- Canada, 4%
- US, 9%
- Japan, –30%
- EU, –1%
- India, 1580%

Figure 2.5 (a) World energy demand, BAU scenario 2050. BP Statistical Review 2011 and World Bank for GDP (b) Energy demand growth 2010–2050, BAU scenario. Energy demand for various energy sources: (c) World, (d) ADV economies, (e) BRIC economies, (f) REST group.

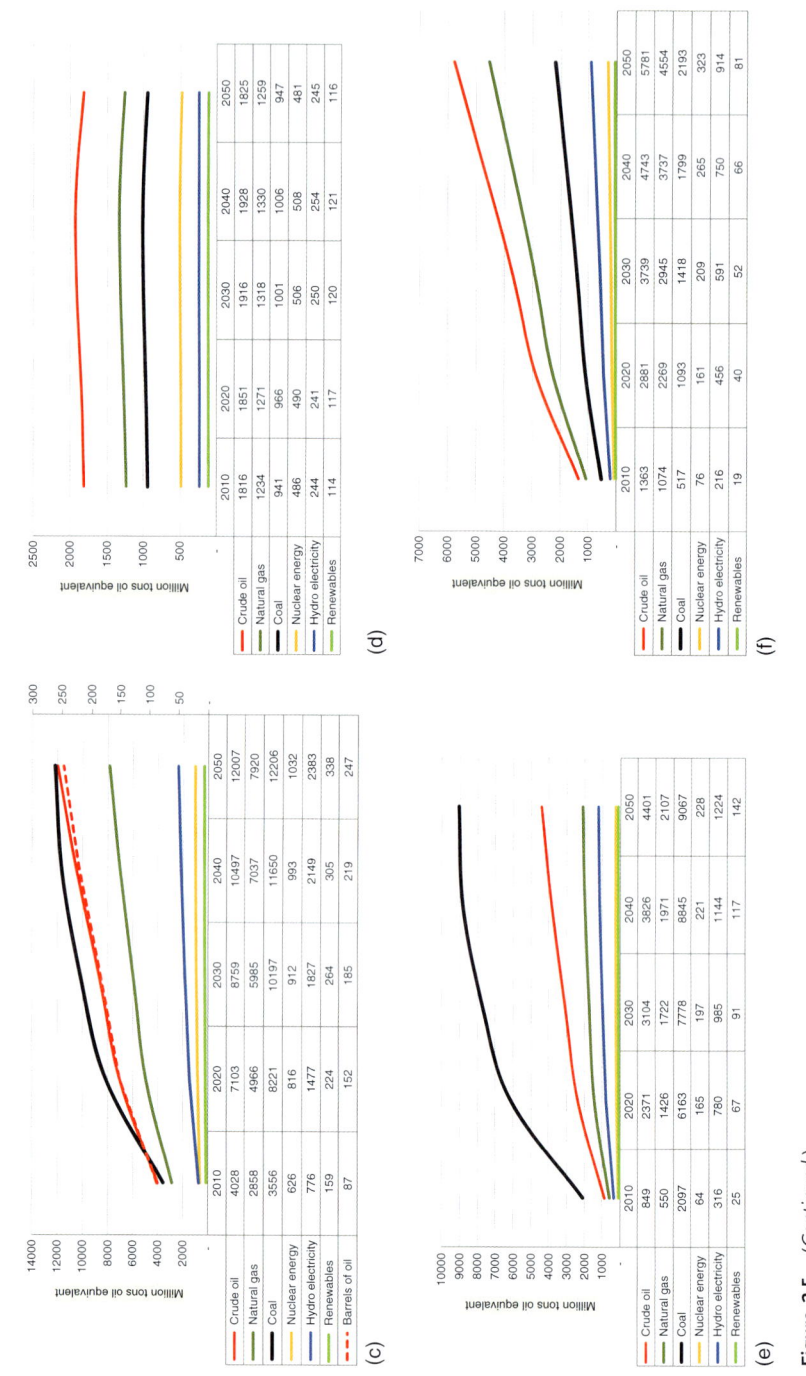

Figure 2.5 (Continued.)

2.3 The World in 2050 | **153**

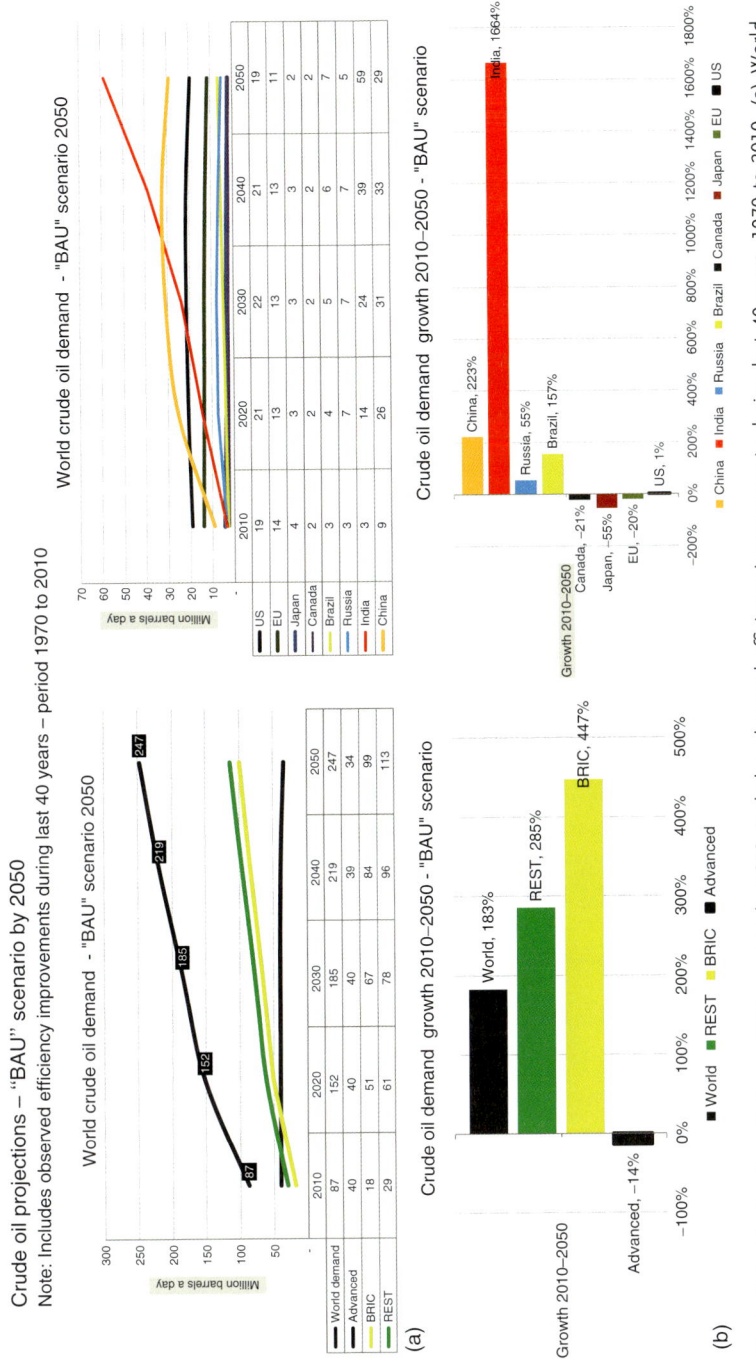

The world by 2050

Crude oil projections – "BAU" scenario by 2050
Note: Includes observed efficiency improvements during last 40 years – period 1970 to 2010

(a) World crude oil demand - "BAU" scenario 2050

	2010	2020	2030	2040	2050
World demand	87	152	185	219	247
Advanced	40	40	40	39	34
BRIC	18	51	67	84	99
REST	29	61	78	96	113

World crude oil demand - "BAU" scenario 2050

	2010	2020	2030	2040	2050
US	19	21	22	21	19
EU	14	13	13	13	11
Japan	4	3	3	3	2
Canada	2	2	2	2	2
Brazil	3	4	5	6	7
Russia	3	7	7	7	5
India	3	14	24	39	59
China	9	26	31	33	29

(b) Crude oil demand growth 2010–2050 - "BAU" scenario

- World, 183%
- REST, 285%
- BRIC, 447%
- Advanced, −14%

Crude oil demand growth 2010–2050 - "BAU" scenario

- China, 223%
- India, 1664%
- Russia, 55%
- Brazil, 157%
- Canada, −21%
- Japan, −55%
- EU, −20%
- US, 1%

Figure 2.6 Crude oil projections, BAU scenario by 2050, includes observed efficiency improvements during last 40 years, 1970 to 2010. (a) World crude oil demand, (b) crude oil demand growth 2010–2050, (c) crude oil per capita demand.

2 The World by 2050

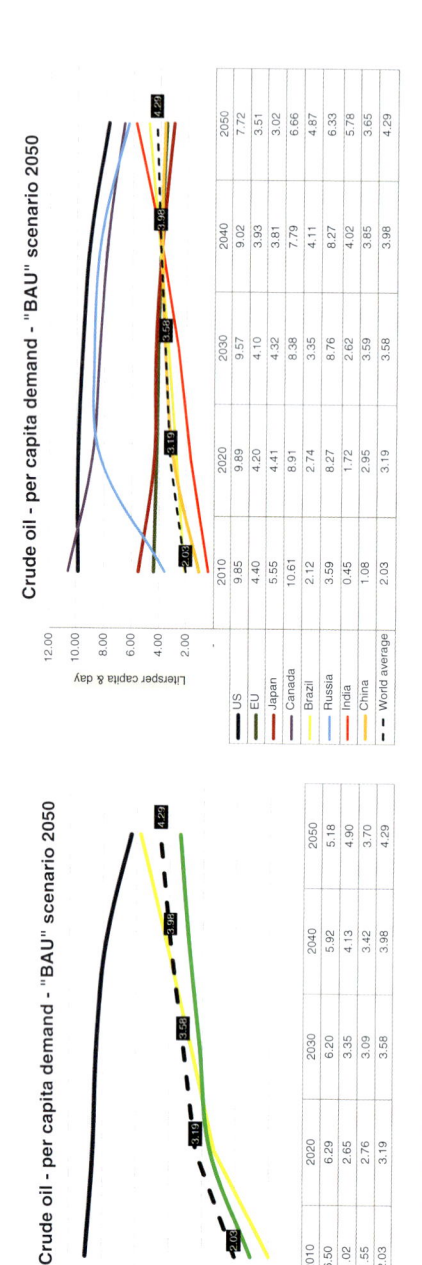

Crude oil - per capita demand - "BAU" scenario 2050

Liter per capita & day	2010	2020	2030	2040	2050
US	9.85	9.89	9.57	9.02	7.72
EU	4.40	4.20	4.10	3.93	3.51
Japan	5.55	4.41	4.32	3.81	3.02
Canada	10.61	8.91	8.38	7.79	6.66
Brazil	2.12	2.74	3.35	4.11	4.87
Russia	3.59	8.27	8.76	8.27	6.33
India	0.45	1.72	2.62	4.02	5.78
China	1.08	2.95	3.59	3.85	3.65
World average	2.03	3.19	3.58	3.98	4.29

Crude oil - per capita demand - "BAU" scenario 2050

Liter per capita & day	2010	2020	2030	2040	2050
Advanced	6.50	6.29	6.20	5.92	5.18
BRIC	1.02	2.65	3.35	4.13	4.90
Rest	1.55	2.76	3.09	3.42	3.70
World average	2.03	3.19	3.58	3.98	4.29

Crude oil	Crude oil Million barrels a day		GDP in $ billion		Ration energy on GDP		Efficiency 1970-2011	Ratio estimated 2050	Ratio final 2050
Rations	1970	2010	1970	2010	1970	2010			
US	14.7	19.1	$ 4,266	$ 14,582	3.45	1.31	2.63	0.50	0.50
EU	12.6	13.9	$ 5,829	$ 16,282	2.17	0.85	2.54	0.34	0.34
Japan	3.9	4.5	$ 1,559	$ 5,497	2.49	0.81	3.07	0.26	0.26
Canada	1.5	2.3	$ 382	$ 1,574	3.85	1.45	2.66	0.54	0.54
Brazil	0.5	2.6	$ 228	$ 2,087	2.29	1.25	1.84	0.68	0.68
Russia	3.6	3.2	$ 506	$ 1,479	7.01	2.16	3.24	0.67	0.67
India	0.4	3.3	$ 150	$ 1,729	2.61	1.92	1.36	1.41	1.41
China	0.6	9.1	$ 118	$ 5,878	4.70	1.54	3.05	0.51	0.51
Advanced	32.7	39.8	12,036	37,935	2.72	1.05	2.59	0.40	0.41
BRIC	5.0	18.2	1,002	11,173	5.01	1.63	3.08	0.53	0.85
Rest	7.7	29.4	2,405	13,940	3.20	2.11	1.51	1.39	1.39
World	45.4	87.4	15,443	63,048	2.94	1.39	2.12	0.35	0.88

Million barrels a day. One barrel equals to 158.987295 liters of crude oil
Source: BP statistical review 2011 and world bank for GDP
* lmtbs review, primary energy comprises commercially traded fuels including modern renewables used to generate electricity
♦ Less than 0.05.
● Less than 0.05%.
Excludes Estonia, Latvia and Lithuania prior to 1985 and Slovenia prior to 1991.
Notes: Oil consumption is measured in million tonnes; other fuels in million tonnes of oil equivalent
Russia GDP in 1970 estimated by author
GDP per capita in 1970 based on world bank historical series - based at 2005 $US dollar
GDP per capita in 2010 based on world bank historical series - based at 2010 $US dollar
(c)

Figure 2.6 (Continued).

In that sense, and if we will keep this average US$ 70 delta to the following billions of crude oil to be producer in the next decades, and assuming producers will be willing to producer at the current margins, we might end up with crude oil prices that could range from US$ 140 to 200 for the most expensive crude oil. As you can imagine, at these price levels we might wonder if crude oil will be still used and competitive versus alternative sources.

According to the EIA, at the current crude oil prices of US$ 110 we could start covering the production cost of some crude oil alternatives, however, we will need higher prices to make these economically attractive for mass production. Additionally, and as reviewed in Chapter 1, when the real price of an energy is equal to or higher that its nominal price at its inception, this energy tends to be not very efficient, fostering the entrance of alternative energy. That golden rule of energy was already broken in 2011 and would be broken again if we consider the potential expected prices for the next billion barrels of oil Figure 2.4(b).

Therefore, with crude oil expected to move in the next decades to at least US$ 140 per barrel and up to US$ 200 per barrel; and knowing that at these high price levels alternatives energies will start to become very appealing for substitution – CO_2 considerations not even factored yet – we wonder for how long crude oil would remain the preference choice of energy and even when that would still remain the case, if that quantities could become available.

Society could perhaps afford still these very high prices but, as discussed before, supply might still not be available due to lack of refining capacity and insufficient crude oil reserves. For all the above reasons and under this BAU scenario we question the sustainability of crude oil as the preferred source of energy.

Additionally to the potential energy and crude oil concerns, if we consider the imperative need to address climate change and reduce the world CO_2 emissions from the projected 94 Gt under the BAU scenario to the sustainable level of just 14 Gt by 2050 according to the IEA; then this BAU scenario certainly seems unsustainable and a new sustainable scenario will be required.

2.3.2
Sustainable Scenario

The "sustainable" scenario will present a "larger, wealthier, healthier, wiser, and sustainable world." This scenario still considers the positive developments of the previous BAU scenario, especially for population, economic growth and life expectancy as projected by the World Bank and PWC, while incorporating the sustainable levels of CO_2 emissions according to the IEA (Figure 2.7).

According to the IEA's 2008 Energy Technology Perspectives and its Scenarios, and Strategies to 2050, if the world wants to avoid climate change and to limit the long term global average temperature increase to 2.0–2.4 °C, the world will need to reduce its CO_2 emissions from the current 33 Gt in 2010 to 14 Gt by 2050. The IEA defined this scenario as the "Blue Map" scenario (Figure 2.8), and under that scenario define several key parameters. First, the world CO_2 emission by 2050 will need to be reduced to just 14 Gt annually. Secondly the level of CO_2 in the atmosphere will need to remain at 450 ppm. Thirdly the average temperature

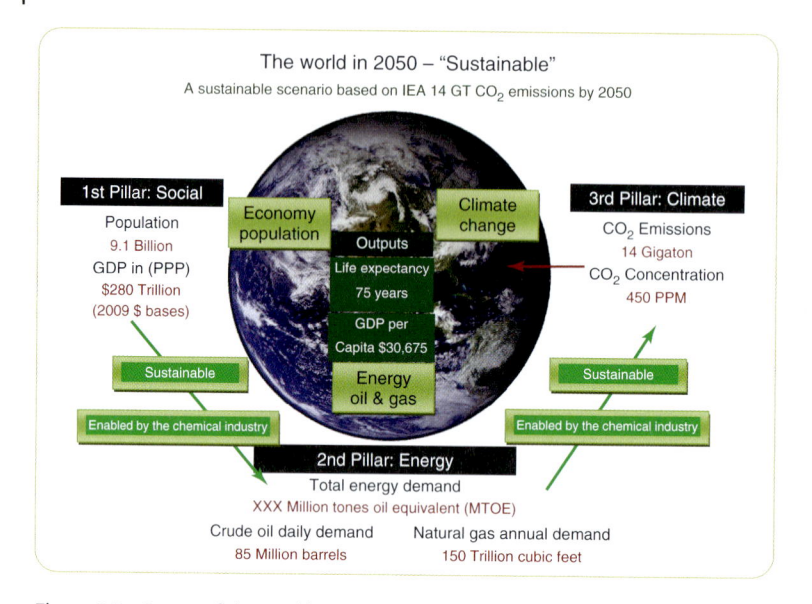

Figure 2.7 Status of the world in 2050 – "sustainable scenario". Source: Own elaboration, based on several sources including World Bank and United Nations Economic and Social projections. PWC economic projections for 2050 and International Energy Agency, Scenarios and Strategies to 2050.

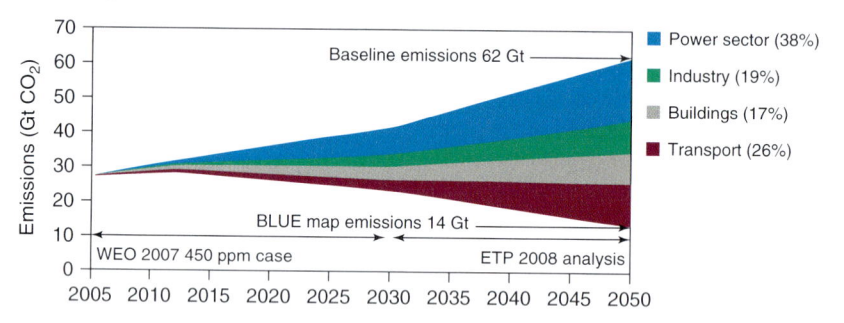

Figure 2.8 Reduction of energy related CO_2 emissions from the baseline scenario in the Blue Map scenario by sector – 2005 to 2050. Source: International Energy Agency – Energy Technology and Perspectives 2008 – in support of the G8 Plan of Actions. Scenarios and Strategies to 2050.

increase must remain between 2.0 and 2.4 °C. We will use the IEA Blue Map scenario as our base for our sustainable scenario.

The IEA also defined a BAU scenario for 2050 that projected world CO_2 emissions of 62 Gt Annually and 550 ppm versus our 94 Gt and 550 to 800 ppm. That large difference in total CO_2 emissions could be explained as follows. First, due to a difference in baseline numbers. The IEA study used the 2005 world CO_2 emissions (28.8 Gt) as their base line, while we used the 2010 figure (33.1 Gt). Indeed if we use 2005 as the baseline for our BAU scenario the difference will be automatically reduced, as we will project 81 Gt instead of 94 Gt. Secondly, our projections do not factor any further potential and logical improvement in energy efficiency, our BAU scenario simply took the observed CO_2 efficiencies during the 1970–2010 period and extrapolated them to the period 2010–2050, this is certainly a simplification as indeed technology might enable further progress. Thirdly the IEA might be using different economic projections and much more complex econometric models that we did. However the differences between the two BAU scenarios still remain very large, almost 19 Gt.

In any case the IEA, as the world experts on the subject, providing regular and detailed updates on the different scenarios and subsequent recommendations should remain the leading authority to be followed and consulted in case of further inquiries.

Therefore, and for this sustainable scenario, we will not spend too much time on it now, but will just share for illustration purposes and as an introduction to this subject the high level conclusions and recommendations from the IEA Energy Technology & Perspective (ETP) 2008; while for further analysis and updates we recommend to contact the IEA directly. As the world keeps changing very rapidly and this is an area subject to tremendous political and economic influence, we will recommend to review regularly the latest IEA analysis.

According to the IEA sustainable scenario, the Blue Map Scenario, the Blue Map scenario, the world will need to be able to "live" with just 14 Gt of CO_2 annually by 2050; that is, it will need to reduce its annual CO_2 by 48 Gt from the IEA projected BAU scenario in 2050 to its sustainable Blue Map scenario. As the IEA BAU scenario was based on 2005 CO_2 emissions data, and in 2010 the amount of CO_2 emissions were already 4.3 Gt higher that in 2005, new updates from the IEA might show slightly higher numbers.

In order to close that gap and according to the IEA Blue Map Scenario most industries will be impacted by severe CO_2 emissions reductions, however, some sectors will be more affected than others. Power generation with 38% emission reduction by 2050 will be the most affected. Transportation will follow with 26% emission reduction, industry will account for 19% and buildings 17% (see Figure 2.9).

The chemical industry represent around 17% of the industrial segment, accounting for 4.1% of the world CO_2 emissions in 2007, that is, around 1.3 Gt of CO_2 emissions annually.

From a technology point of view the IEA Blue Map scenario also defined a clear roadmap with different ways to achieve this emission reduction (Figure 2.9). According to the IEA the following actions will need to take place in order to achieve the desired emissions reductions:

- 24% reduction through increases in fuel efficiency, accounting for 10 Gt
- 21% reduction through the increase in renewable energies, accounting for 9 Gt
- 10% carbon dioxide capture and storage (CCS) in power generation, accounting for 4.2 Gt
- 9% CCS in industry and transformation, accounting for 3.8 Gt
- 7% power generation efficiency and fuel switching, accounting for 3 Gt
- 6% increase in nuclear power, accounting for 2.5 Gt.

158 | *2 The World by 2050*

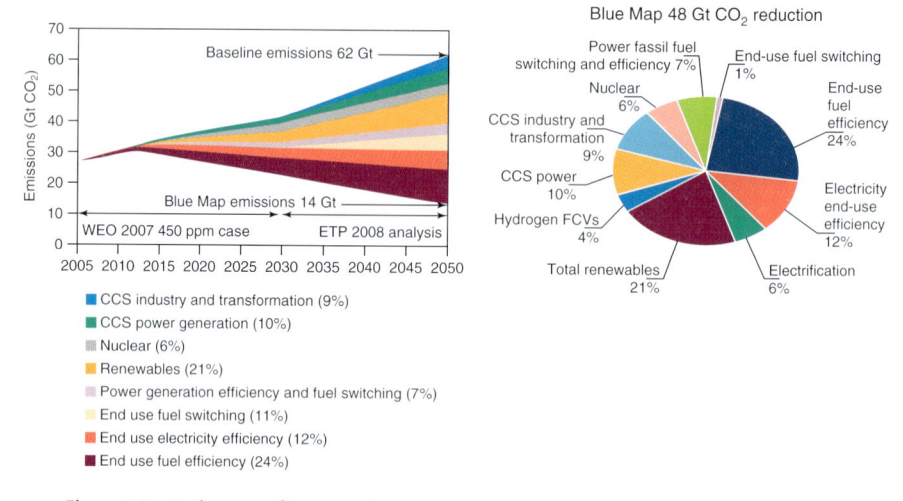

Figure 2.9 Reduction of energy related CO_2 emissions from the baseline in the Blue Map scenario by technology – 2005 to 2050. Source: International Energy Agency – Energy Technology and Perspectives 2008 – in support of the G8 Plan of Actions. Scenarios and Strategies to 2050.

As we can see, none of these targets will be simple to achieve nor inexpensive. First in the ranking of potential solutions, and accounting for 24% of the world emissions reduction, we have improvements in fuel efficiency. These improvements will need to come on top of those regularly observed in our economy, implying a significant increase. In second position, accounting for 21% of emissions reduction, we have the increased use of renewable energies.

This large increase in the use of renewable energies will lead to significant changes in our world energy mix, including increases in the supply of different renewable energies like wind, solar, nuclear, biomass, and so on, but also natural gas. Natural gas demand under this scenario is expected to increase by an impressive 34% by 2050, while nuclear, biomass and other renewables are expected to increase by much larger percentages. Crude oil and coal are expected to be reduced by 27 and 18%, respectively.

All these massive changes in the world energy mix will imply significant investment in very short periods of time. Investment that without the right sense of urgency and collaboration across governments, industries, and companies might not happen in time. See Figure 2.10 and please note that the Baseline and Act Map scenarios to 2050, are IEA scenarios not explained or used by us in this book.

In third position and with a critical role to be played, the EIA recommend a massive increase in CCS, with 19% of the total reduction, equivalent to 8 Gt. These reductions will be equally distributed between CCS for power generation with 10% of the total reduction and CCS for industry and transformation with the remaining 9%.

Disregarding at this time the potential discussion about the appropriateness and readiness of the CCS technology as the world cannot afford not to use it; we can

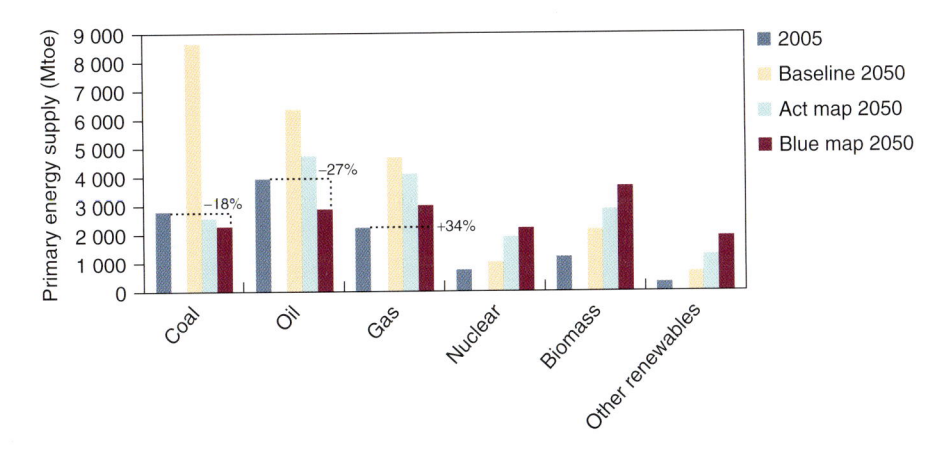

Figure 2.10 World fuel supply for Baseline, and Blue Map 2050. Source: IEA – Energy Technology and Perspectives 2008. In support of the G8 Plan of actions. Scenarios and strategies to 2050.

expect humungous investments in this area with thousands of new plants required around the world.

Let us illustrate the potential cost and size of the required investments on a global basis by reflecting on the figures and economics of perhaps the largest CCS project in the world. The Gorgon project in Barrow Island in Australia is expected to be able to store up to 4 Mt of CO_2 per year for an initial estimated cost up to US$ 45 billion. The Gorgon CCS Project is a joint venture between Chevron Australia, ExxonMobil, Shell, Tokyo Gas, Osaka Gas, and Chubu Electric. Operation is expected to start in 2015 and last for approximately 40 years.

Considering the economics of the Gordon project as a first benchmark, we can observe that in order to store 1 Mt of CO_2 per year, the industry will need to make an initial investment of US$ 11 billion. Considering that according to the IEA Blue Map scenario 8 Gt will need to be stored by 2050, we can observe the magnitude of the investment required.

So under this sustainable scenario, and with the economics of the Gordon project in mind, the world will require at least between 2000 and 2500 CCS units around the world with a potential cost of more than US$ 90 trillion by 2050. If we were to do this analysis under our BAU scenario, the investment would be much greater than that.

Considering now all the different actions that need to be taken, several things can be concluded. The expected emission reduction under this sustainable scenario will not only be extremely challenging from a technology point of view, but also the amount of investment required would be enormous and might be physically impossible to deliver in a limited amount of time.

Living in a world of 14 Gt or 4000 grams of CO_2 per capita per day (a 30 km drive in a car with an efficiency of 130 g km^{-1}) will be certainly very challenging, but much less so than suffering the dangers of climate change. Let us illustrate the challenge with the following comparison of world CO_2 emissions per capita per day.

Looking at recent experiences we might be positively surprised to know that during the last 40 years from 1970 to 2010, the world already achieved significant CO_2 reductions on a GDP per capita basis. Indeed, during this period emissions measured versus GDP per capita have been reduced by 50%. In 1970, the ratio between the world GDP (US$ 15 443 billion) and world CO_2 emissions (14.993 billion tons) was one to one. In 2010, although the world economy increased by a factor of four to US$ 63 048 billion, the world CO_2 emissions only doubled, to 33 158 billion annually, highlighting the tremendous achievements made in energy and emissions efficiency. For the next 40 years until 2050 under the sustainable scenario, the challenge would be even greater. According to PWC the world economy is expected to grow to US$ 280 trillion, while CO_2 emissions, according to the EIA sustainable scenario, should remain at 14 Gt annually. That would imply that the ratio of CO_2 emissions to GDP will be reduced to 0.05. See Figure 2.11. In other words in order to avoid climate change the world will need to accomplish a 90% reduction in CO_2 emissions versus GDP.

On the positive side we know that during the last 40 years the world did manage to reduce by 50% its CO_2 emissions versus GDP. This time the world will need to figure out how to reproduce the achievements of the last 40 years, while achieving an extra 40% of reductions. The task appears gigantic and tremendously complex, but not impossible.

Under this sustainable scenario the world can have a clear roadmap to avoid climate change by 2050. The challenges would certainly be enormous but the rewards will be equally impressive. The world has a unique opportunity to host the wealthiest, healthiest, and wisest society in human history, and this is a target that the world must not fail to meet. A strong partnership and open collaboration among the different stakeholders across society and the value chain, including a pivotal and decisive role for the chemical industry, will be critical. The stakes are certainly high, but the rewards are so huge that we are confident that the world will get to work closely together to ensure this major issue is solved. We are positive that the chemical industry, as a key enabling industry, is poised to play an even much larger role in the decades to come.

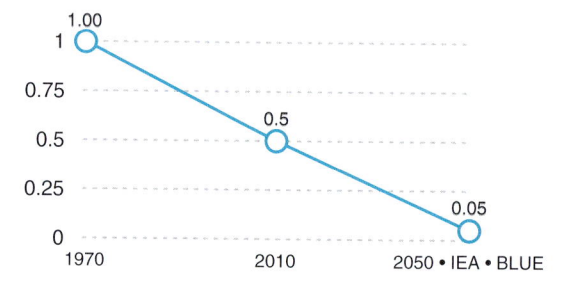

Figure 2.11 Ratio World CO_2 emissions per World GDP. Source: Own elaboration, compilation of several sources including GDP data from the World Bank and CO_2 emissions and projections from the International Energy Agency.

2.3. The World in 2050 | 161

Appendix: Roadmaps to a World of 4000 g of CO_2 per Capita per Day

The world in 1970

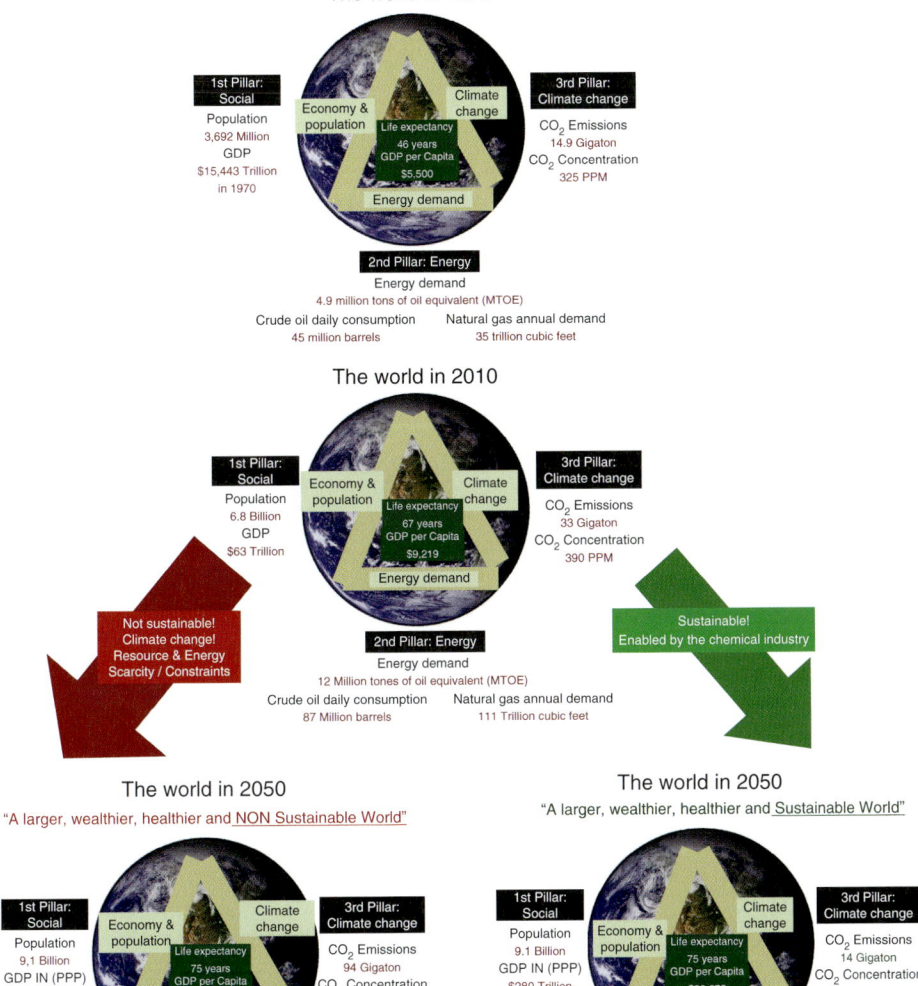

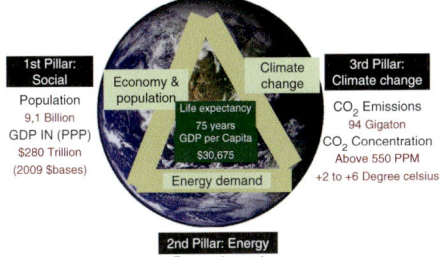

1st Pillar: Social
Population
3,692 Million
GDP
$15,443 Trillion
in 1970

Economy & population
Life expectancy 46 years
GDP per Capita $5,500
Energy demand

3rd Pillar: Climate change
CO_2 Emissions
14.9 Gigaton
CO_2 Concentration
325 PPM

2nd Pillar: Energy
Energy demand
4.9 million tons of oil equivalent (MTOE)
Crude oil daily consumption
45 million barrels
Natural gas annual demand
35 trillion cubic feet

The world in 2010

1st Pillar: Social
Population
6.8 Billion
GDP
$63 Trillion

Economy & population
Life expectancy 67 years
GDP per Capita $9,219
Energy demand

3rd Pillar: Climate change
CO_2 Emissions
33 Gigaton
CO_2 Concentration
390 PPM

2nd Pillar: Energy
Energy demand
12 Million tones of oil equivalent (MTOE)
Crude oil daily consumption
87 Million barrels
Natural gas annual demand
111 Trillion cubic feet

→ **Not sustainable!** Climate change! Resource & Energy Scarcity / Constraints

→ **Sustainable!** Enabled by the chemical industry

The world in 2050
"A larger, wealthier, healthier and **NON Sustainable World**"

1st Pillar: Social
Population
9,1 Billion
GDP IN (PPP)
$280 Trillion
(2009 $bases)

Economy & population
Life expectancy 75 years
GDP per Capita $30,675
Energy demand

3rd Pillar: Climate change
CO_2 Emissions
94 Gigaton
CO_2 Concentration
Above 550 PPM
+2 to +6 Degree celsius

2nd Pillar: Energy
Energy demand
35,887 Million tones oil equivalent (MTOE)
Crude oil daily consumption
247 Million barrels a day
12,007 MTOE annually
Natural gas annual demand
307 Trillion cubic feet annually
7,920 MTOE annually

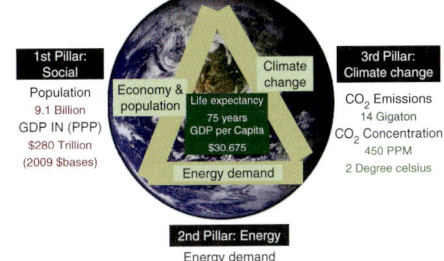

The world in 2050
"A larger, wealthier, healthier and **Sustainable World**"

1st Pillar: Social
Population
9.1 Billion
GDP IN (PPP)
$280 Trillion
(2009 $bases)

Economy & population
Life expectancy 75 years
GDP per Capita $30,675
Energy demand

3rd Pillar: Climate change
CO_2 Emissions
14 Gigaton
CO_2 Concentration
450 PPM
2 Degree celsius

2nd Pillar: Energy
Energy demand
Sustainable "X" million tones oil equivalent (MTOE)
Crude oil daily consumption
85.5 Million barrels
Natural gas annual demand
150 Trillion cubic feet

3
The Chemical Industry in 2010

> *Study the past if you would like to divine the future.*
>
> Confucius

The aim of this chapter is to provide a high level and comprehensive overview of the chemical industry as we know it today, defining the key features and characteristics of the industry. The selected features will be used later in the analysis and future projections for the industry.

For this purpose, we will use extensively data from the CEFIC Facts and Figures Report for 2011 on the industry as well as some generic overview of the industry from other authoritative sources.

3.1
Chemical Industry: Economic Relevance

The broad chemical industry with US$ 3996 billion sales in 2010 and 6.3% of the world gross domestic product (GDP) is one of the world's largest industries. (See Table 3.1)

The pharmaceutical industry with an estimated US$ 875 billion accounted for 1.4% of the global GDP and 21% of the broad chemical industry. The chemical industry with estimated US$ 3121 accounted for 5% of the world GDP and the majority of the broad chemical industry (Figure 3.1).

Within the chemical industry, the ADV economies have the largest share of the chemical industry with 47%. However, China with 24% of global chemical sales in 2010 has the largest chemical market. The European Union 27 appears to be the second largest market with 21%, followed by the United States with 17% and Japan with 7%. The large size of the Chinese chemical market explains why the BRIC group already accounted for almost 30% of the world chemical sales in 2010.

The enormous growth in chemicals demand in China and the Asia Pacific region has been supported by a massive investment in chemicals, including pharmaceuticals, in that region. In 2010 alone more than US$ 300 billion was invested in that region, and indeed if we look further during the last decade, the bulk of the investments in the chemical industry has been made in China and the Asia Pacific region, excluding Japan (Figure 3.2).

The Future of the Chemical Industry by 2050, First Edition. Rafael Cayuela Valencia.
© 2013 Wiley-VCH Verlag GmbH & Co. KGaA. Published 2013 by Wiley-VCH Verlag GmbH & Co. KGaA.

3 The Chemical Industry in 2010

Table 3.1 Chemical industry in 2010.

REVENUE	In $ US	in €	In % GDP
Chemicals	3121	2353	1.4
Pharmaceuticals	875	660	5.0
TOTAL	3996	3013	6.4

Exchange rate : 1,326

Author Elaborated based on European Chemical Industry Council – CEFIC – Facts and figures 2011 and the European Federation of Pharmaceutical Industries and Association – EFPIA – Pharmaceutical Industry in Figures for 2010.

Chemicals sales 2010
- Advance 47%
- BRIC 30%
- Rest 23%

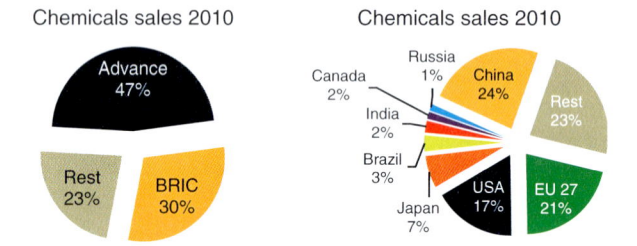

Chemicals sales 2010
- Canada 2%
- Russia 1%
- China 24%
- India 2%
- Brazil 3%
- Japan 7%
- USA 17%
- EU 27 21%
- Rest 23%

Figure 3.1 Chemical sales 2010. Author elaborated based on several sources including CEFIC 2011 Facts and Figures.

The massive growth in chemicals demand in China and the rest of Asia, excluding Japan, and the BRIC in general has resulted in an impressive 63% growth for the chemical industry, excluding pharmaceuticals, during the last decade.

Sales in the chemical industry, excluding pharmaceuticals, grew from the €1437 billion in 2000 to the €2353 billion in 2010 (Figure 3.3).

The massive growth in chemical demand in China and Asia Pacific, excluding Japan, has resulted in a massive relocation of sales and focus of the industry from the ADV economies to the rest of the world. In 2000 the ADV economies accounted for almost 70% of the chemical sales but by 2010 that was reduced to just 47% (Figures 3.1 and 3.2).

After this short term transition, China became the world's largest chemical market, increasing its market share from 6.4% in 2000 to an impressive 24.4% in 2010, a 524% increase. Its market increased from €92 billion to €575 billion euro, an extra €484 billion market added into the world chemical industry in just one decade. Let us keep good note of this figure as it will be useful to us when reviewing the projections for the industry by 2050 (Figure 3.2).

In the pharmaceutical industry this sort of transition has not been seen yet. Indeed in 2010 the ADV economies accounted for almost 75% of the global pharmaceutical sales. The REST had 15% and the BRIC, with already 42% of the world population and 18% of the world GDP, only accounted for 10% of the pharmaceutical sales (Figure 3.4).

The United States was the largest pharmaceutical market in the world with 37% of the global demand, followed by Europe with 28%, and the REST with 15%.

3.1 Chemical Industry: Economic Relevance | 165

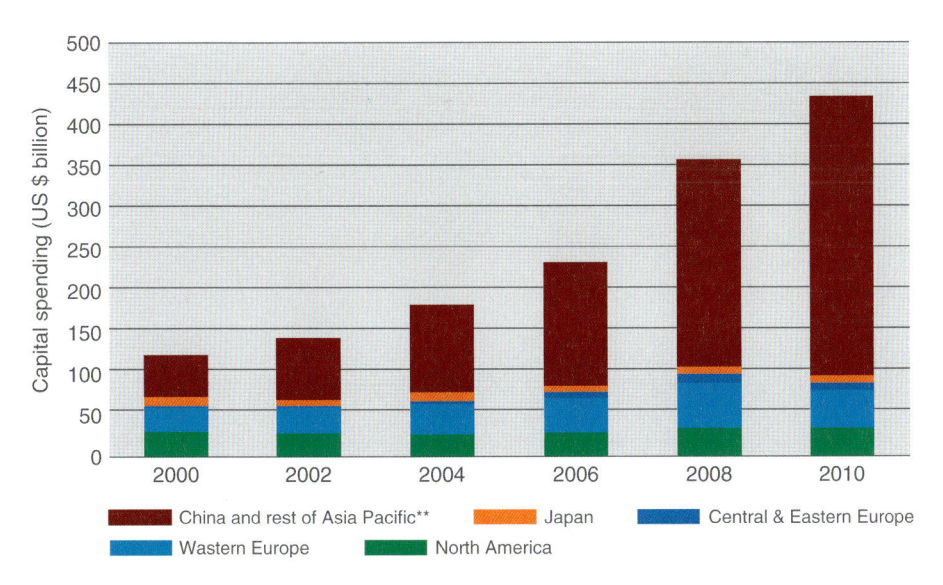

Figure 3.2 China and the rest of Asia-Pacific attract the bulk of chemical investment.
* Including pharmaceuticals. ** Excluding Japan. American Chemistry Council (ACC).
Source: CEFIC Chemdata international – CEFIC Facts & Figures 201.

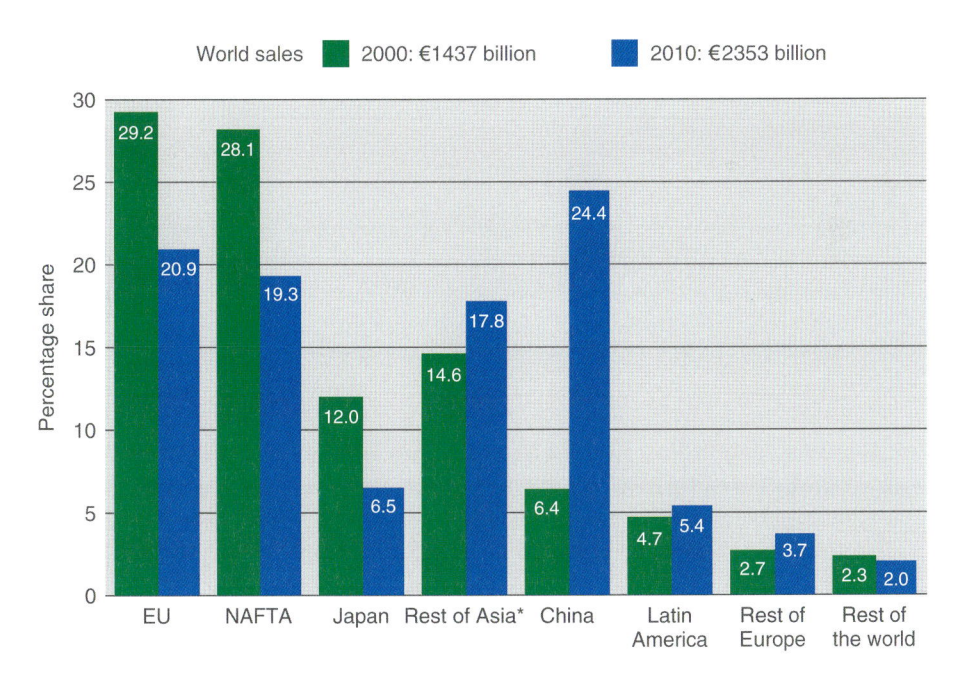

Figure 3.3 Sales of chemicals excluding pharmaceuticals. Source: CEFIC Chemdata international – CEFIC Facts & Figures 201.

Pharmaceuticals 2010

- Advance 75%
- BRIC 10%
- Rest 15%

Pharmaceuticals 2010

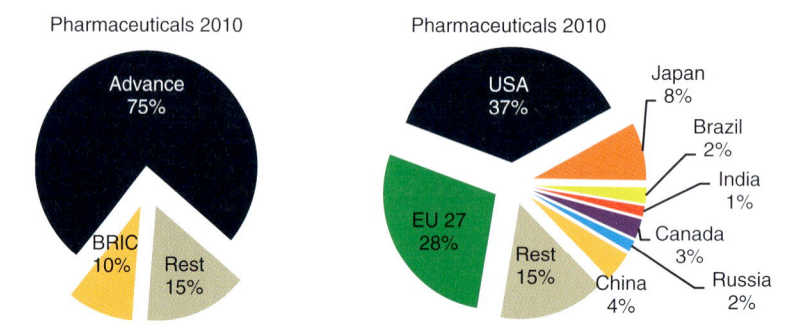

- USA 37%
- EU 27: 28%
- Rest 15%
- Japan 8%
- China 4%
- Canada 3%
- Brazil 2%
- Russia 2%
- India 1%

Figure 3.4 Pharmaceuticals sales 2010. Source: The Pharmaceutical Industry in Figures – 2010 edition. European Federation of Pharmaceutical Industries and Associations.

Within the BRIC, China remains as the largest pharmaceutical market by sales with 4% of the global demand, followed by Russia and Brazil with 2% of the world demand, and India with just 1%.

These large differences in the pharmaceutical market share are matched by equally large differences in per capita demand. In 2010, the ADV economies consumed on average US$ 678 of pharmaceuticals per capita annually, while the REST group consumed US$ 44, and the BRIC only US$ 30. At the country level the disparities can also be enormous. The United States with US$ 1000 per capita demand for pharmaceuticals annually remain among the highest, while India with just US$ 11 remain among the lowest. For further data and analysis please visit Chapter 5, Segment 2.1 – Economic Relevance of the industry.

3.2
Chemical Industry: Technological Relevance

The technological relevance of the industry is certainly one of its major characteristics. The chemical industry, led by the pharmaceutical industry, appears as one of the industries with the highest R&D investment versus revenue. Indeed the pharmaceutical industry with 16.5% R&D expenses versus revenue appeared to be the industry with the highest expenditure in R&D across all industries in 2010 (Figure 3.5).

On the other hand, the chemical industry, excluding the pharmaceutical industry, has 3% R&D expenses versus revenue, much lower than for the pharmaceutical industry but still close to the all sectors average.

The chemical industry also presents a very broad range of R&D ratios, depending on the sectors of the industry and the regions where the industry operates. Indeed the R&D intensity in Japan appears to have been much higher, sometimes double or triple, than that in the USA or Europe during the last two decades. At a product level there are also significant differences, and we will discuss this further when reviewing the industry outlook by 2050.

Additionally, the enormous growth the chemical industry has been facing during the last decades has somehow "forced" companies to shift the focus of investments

3.2 *Chemical Industry: Technological Relevance* | 167

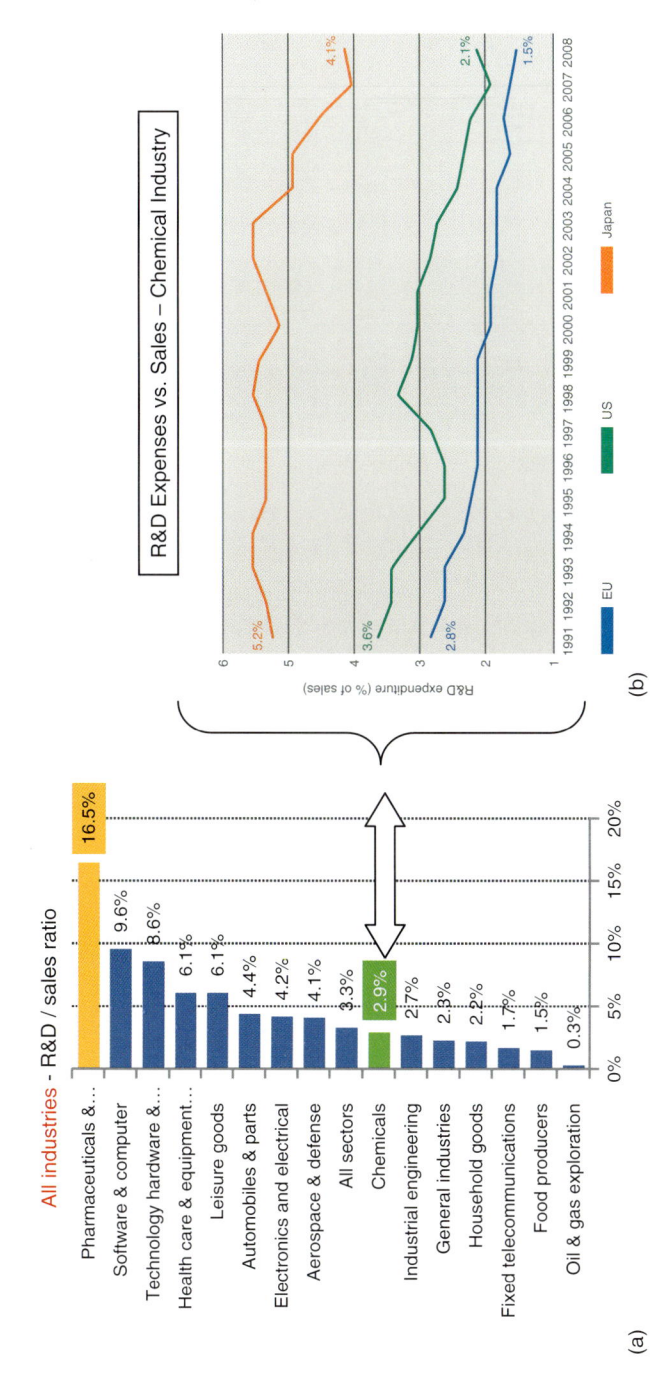

Figure 3.5 (a) R&D expenditure/sales 2010. Source: 2009 EU industrial; R&D Investment Scorecard, EU Commission, JRC/Dg Research. (b) R&D expenses versus sales in the chemical industry from 1991 to 2008. Source: CEFIC Chemdata International – CEFIC Facts & Figures 201.

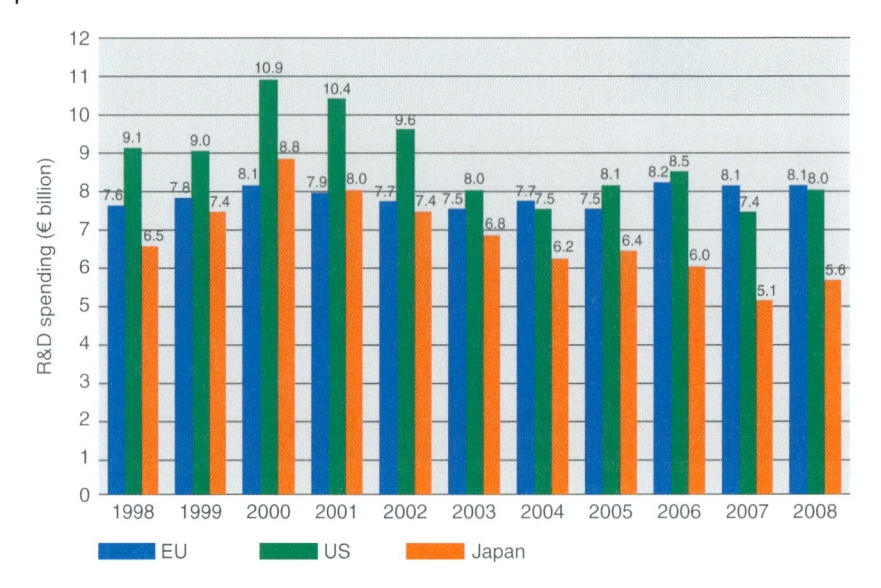

Figure 3.6 R&D spending in the chemical industry, excluding pharmaceuticals. Source: CEFIC Chemdata international – CEFIC Facts & Figures 2011.

from the qualitative or technological to the quantitative point of view. In other words, companies have been more focused on growing their supply globally, in order to support the enormous demand coming from the emerging geographies, and more specifically from the BRIC economies, than investing in technology and innovation.

Indeed the absolute amount of R&D expenditure has remained more or less stable during the last decades, however the percentage of R&D on sales has been reduced drastically in every region and specially in USA and Europe (Figures 3.5 and 3.6). Meanwhile the chemical industry tripled its capital expenditure from US$ 100 billion in 2000 to almost US$ 450 billion in 2010; in order to enable its massive growth in Asia Pacific and more specific into China (Figure 3.2).

3.3
Industry Relevance: Profitability

Looking into the profitability of the industry the first thing to be noticed is the large differences existing between the profitability of the chemical and pharmaceutical sides of the industry. Using as a benchmark and approximation the percentage earnings before interest, tax, depreciation and amortization (EBITDA) on sales of the top 10 pharmaceutical companies in 2010 versus the average percentage operating margin on sales of the top 100 chemical companies in 2010; we can observe how the profitability of the pharmaceutical industry (17%) was almost double that of the chemical industry (9%).

Within the chemical industry, we can observe large deviations in profitability, with certain sectors like fertilizers having a 22% operating margin on sales in 2010,

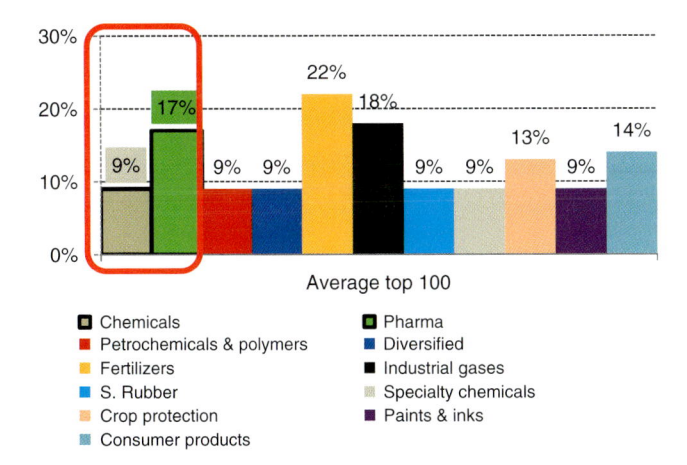

Figure 3.7 2010 %EBITDA on revenue for the top chemical companies including pharmaceuticals. Source: Based on publically available data for the top 100 chemical companies and the top 10 pharmaceutical companies in 2010.

industrial gases 18%, consumer products 14% and crop protection 13%, all above the chemicals average, while others like petrochemicals and polymers, diversified, and so on, were at the average or below (Figure 3.7).

The results from this analysis have to be treated prudently and placed in the right context. The sample we took is an oversimplification of reality as we only considered data for 2010, and took data from the top 100 "Publically Listed" chemical companies. We also assigned some companies completely to one of the sectors, when that might not always be 100% the case. Furthermore, in some sectors, like consumer products, we only considered one company among the top 100. Having said this and for illustration purposes these results could serve us to illustrate the fact that indeed there are significant differences within the two major sectors of the chemical industry and also within all the different sectors of the chemical industry, and these differences have to be a factor when reflecting about the industry and its future.

Looking now at profitability from a broader perspective including net margin on sales and price/earnings (P/E) ratios for all industries in the USA in 2010, we can find similar results. In terms of net margin, it is interesting to see that the average margin for pharmaceuticals (drugs) in the USA in 2010 was also 16.64%, and 13.94% for biotechnology. Within the chemical sector, basic chemicals with 9.34% net margin lead, followed by diversified chemicals at 6.92% and specialty chemicals at 6.26% (Figure 3.8a).

In terms of P/E ratios, the chemical industry in 2010 appears to have very healthy ratios, compared to other industries. The chemical industry as a cyclical industry, especially on its chemical side, tends to have P/E ratios quite linked to the economic cycle and growth potential. It is interesting to see that in 2010 the highest P/E of the chemical industry was for the biotechnology side (27.8) and for chemicals (24.3), while specialty chemicals (19.5), drugs/pharmaceuticals (18.8) and basic chemicals (17.1) appeared to be much lower in the USA ranking (Figure 3.8b).

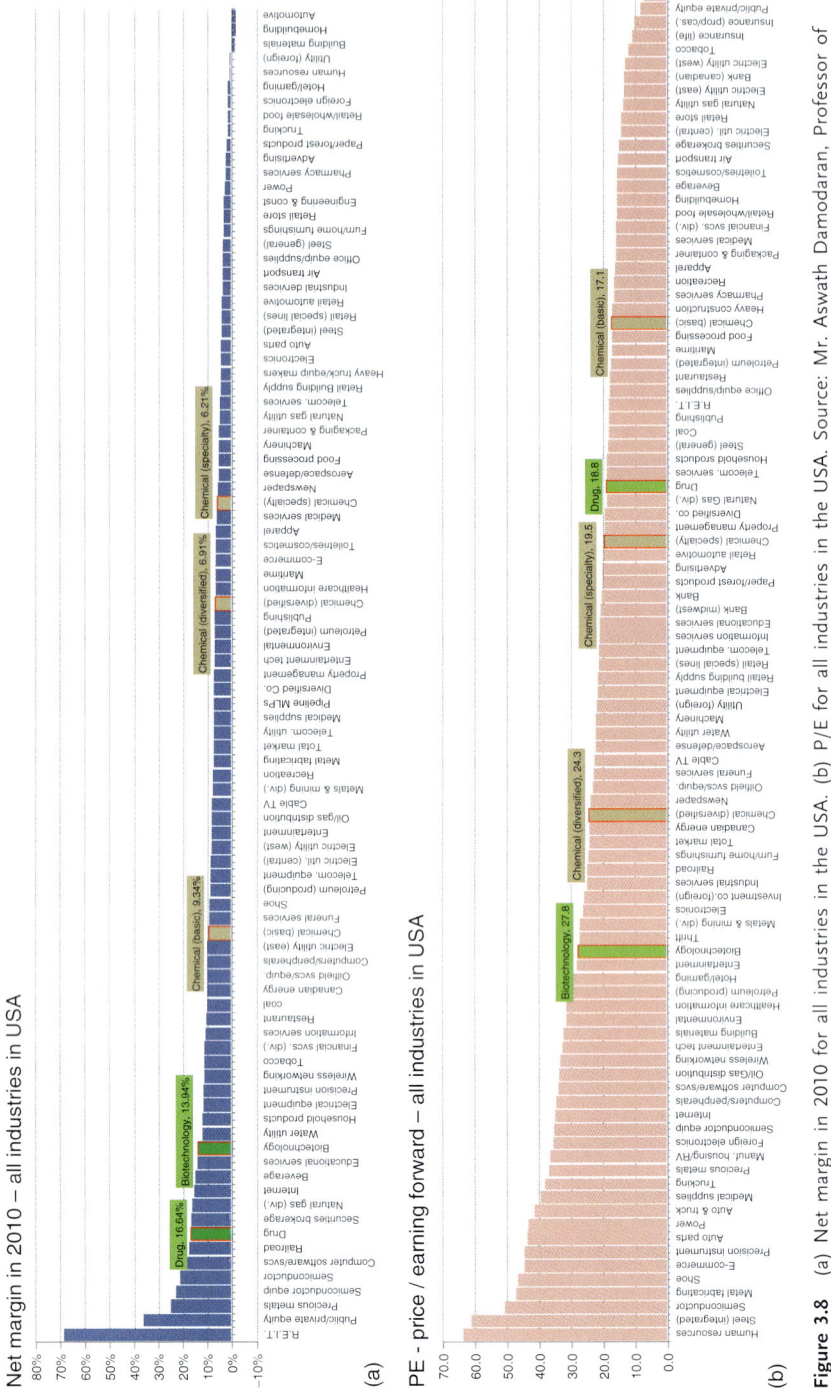

Figure 3.8 (a) Net margin in 2010 for all industries in the USA. (b) P/E for all industries in the USA. Source: Mr. Aswath Damodaran, Professor of Finance at the Stern School of Business at New York University. http://pages.stern.nyu.edu/~adamodar/

The chemical industry tends to have a very broad range of profitability levels – measured by margin and P/E ratios – with the pharmaceutical normally on the higher side. However, margins and P/E ratios in the industry tend to be cyclical, so profitability analysis should be done considering these factors while also observing other long term or more structural ratios.

3.4
Feedstocks and Energy

The chemical industry as one of the largest industries in the world is also a large energy consumer, not only because it uses large amounts of energy to manufacture its goods but also because it requires even larger quantities of oil and gas to be used as feedstock for its final products. In 2006 almost 100% of the feedstock used in chemical industry was based on oil and gas, amounting to almost 81 million tons of oil equivalent (MTOE) by 2006 (Figure 3.9).

From an energy perspective the industry consumed almost 60 MTOE. Gas and electricity appear to be the preferred energy sources, although others like oil, coal, and others were also used. During the last decades the industry has been making tremendous efforts to reduce its energy consumption and efficiency across the world, Europe being one of the leading regions in this area.

During the last two decades, the European chemical industry has been able to decouple the production of chemicals from its energy consumption or greenhouse emissions, increasing production while decreasing energy consumption. Indeed, during this period the European chemical industry, including pharmaceuticals increased its production by 60%, while energy demand declined by almost 30% and CO_2 emissions by almost 50% (Figure 3.10).

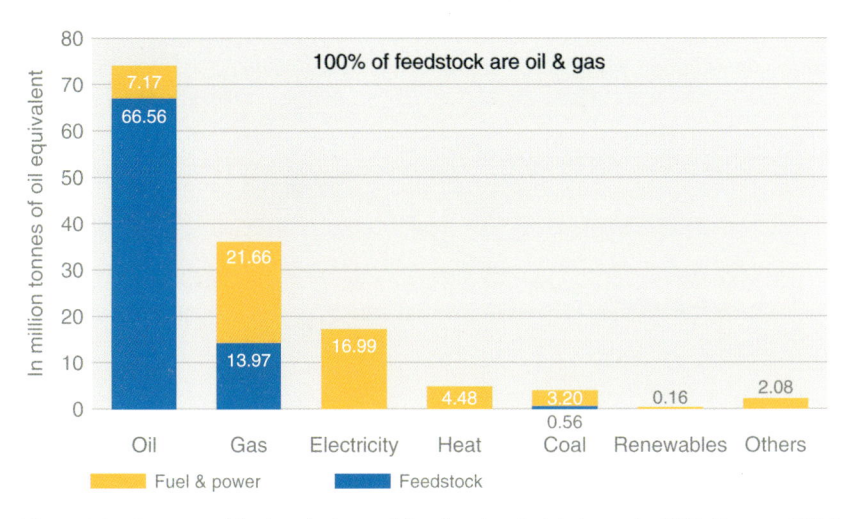

Figure 3.9 Energy and feedstock demand for the chemical industry in 2006. Source: CEFIC Chemdata international – CEFIC Facts & Figures 2011.

3 The Chemical Industry in 2010

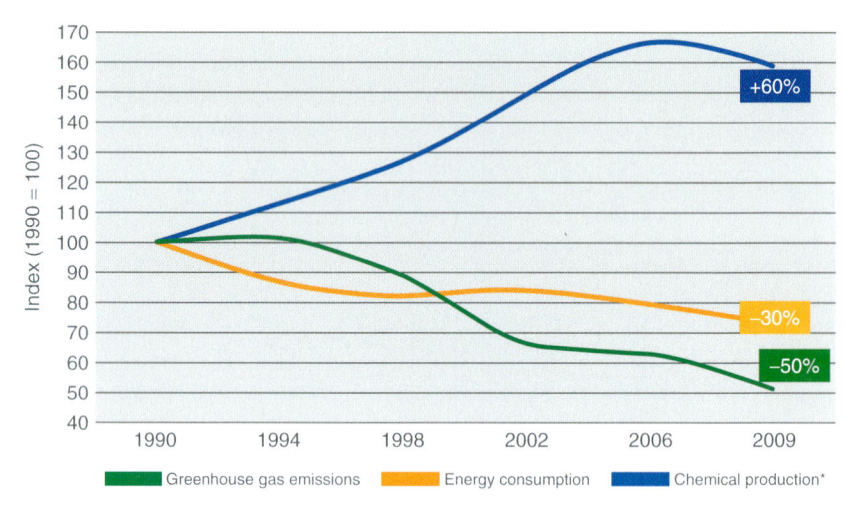

Figure 3.10 Chemical production decoupled from energy and greenhouse emissions. * Including pharmaceuticals. Source: CEFIC Chemdata international and European Environmental Agency (EEA) – CEFIC Facts & Figures 2011.

The chemical industry in general, and especially in the ADV economies has been making tremendous effort and progress in reducing greenhouse emissions while increasing production. In the case of Europe that has been particularly notable, especially when comparing with the performance of the overall European economy during the same period.

During the last 20 years, the European Union CO_2 emissions declined by just 8% (Figure 3.10), while the European chemical industry reduced its overall greenhouse emissions by an impressive 50% (Figure 3.10). This extraordinary performance confirms the commitment of the industry to reduced emissions and energy consumption while increasing chemicals production.

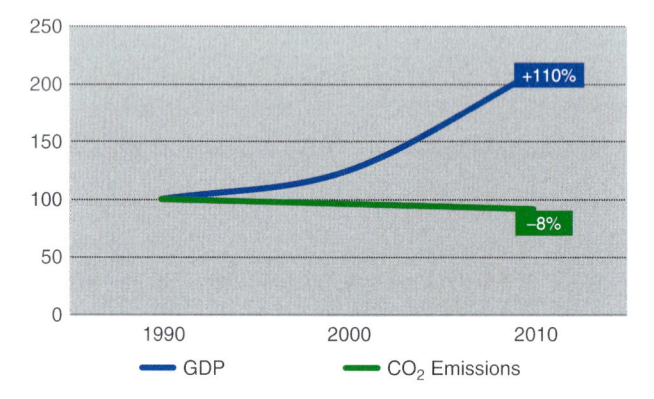

Figure 3.11 European Union 27 CO_2 emissions. Source: Author elaborated based on British Petroleum Statistical Energy Review 2011. GDP at 2010 dollars based.

3.5
Major Sectors and Products of the Chemical Industry

The broad chemical industry, with thousands of products across multiple diverse sectors, appears to be very complex. The most notable example comes when comparing the two largest sectors – pharmaceuticals and chemicals.

These two sectors are very different, with different customers, products, margins, R&D spending, labor intensity, and even geographical presence.

Within the chemicals industry there are also significant differences. The industry dynamics, customers, products, profitability, energy or labor intensity for petrochemicals, and polymers is completely different than for fertilizers, crop protection, industrial gases, and so on.

Therefore, the chemical industry is normally referred to as an *industry of industries*, all with very different dynamics and characteristics, and even though comparing smaller segments than when comparing chemicals to pharmaceuticals, we should keep this in mind when reviewing the chemical industry.

In terms of the major segments and products of the industry from a sales perspective, pharmaceuticals was the largest segment in 2010 with 22% of the global sales and US$ 875 billion, while chemicals accounted for the remaining 78% of the industry (Table 3.2).

Specialty chemicals, petrochemicals, and polymers all had a similar market share around 18%, followed in the distance by basic inorganic and consumer products with 10%. Finally, excluding pharmaceuticals the single largest sub-segment is plastics with 16% of the global demand for chemicals. Figure 3.12 shows the relative sizes of the industry segments in 2010, and also the major petrochemical feedstocks used and the geographical split of the chemicals and pharmaceutical industries.

3.6
Industry Structure and Companies

The large complexity of the chemical industry with multiple segments and products in combination with the global nature of the industry has resulted in a high level of complexity and fragmentation within the industry, with thousands of companies around the world.

That high level of fragmentation is especially obvious on the chemical side of the industry, where the top 10 chemical companies in 2010 only accounted for 15% of the global chemical sales, and the top 100 chemical companies (Table 3.3) only represent 40% of the industry. Even the largest companies, giants, and diversified chemical companies like BASF or Dow Chemical, only participate in limited segments of the industry but not in all of them.

On the pharmaceutical side the level of concentration is certainly much higher with 51% of the global sales conducted by the top 10 companies. That lower level of fragmentation is a result of the large number of sales conducted in the ADV economies. Indeed in 2010, more than 70% of the pharmaceutical sales occurred in the ADV economies compared to 47% for the chemical industry.

Table 3.2 Major segments and products of the chemical industry.

Chemical Industry			2010
SEGMENT	$US Billion	% Industry	% GDP
PETROCHEMICALS	749	18.74	1.2
C2 - Ethylene	210	5.26	0.3
C3 - Propylene	124	3.09	0.2
C4 - Butadiene	45	1.13	0.1
C5-C8 & Others	370	9.27	0.6
BASIC INORGANICS	424	10.62	0.7
Fertilizers	157	3.92	0.2
Industrial Gases	86	2.15	0.1
Other Inorganics	182	4.56	0.3
POLYMERS	749	18.74	1.2
Plastics	640	16.01	1.0
Synthetic Rubber	56	1.41	0.1
Man-Made Fibers	53	1.33	0.1
SPECIALTY CHEMICALS	799	19.99	1.3
Dyes & pigments	71	1.78	0.1
Crop Protection	54	1.34	0.1
Paints & Inks	252	6.30	0.4
Auxiliaries for industry	423	10.58	0.7
CONSUMER PRODUCTS	399	10.00	0.6
PHARMACEUTICALS	875		1.4
CHEMICAL INDUSTRY	3121	78	4.95
PHARMACEUTICALS	875	22	1.39
BROAD CHEMICAL INDUSTRY	3996		6.33

Source: CEFIC – Facts and figures 2011 and own estimations for some of the sub segments; and EFPIA – Pharmaceutical Industry in Figures for 2010.

With regard to industry structure, we can already notice large differences between the two major segments – chemicals and pharmaceuticals. Differences are also noticeable within the chemicals sector as we will see later.

This low level of concentration in the chemical industry is clearly in contrast with the high level of concentration observed in some of the industries most critical for it, such as the oil industry where the top 10 producers accounted for 67% in 2010; or the automotive industry, where the top 10 producers accounted for 61% (Figure 3.13). The lower concentration of the chemical industry may have resulted in a lower or reduced "selling" or "purchasing" power for the chemical industry or,

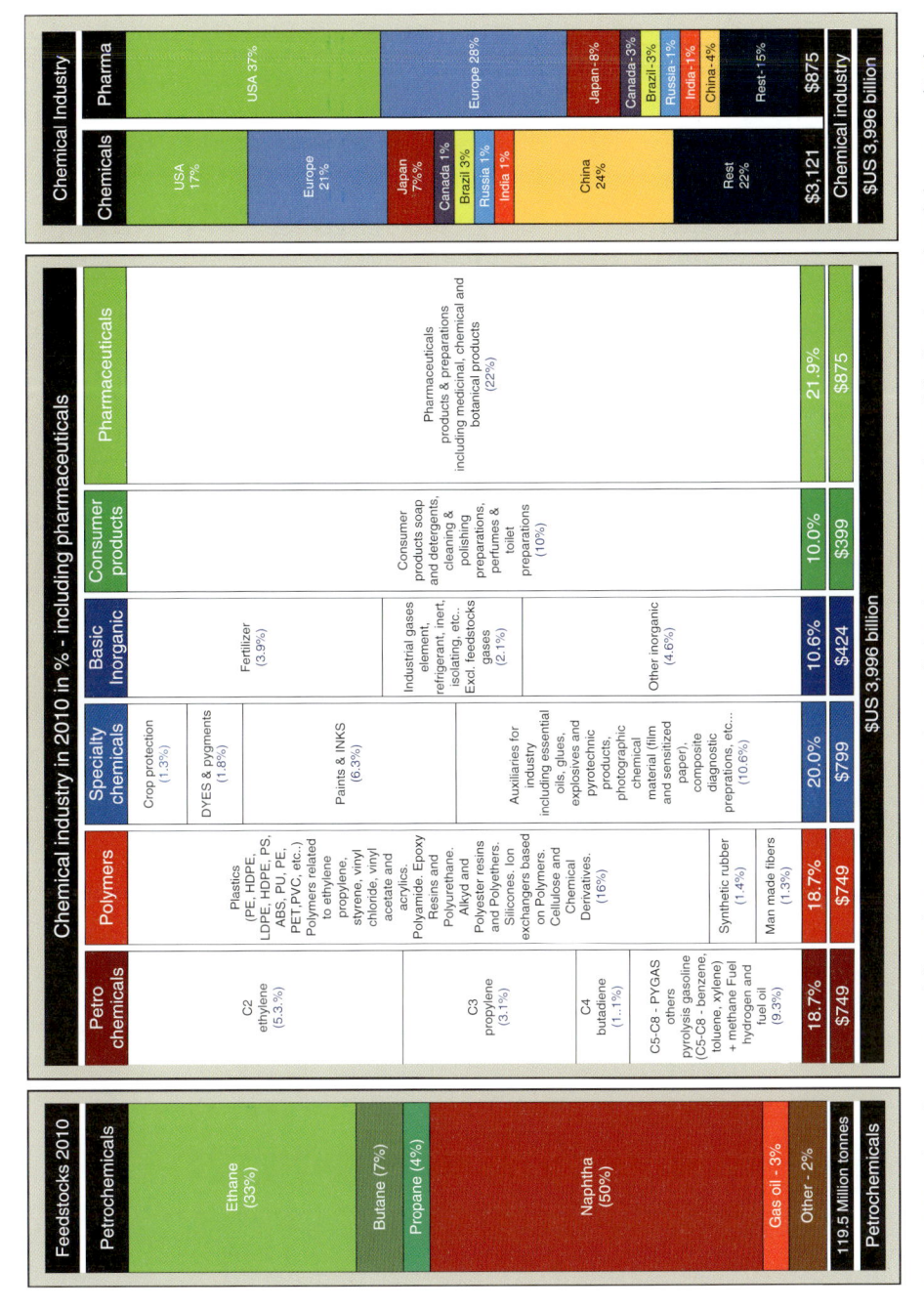

Figure 3.12 The relative sizes of the industry segments in 2010, and also the major petrochemical feedstocks used and the geographical split of the chemicals and pharmaceutical industries. Own elaboration based on CEFIC2011 Facts Report and additional calculations and Estimations.

Table 3.3 The ICIS top 100.

ICIS TOP 100 ANALYSIS

THE ICIS TOP 100

Rank 2010	Company	Sales $m	Sales % Change Reporting currency	Sales % Change in $	Operating profit $m 2010	Operating profit $m 2009	Net profit $m 2010	Net profit $m 2009	Total assets $m	Total assets % Change	R&D $m	R&D % Change	Capital spending $m	Capital spending % Change	Employees Numbers	Employees % Change
1	BASF [a]	84,651	26.0	16.5	10,286	5,270	6,039	2,021	78,714	15.8	1,977	6.7	3,377	1.6	109,140	4.2
2	Dow Chemical	53,674	19.6	19.6	2,802	469	1,970	336	69,588	5.4	1,660	11.3	2,130	26.6	49,505	−5.2
3	ExxonMobil [b]	53,636	30.9	30.9	3,392	2,408	4,913	2,309	26,235	8.6			2,215	−29.6		
4	Sinopec [b]	48,725	50.2	55.3	2,281	2,025			19,130	−0.8			1,956	−48.8	65,623	−4.9
5	LyondellBasell Industries [c]	41,151	33.5	33.5	2,944	317	10,151	−2,865	25,494	−8.2	154	6.2	692	−11.2	14,000	−5.8
6	SABIC	40,525	47.4	47.4	10,105	5,014	5,741	2,420	84,688	7.0	174	23.2	4,293	−32.9	33,000	0.0
7	Shell [b]	39,629	43.8	43.8			1,511	316					809	−59.3		
8	Mitsubishi Chemical [d/x]	38,241	25.9	40.9	2,735	716	1,009	138	39,778	−1.8	1,580	−4.4	1,423	−1.0	53,882	0.0
9	INEOS [o]	34,561	37.2	26.8	1,528	838	−32	−937	17,925						13,682	
10	DuPont	31,505	20.7	20.7	3,711	2,184	3,031	1,755	40,410	5.8	1,651	19.8	1,508	15.3	60,000	3.4
11	Total [b]	24,480	25.4	16.0	1,278	357	1,209	390							41,658	−6.7
12	Bayer [e]	23,983	20.5	11.4	1,152	525			33,868	7.5	1,308	9.8	1,347	−8.0	55,200	−0.9
13	Sumitomo Chemical [d]	23,939	22.3	36.9	1,062	555	295	159	28,587	−0.7	1,668	17.7	1,215	−15.8	29,382	5.6
14	AkzoNobel	19,402	12.4	3.9	1,616	1,225	999	408	26,631	6.4	443	2.1	708	4.1	55,590	1.6
15	Braskem [f]	19,004	60.1	60.1	1,937	846	1,138	240	19,072	43.2			1,053	56.4	6,799	48.8
16	Toray [d]	18,593	13.2	26.7	1,209	433	699	−153	18,929	0.7	563	0.9	669	2.3	38,740	2.1
17	Air Liquide	17,876	12.6	4.1	2,987	2,808	1,861	1,763	29,870	9.3	311	7.8	2,303	20.7	43,600	3.1
18	Linde [g]	17,054	14.8	6.1	3,877	3,418	1,332	847	35,635	10.3	125	5.6	1,580	8.0	48,430	1.5
19	Evonik Industries [b/h]	17,053	29.0	19.2									798	19.9	31,061	2.5
20	Mitsui Chemicals [d]	16,806	15.2	29.0	490	−102	300	−302	15,646	4.6	460	−6.2	544	−8.2	12,892	−0.6
21	Johnson Matthey [d]	15,153	27.4	35.2	556	389	277	235	4,935	12.5	167	19.7	182	4.3	9,742	8.9
22	LG Chem	15,053	23.0	27.9	2,230	1,594	1,761	1,190	5,184	−38.5			1,181	40.2	13,000	30.0
23	SK Energy [i]	14,279	26.8	31.8	346	536										
24	Reliance Industries [d/i]	14,058	12.9	27.6	2,121	1,699			10,160	−0.2			122	6.9		
25	Asahi Kasei [d/j]	12,955	15.3	29.0	959	305							632	−24.0		
26	Shin–Etsu [d]	12,779	15.4	29.2	1,802	1,265	1,209	905	21,545	0.8	451	11.2	1,589	12.0	19,770	16.6
27	Merck KGaA	12,313	19.9	10.9	1,476	930	838	525	29,671	34.0	1,851	3.9	525	−15.2	40,562	22.7
28	DSM	11,994	15.1	6.4	1,063	767	672	483	13,889	9.0	429	6.2	551	−9.0	21,911	−3.6
29	Syngenta	11,641	5.9	5.9	1,793	1,819	1,397	1,408	17,285	7.2	1,032	8.4	396	−39.3	26,000	4.0
30	PPG Industries [k]	11,297	8.8	8.8	1,341	990										
31	Yara International	11,227	6.4	7.9	1,282	215	1,499	641	11,243	6.2	18	15.9	531	−27.5	7,348	−3.7
32	Chevron Phillips Chemical	11,204	33.3	33.3	1,501	707	1,388	615	8,016	8.1	41	7.9	186	13.4	4,600	−4.2
33	Sekisui Chemical [d]	11,055	6.6	19.3	596	388	285	125	9,542	0.4	298	2.8	256	−42.7	19,770	0.0
34	Honam Petrochemical	10,727	59.3	59.3	1,093	775	774	698	7,562	33.9	19	35.7	1,654	908.5	1,557	1.1
35	Agrium	10,520	15.2	15.2	1,103	580	714	366	12,717	30.0			441	40.9	14,150	26.9
36	Praxair	10,116	13.0	13.0	2,082	1,575	1,195	1,254	15,274	6.7	79	6.8	1,388	2.7	26,261	0.4
37	Mosaic [r]	9,938	47.0	47.0	2,664	1,271	2,515	827	15,787	24.2			1,263	38.7		
38	Teijin [d]	9,850	6.5	19.2	586	145	304	−385	9,196	−7.5	380	−5.7	307	−25.4	17,542	−6.6
39	NPC (Iran) [d/w]	9,810	36.4	36.4	736	382	553	166	22,554	−8.2	3	−72.1	907	−47.3	18,187	33.2
40	Formosa Chemicals & Fibre (Taiwan) [m]	9,719	41.7	41.7			1,729	919	12,969	22.0					4,906	−0.5
41	Henkel (Adhesive segment) [n]	9,683	17.4	8.5							306	2.2	159	−37.8	23,927	−0.9
42	LANXESS	9,436	40.8	30.2	804	214	502	57	7,509	11.8	154	14.9	664	82.2	14,648	2.2
43	Solvay [o]	9,422	−16.2	−22.5	404	1,238	2,355	740	18,573	11.7	180	−2.2	713	−5.1	16,785	−11.6
44	DIC [d]	9,407	2.8	15.0	449	300	190	27	8,498	−6.1			241	−10.5		
45	Huntsman	9,049	19.6	19.6	410	13	27	114	8,714	1.0	151	4.1	236	24.9	13,000	18.2
46	Air Products	9,026	9.3	9.3	1,389	846	1,029	631	13,506	3.7	115	−0.9	1,134	−8.3	18,300	−3.2
47	BP [q]	8,900	30.9	30.9												
48	Sasol [l]	8,572	−13.2	−11.9	721	−290							439	−1.4	11,936	−4.8
49	Borealis	8,308	33.0	23.0	463	34	439	53	7,461	16.9	111	6.3	129	−68.5	5,075	−2.7
50	Tosoh [d]	8,265	8.9	21.8	405	141	121	74	8,766	−1.9			335	−4.6	11,089	0.0

Table 3.3 (continued.)

Rank 2010	Company	Sales $m	Sales % Change Reporting currency	Sales % Change in $	Operating profit $m 2010	Operating profit $m 2009	Net profit $m 2010	Net profit $m 2009	Total assets $m	Total assets % Change	R&D $m	R&D % Change	Capital spending $m	Capital spending % Change	Employees Numbers	Employees % Change
51	Polimeri Europa (ENI)[i]	8,139	46.1	35.1	-114	-967	-113	-487	4,077	19.1			333	73.1	5,972	-1.6
52	Arkema	7,826	32.9	22.9	644	-99	460	-247	6,385	22.1	184	2.2	417	4.7	13,903	0.7
53	Sherwin-Williams	7,776	9.6	9.6	749	695	463	436	5,169	19.6	40	-1.2	125	37.1	32,000	9.5
54	Momentive Performance Materials Holdings[p]	7,406	26.9	26.9	533	15			8,777	164.7	131	9.2	140	81.8	11,000	
55	Rhodia	6,926	29.6	19.9	798	229	343	-189	6,800	20.3	109	12.3	310	40.1	14,000	2.9
56	Formosa Plastics[m]	6,675	36.2	36.2			1,758	922	11,722	20.7					5,033	0.9
57	Clariant	6,650	7.7	4.3	342	-19	168	-198	5,530	-2.8	126	-10.0	209	65.9	16,200	-7.6
58	K+S	6,618	39.7	29.2	969	347	595	138	7,387	6.8	21	-17.1	266	13.2	15,241	0.2
59	PotashCorp	6,539	64.4	64.4	2,548	1,181	1,806	981	15,619	20.9			1,978	12.1	5,400	5.1
60	Wacker Chemie	6,293	27.7	18.0	1,064	221	650	-101	7,291	21.1	219	0.7	818	-19.9	16,314	4.5
61	Dow Corning	5,997	17.8	17.8			866	598	12,648	16.9					9,000	0.0
62	Celanese	5,918	16.5	16.5	503	290	377	498	8,281	-1.6	70	0.0	201	14.2	7,250	-2.0
63	Eastman Chemical	5,842	32.9	32.9	862	345	438	136	5,986	8.5	152	22.6	243	-21.6	10,000	0.0
64	Taiyo Nippon Sanso[d]	5,840	11.6	24.9	428	297	154	170	7,459	0.1						
65	ICL	5,692	25.0	25.0	1,346	938	1,025	770	6,388	8.1	64	19.1	334	-3.6	11,035	3.9
66	Orica[s]	5,602	-10.2	-1.8	900	736	1,271	478	6,737	-5.0	37	14.2	508	47.5	14,000	-7.5
67	Ashland[s/t]	5,593	10.0	10.0	511	338			6,549	0.5			175	21.5		
68	Kaneka[d]	5,480	10.0	23.1	256	189	140	91	5,496	5.1			319	10.4		
69	Mitsubishi Gas Chemical[d]	5,447	17.3	31.3	281	44	438	80	6,968	7.0	198	1.2	427	28.4	4,979	1.2
70	Lubrizol	5,418	18.1	18.1	1,110	859	732	501	4,967	4.1	226	6.5	176	25.7	6,896	2.5
71	Styron[z]	4,967	44.0	44.0	181	153	57	64	2,676	58.2			9	-63.2		
72	ALPEK (Grupo Alfa)	4,957	13.7	20.4	387	322	203	142	3,217	-2.4					4,076	1.9
73	Cementhai Chemicals	4,811	42.7	58.7	271	417	754	376	5,503	-0.5						
74	Honeywell[u]	4,726	14.0	14.0	749	605			4,938	6.0			188	22.9		
75	PKN Orlen[b/i]	4,608	4.2	1.6	165	-68			4,492	8.5			592	-25.8		
76	NOVA Chemicals	4,576	54.8	54.8	590	-112	259	-241	5,670	1.3	35	2.9	126	40.0	2,445	-2.2
77	Showa Denko[b/i]	4,392	9.7	22.8	102	91			4,363	1.4	43	3.4	336	148.2		
78	Kuraray[d]	4,386	9.1	22.1	641	329	347	176	4,188	2.7	18	-3.2	226	-13.6	6,544	-1.3
79	Denki Kagaku Kogyo (Denka)[d]	4,322	10.5	23.7	297	234	173	113	4,859	0.5						
80	Daicel Chemical Industries[d]	4,271	10.4	23.6	395	225	203	119	4,964	-4.0	145	5.8	138	-36.1	1,948	-1.2
81	Airgas[d]	4,252	9.7	9.7	468	400	250	196	4,936	9.8			256	1.3	14,000	0.0
82	Nalco	4,251	13.4	13.4	578	404	196	61	5,224	5.2	80	8.1	156	53.2	12,400	7.0
83	JSR[d]	4,114	9.8	22.9	472	218	333	147	4,717	4.6	208	-6.0	161	-29.9	5,259	0.9
84	Occidental Chemical[b]	4,016	24.5	24.5	438	389							237	15.6		
85	CF Industries[v]	3,965	52.0	52.0	896	680	349	366	8,759	251.1			258	9.5	2,400	60.0
86	Givaudan	3,959	7.1	3.8	519	443	318	192	6,466	-2.3	314	3.1	112	26.3	8,618	1.4
87	PETRONAS[i]	3,868	60.2	72.7												
88	Nippon Shokubai[d]	3,482	18.0	32.1	360	150	255	117	3,977	5.9	126	4.6	185	-36.3	3,576	4.3
89	PTT Chemical	3,440	18.5	31.8	479	356	343	204	5,601	4.8			224	-61.3		
90	CEPSA[i]	3,427	47.7	29.5	177	87							39	-5.9		
91	RPM[r]	3,382	-0.9	-0.9	345	320	189	180	3,515	17.0	41	-1.0	40	71.6	9,000	0.0
92	PEMEX[i]	3,348	-17.3	-12.8	-1,245	-1,565	-1,225	-1,536	7,254	3.0						
93	Ube[b/d]	3,299	21.3	35.7	347	85										
94	Indorama Ventures	3,292	22.1	35.8	371	245	352	145	2,600	5.0	2	0.0	81	-36.1	4,083	17.5
95	Zeon[d]	3,265	19.7	34.0	426	101	221	54	3,509	3.4			472	19.0		
96	Hanwha Chemical	3,241	19.6	24.3	434	353	357	295	4,843	12.5						
97	Valspar	3,227	12.1	12.1	376	281	222	150	3,868	10.2	100	9.7	68	16.9	10,180	15.8
98	Tessenderlo Group	3,217	15.9	7.2	78	-74	27	-239	2,248	7.6	18	-3.6	155	4.2	8,262	-0.7
99	EuroChem	3,210	32.9	32.0	854	354	656	367	4,906	16.1			670	9.8	19,614	-2.1
100	Rockwood Holdings	3,192	15.3	15.3	358	198	239	21	4,724	-1.3	49	12.3	180	19.0	9,600	1.1

Source: ICIS. Full list available at http://www.icis.com/Articles/2010/09/13/9392596/the-icis-top-100-chemical-companies-analysis.html

Industry concentration: Comparatively low

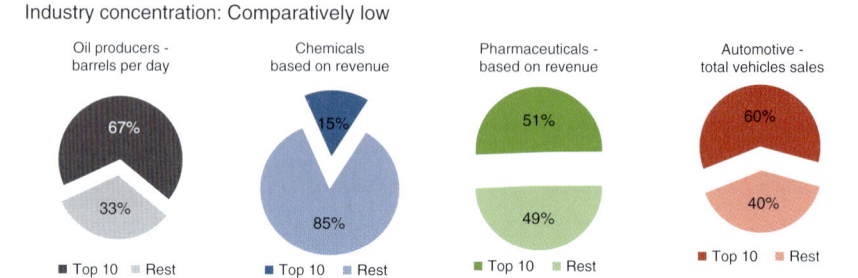

Figure 3.13 Industry concentration for oil producers, chemicals, pharmaceuticals and the automotive industry. Source 1: OPEC for largest oil producers in 2010. Source 2: ICIS top 100 Chemical Companies – September 2011, ICIS Chemical Business Magazine, p. 29 Source 3: IMAP – Pharmaceutical and Biotech Industry Global Report 2011 Source 4: Organization International des Constructeurs d'Automobiles. World Ranking of Manufactures in 2010.

in other words, in reduced margins versus other industries; however, that situation has to be evaluated segment by segment and area by area.

Indeed, in certain segments like fertilizers, crop protection, or industrial gases, the level of concentration can be higher much higher. In some of these segments we can find companies like Air Liquide, Linde, Syngenta, Yara, Mosaic, Praxair, Agrium, or Monsanto, just to mention a few, that completely specialize in one of these fields and can have a significant market share.

Therefore, the low level of concentration in the industry is certainly a characteristic of the chemical industry at the high level, and in some cases can even be a disadvantage, especially when comparing and doing business with other industries but that statement has to be properly qualified by looking individually at each company and segment.

In terms of the top chemical companies by sales in 2010, we can see a large mix of companies, with some oil companies like Exxon or Shell, some companies backward integrated into crude oil like for instance BASF, some diversified chemicals like Dow and some specialty chemicals like DuPont (Figure 3.14). However, it is the large petrochemical and polymers chemical companies who top this ranking. Indeed in 2010 the largest chemical company in the world, much larger than the rest, was BASF with US$ 84.7 billion. Dow Chemical and ExxonMobil appear in second position with similar sales at US$ 53.6 billion, Sinopec appears in fourth position with US$ 48,7 billion and then the rest. Perhaps it is interesting to consider that in 2010, 80% of the top 10 chemical companies were based in the ADV economies, only one in the BRIC (Sinopec) and only one in the rest (SABIC). We can wonder how different that split will look by 2050.

In terms of the top pharmaceuticals companies, Pfizer appears to be the largest with $67.8 billion sales in 2010, followed closely by Johnson & Johnson with $61.6 billion (Figure 3.15). Considering that 75% of the world pharmaceutical sales in 2010 took place in the ADV economies, it is not surprising to find that 80% of the top 10 pharmaceutical companies were located in the ADV economies,

3.6 Industry Structure and Companies

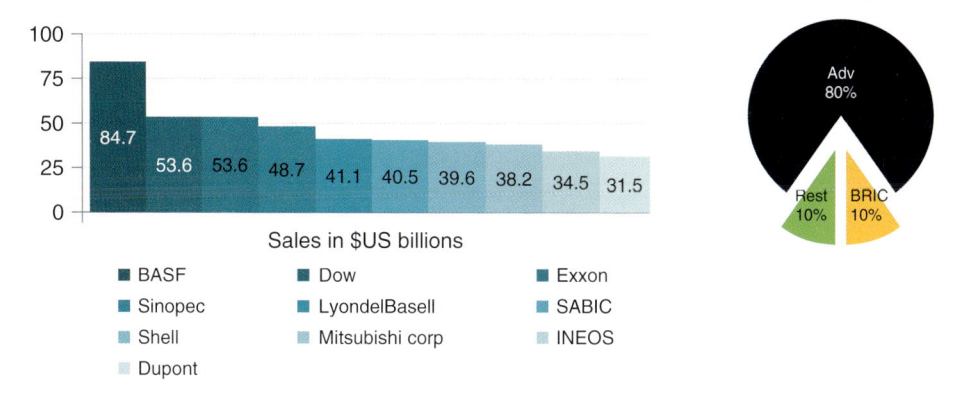

Figure 3.14 Top 10 chemical companies in 2010. Source: ICIS Top 100 Chemical Companies in 2010.

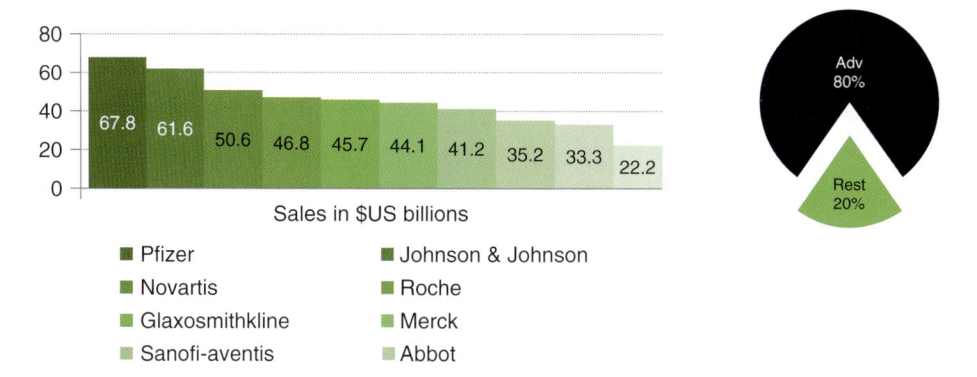

Figure 3.15 Top 10 pharmaceutical companies in 2010. Source: IMAP Health care report – Pharmaceuticals and Biotechnology Industry Global report 2011.

and the remaining 20% were based in Switzerland (REST). Among the top 50 pharmaceutical companies in 2010, 45 were based in the ADV economies, 3 in Switzerland (REST), 1 in Australia (REST) and 1 in Iceland (REST). Considering the large populations of the BRIC and its expected growth in pharmaceuticals, we wonder how the structure of the industry will change by 2050.

Another key characteristic of the industry is the clear correlation among the following variables: profitability, R&D spending, sales per employee, and P/E ratios. Looking in detail at these different comparisons we can consider individually the different segments of the industry, but more importantly several key features of the current structure of the industry:

- **Profitability versus Sales per Employee**
  Looking at this first correlation within the broad chemical industry and then within the chemicals side of the industry, once things appear to be clear and somehow logical: *"segments and companies with higher margins tend to have much lower sales per employee than those with lower margins"*.

Ranking the top 100 chemical companies and the top 10 pharmaceutical companies for 2010 by their profitability in terms of sales per employee, and with the clear exception of fertilizers, we can observe how the segments with the highest profitability like pharmaceuticals (17%), industrial gases (16%), consumer products (14%), and crop protection (12%) had some of the lowest ratio of sales per employee, fluctuating around the US$500 000 sales per employee. In contrast, petrochemicals and polymers with lower margins among the top 100 chemical companies tend to have much higher sales per employee with almost US$1.6 million sales per employee (Figure 3.16). Looking briefly at the chemicals side of the industry and its top chemicals producers, we can validate that correlation also applies among its top companies; with two major players, LyondellBasell and INEOS having the highest sales per employee in 2010, with US$3 and 2.5 million, respectively (Figure 3.17).

- **Profitability versus R&D Spending**

The second correlation tends also to be straightforward and a key feature of the industry. This feature holds very well during the economic cycle, and across the whole industry; a correlation that basically states that *"segments and companies with the highest margins tend to have the highest R&D ratios and vice versa"*.

Looking now at the selected group of segments and companies for this exercise, and with the exception of fertilizers, it seems that this correlation also hold very well. The pharmaceutical segment had the highest profitability of the industry but also the highest R&D spending relative to sales, while for the rest of the segments similar patterns could be observed (Figure 3.18).

The philosophical question that traditionally arises at this point is what triggers what? Do high R&D spending ratios enable high margins or vice versa? This is certainly a very fair, interesting, and critical question. A question that cannot be answered simply, but one that managers certainly get asked all the time.

In any case this is a very simple analysis based on a small sample and period of time, so more rigorous and lengthy analysis should and could be done to extract the real value of these correlations, however, at this stage we just want to highlight this as one of the major features of the industry structure. A feature, that we can easily extrapolate to other similar industries, and certainly can explain certain industry and companies behaviors.

- **Profitability versus P/E Ratio**

Finally, and perhaps the most obvious of all the observed correlations is that between higher margin and higher P/E ratios for the stock value. Companies and industries with higher margin or higher margin expectation tend to have higher valuation.

As large parts of the chemical industry tend to be cyclical, P/E ratios can fluctuate significantly during the cycle and these analyses should be done on a long term basis and considering in which part of the cycle a certain industry segment or company is. Please see Figure 3.19 as an illustration and old example of the potential correlation between P/E ratios and margin of some selected chemical and pharmaceutical companies in 2006. As can be observed, and not surprisingly, there is a large correlation between margin and P/E ratios.

3.6 Industry Structure and Companies | 181

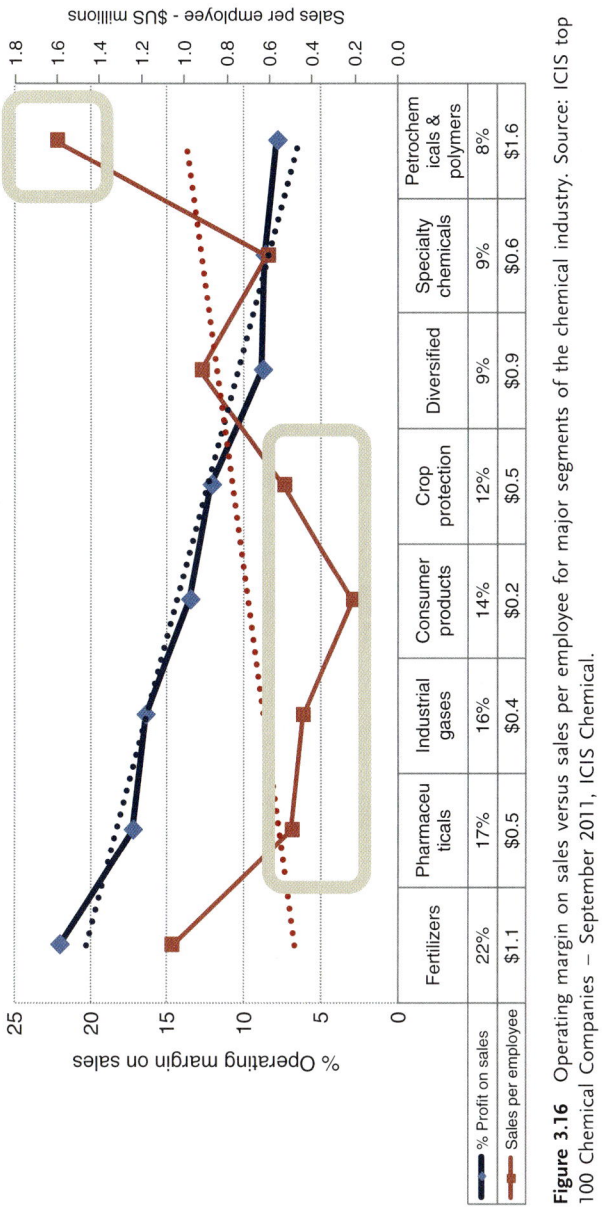

	Fertilizers	Pharmaceuticals	Industrial gases	Consumer products	Crop protection	Diversified	Specialty chemicals	Petrochemicals & polymers
% Profit on sales	22%	17%	16%	14%	12%	9%	9%	8%
Sales per employee	$1.1	$0.5	$0.4	$0.2	$0.5	$0.9	$0.6	$1.6

Figure 3.16 Operating margin on sales versus sales per employee for major segments of the chemical industry. Source: ICIS top 100 Chemical Companies – September 2011, ICIS Chemical.

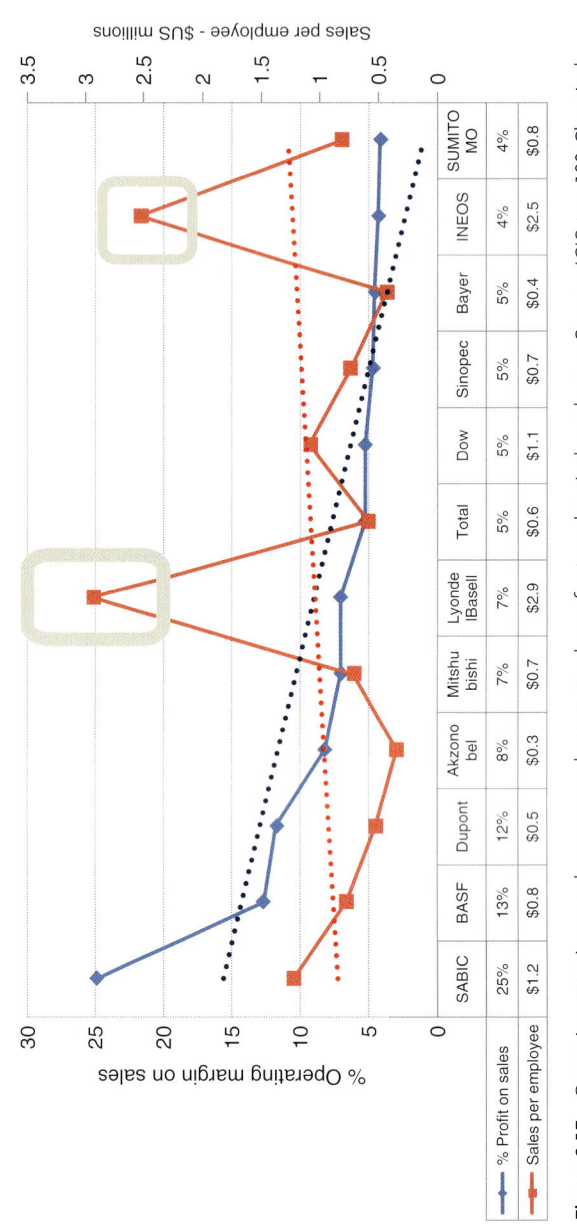

Figure 3.17 Operating margin on sales versus sales per employee for top chemical producers. Source: ICIS top 100 Chemical Companies – September 2011, ICIS Chemical.

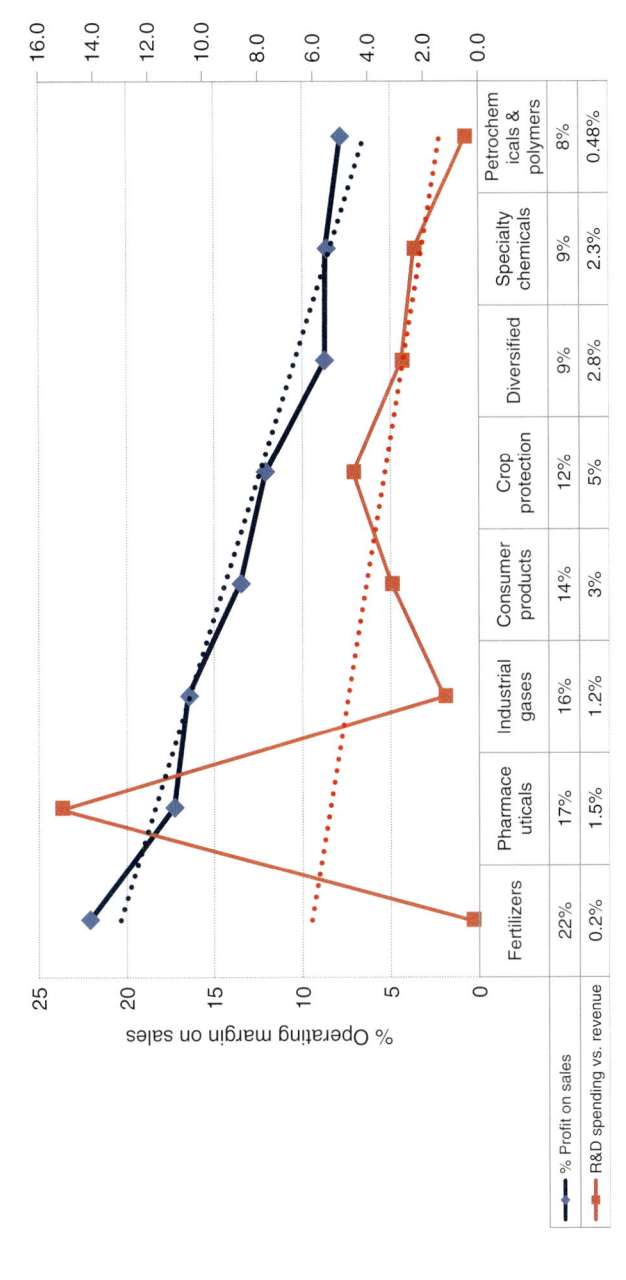

Figure 3.18 Operating margin on sales versus R&D expenditure on sales for major segments of the chemical industry. Source: ICIS top 100 Chemical Companies – September 2011, ICIS Chemical.

	Fertilizers	Pharmaceuticals	Industrial gases	Consumer products	Crop protection	Diversified	Specialty chemicals	Petrochemicals & polymers
% Profit on sales	22%	17%	16%	14%	12%	9%	9%	8%
R&D spending vs. revenue	0.2%	1.5%	1.2%	3%	5%	2.8%	2.3%	0.48%

184 | *3 The Chemical Industry in 2010*

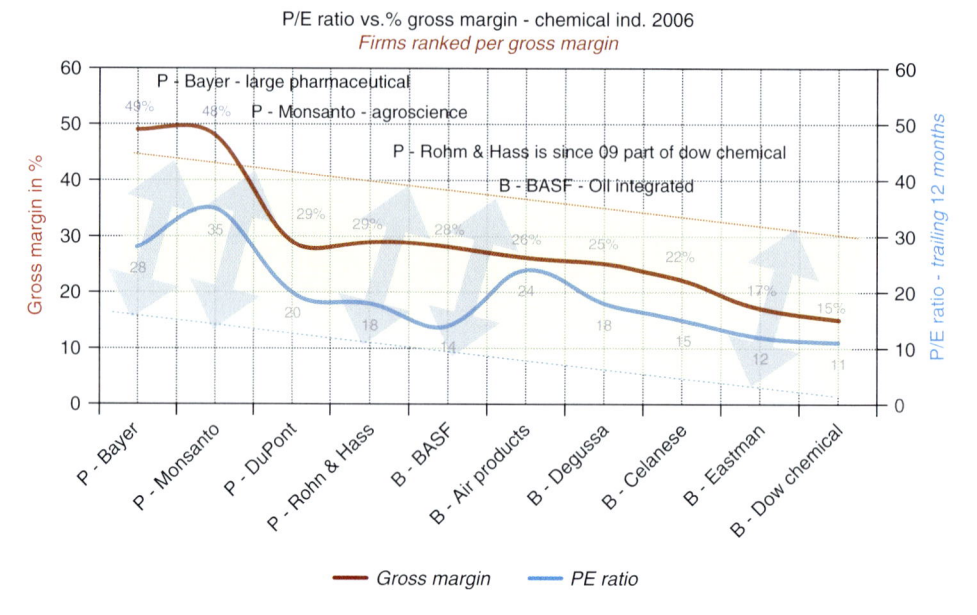

Figure 3.19 P/E versus % gross margin for the chemical industry in 2006. Source: ICIS top 100 Chemical Companies – September 2007, ICIS Chemical.

This correlation can be very useful when trying to estimate which segments of the industry and companies, and from which part of the world, might enjoy higher margins and P/E ratios in the future.

3.7
Safety

Perhaps the most important, remarkable, and noticeable feature across the whole chemical industry, including pharmaceuticals, is its constant focus on safety and security, in combination with the high levels of regulation.

Among the many examples we could use to illustrate this point we have chosen the "incident rate" for the European chemical industry (Figure 3.20), while for other regions and for similar safety metrics we should expect very similar performance.

Looking at the rate of accidents at work with the European manufacturing sector, the chemical industry not only appears to have one of the lowest accident ratios among the top 10 manufacturing sectors in EU 15, but actually that has been the case for the last decades.

Indeed, since 1995, as in Figure 3.21, we can observe how the chemical industry has been constantly outperforming the European manufacturing performance while improving this performance year after year.

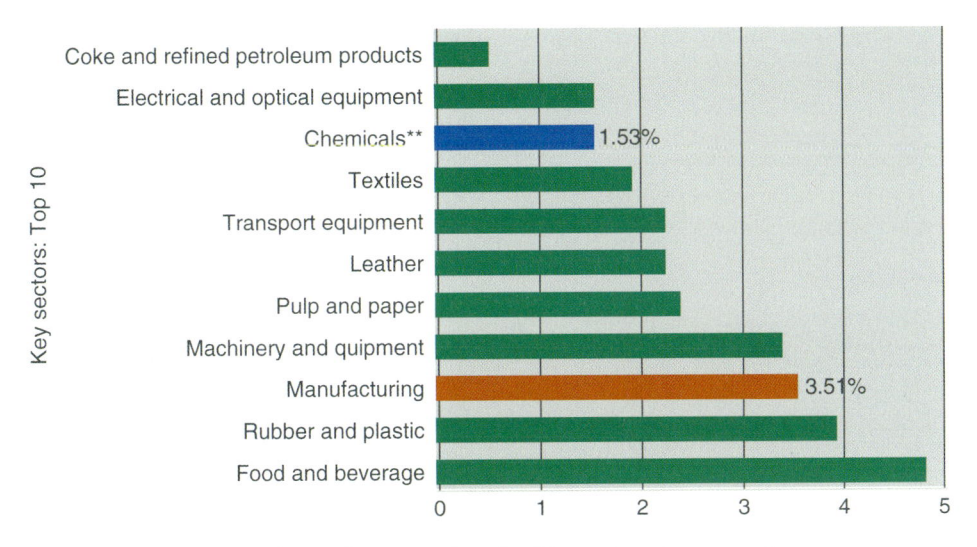

Figure 3.20 Incident rate* for accidents at work in the EU manufacturing sector. Source: Eurostat and Cefic analysis. *Incidence rate of accidents at work per 100 employee (more than three days lost). **Including pharmaceuticals.

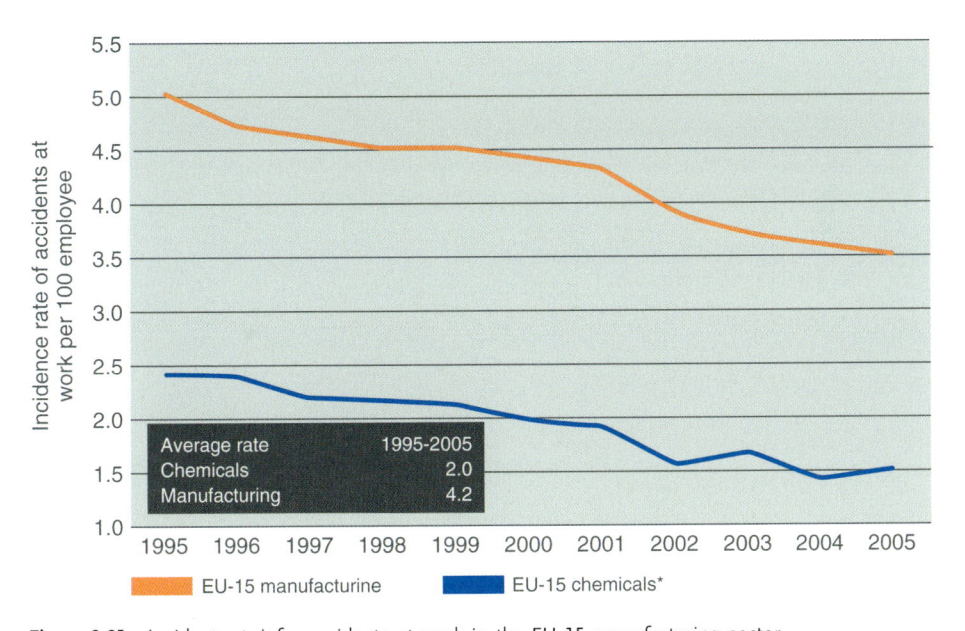

Figure 3.21 Incident rate* for accidents at work in the EU 15 manufacturing sector (more than 3 days lost) for the years 1995 to 2005. Source: Eurostat (Health_Safety_Work) Database and Cefic Analysis. *Including pharmaceuticals.

The enormous focus and efforts of the industry on safety and compliance with regulation and safety standards is something that can be seen across the whole industry.

Safety is certainly one of the key features of the industry, a feature that will always remain core for the industry, even when looking into 2050 and beyond.

3.8
Background

3.8.1
Recent History of the Chemical Industry Excluding Pharmaceuticals

This summary is certainly a huge over simplification of the recent history of the chemical industry but we could not move forward into the future of the industry without having a high level view and understanding of the recent history of the industry, its bases, its relevance and its contribution to other industries, society, and mankind.

The summary will serve to introduce the passionate, intriguing, and thrilling history of the chemical industry. This is the history of an industry that has been and still remains today at the core of innovation and human progress. An industry that has distinguished itself for being the industry of last resort, enabling many other industries through the last centuries, solving some of the most critical issues for nations, industries, and mankind.

The following pages will be devoted to this purpose, and even when we exclude the invaluable contributions of the pharmaceutical industry to human progress, we hope we will still be able to convey part of the excitement and pride in these contributions to our society over the last centuries.

- 1750 Industrial revolution and inorganic chemistry
- 1850 Synthetic dyes from coal for textiles, and chlorine bleach
- 1869 Plastics/celluloid
- 1900 Electrolysis of brine
- 1909 Synthetic fertilizers
- 1914 Rayon from wood fibers
- 1928 Nylon (DuPont)
- 1910 Steam cracker (ethylene, propylene, butadiene)
- 1920 Styrene cracking (ethylene, benzene)
- 1930s Petrochemicals and polymers (polyethylene (PE), others)
- 1930s Synthetic rubber (SBR, polybutadiene rubber PBR, others)
- 1950s Plastics demand explodes
- 1960s Internationalization
- 2010–2050 The chemical industry leads the revolution against climate change.

3.8.1.1   1750–1850 Industrial Revolution and Inorganic Chemistry

With the industrial revolution and mechanization of many industries, from textiles to agriculture, manufacturing, mining, transportation, and many others, profound changes occurred in most of these industries as well as in the society and economy. All these changes created new demands for the incipient chemical industry, being the first time the industry became an enabler of change. The chemical industry started to grow fearlessly, especially in the UK, on the back of growing demand from the textile, glass, metal, coal, and soap industries. Indeed by 1850, the UK had the largest chemical industry in the world, and it seemed the growing demand for chemicals was endless. Inorganic chemicals, such as sulfuric acid, soda ash, and chlorinated lime were in high demand, and as society kept progressing, demand kept increasing.

3.8.1.2   1850 Synthetic Dyes from Coal for Textiles, and Chlorine Bleach

The fast developments of the textile industry started to create additional needs for the chemical industry. Soda ash was the base of the successful British textile industry and the glass industry. The large use and commercialization of soda ash increased the affordability of soap, improving the hygiene conditions of millions of people, reducing infections and illness around the world, and making soda one of the first globally traded commodities. As the textile industry developed, new improvements were required, including faster processing, and synthetic dyes as the natural ones started to be depleted. These two constraints pushed the industrial development of chlorine bleach and the creation of synthetic dyes from coal. Chlorine led to the creation of bleach for disinfectant uses, whitening and as an essential precursor in the chemical industry.

Synthetic dyes extracted from coal were another important milestone for the chemical industry, not only because thanks to that we can all enjoy the current colors in our textiles, but also because they were the basis for some of the largest chemical companies in the world. Bayer was created in 1863 in Germany for manufacturing and selling synthetic dyes, opening its famous pharmaceutical department only in 1881.

BASF was also created around the same time, actually in 1865, and also based on the production and manufacturing of synthetic pigments and still today its name indicates its origins. BASF in German stands for "Badische Anilin und Soda Fabrik." Badische, or Baden in English, refers to the region where BASF started, Anilin to the major component used for synthetic dyes, although aniline is largely used today for methylene diphenyl diisocyanate (MDI) and polyurethane (PU) as well as a precursor in the chemical industry. Soda refers to all the chemistry developed around that, especially for the textile, water, and glass industries, while Fabrik in English stands for factory.

3.8.1.3   1870 Celluloid

Traditionally considered as the first plastic and thermoplastic, celluloid is a compound created from nitrocellulose and camphor, plus dyes and other agents. It was used in the manufacture of dressing table sets, dolls and toys, picture frames, kitchen utensils, cheap jewelry, and many items that would earlier have been manufactured from ivory, horn, wood, or other animal products. However, its use as a photographic film and film for movies was perhaps its most famous use. Indeed, the film industry is often referred as the *"celluloid"*. In 1885, George Eastman Kodak patented a machine for producing continuous photographic film based on cellulose nitrate. However, the fragility, and more importantly the flammability of the material, led to its replacement by polyester (PES) in the film industry during the 1940s, and in many other applications of other plastics through the following decades. Celluloid is still considered as the first plastic.

3.8.1.4 1880 Rayon from Wood Fibers

Rayon was the first man-made fiber to be developed; it is made from wood and was first known as *artificial silk*. It was discovered by the Frenchman, Count Hillaire de Chardonnet in the 1880s but at the same time the English inventors Charles Frederick Cross and Edward John Bevan invented a more effective process and product called *viscose*. After the generic name "rayon" was coined in 1924, the enterprise was renamed the DuPont Rayon Company.

Rayon has been traditionally a subject of large confusion as it is not a natural fiber nor a synthetic one. It is a fiber on its own. Normally associated with the replacement of silk, cotton, or linen, and with properties indeed closer to the natural cellulosic fibers rather than those of thermoplastic and petroleum-based ones, such as nylon or polyester, the reality is its properties are somehow unique and different to all of them.

Although the process to extract rayon had been known for centuries, it was only in the early nineteenth century when its industrial production started to be considered. Rayon has a silk esthetic with excellent feel and drape and is able to retain very bright colors. It is also very versatile, allowing it to blend very easily with many other fibers. With all these merits, rayon can be found today in yarns, fabrics (including gabardines, suiting, coats, and outwear), apparel (including blouses, dresses jackets, lingerie, etc.), domestic textiles (including blankets, curtains, draperies, covers, tablecloths, etc.) or in various applications like cellophane. Despite the fact the major raw material come from the cellulose of trees, and therefore has a renewable base, the high energy and water requirements to process the materials have been subject to criticism. However, the recent developments on the cotton prices have stimulated the demand for rayon again.

3.8.1.5 1900 Electrolysis of Brine (Chlorine)

During this process caustic soda is created. Caustic soda is used in several industries, including textiles, pulp and paper products, soaps, detergents, polyvinyl chloride (PVC), packaging, petroleum refining, gas treatment, and other chemical processing. Chlorine is used as a bleach, an oxidizing agent, a disinfectant in water purification, in agrochemicals, solvents, pharmaceuticals, plastics and polymers, and as a chemical reagent. Around the chlorine chemistry, The Dow Chemical Company was founded by Herbert H. Dow in Midland, Michigan, USA. Traditionally perceived as the largest USA chemical company and the second largest chemical company in the world, the Dow Chemical Company started its journey in the chemical industry with the production and commercialization of bleach in 1989.

3.8.1.6 1913 Synthetic Fertilizers

Fertilizers have been used and known for centuries, first in the form of organic fertilizers (composed of enriched organic matter – based on plants or animals) and then during the industrial revolution with the use of inorganic fertilizers (composed of synthetic chemicals and/or minerals). They are added to the soil to supply additional nutrients essential for the growth of plants. In general they tend to complement the soil with several nutrients including nitrogen (N), phosphorus (P), magnesium (K), and calcium (Ca). At the end of the nineteenth century with the outlook of increasing populations and limited agricultural resources, chemical companies were confronted with the challenge of how to increase the supply capacity of agricultural products, with better and more efficient fertilizers, possibly inorganic fertilizers.

However, that challenge became even bigger during World War I. At the beginning of the War the importance of nitrogen increased dramatically due to its use for the production of explosives and, with that in mind, Germany was prevented from importing nitrogen from Chile which had the only commercial viable deposits of sodium nitrate in the world.

Inorganic fertilizers are often associated with BASF and synthesized by the Haber–Bosch process that involve the fixation of atmospheric nitrogen to produce synthetic ammonia. In 1913, BASF started the first ammonia synthesis plant and began producing nitrogen fertilizers, although, during the war periods, part of this production was also used for military purposes.

Carl Bosch and Friedrich Bergius, employees of BASF, won the Nobel Prize for the development of high-pressure technology for ammonia synthesis and coal hydrogenation.

Fertilizers have played a significant role in our societies, originally during the British Agricultural revolution. They enabled significant increases in absolute production and productivity during the industrial revolution, raising living standards across the UK and globally. Most recently, fertilizers have certainly enabled the spectacular growth in population our world has experienced during the last centuries. Indeed, during the last century, our world population grew by a staggering 368% from 1.6 billion in 1900 to 5.9 billion in 2000.

Considering the expected economic and population growth in the next decades until 2050, we could expect a bright future for inorganic fertilizers. However, they will need to be able to counterbalance the claims that ammonia-based fertilizers increase the methane emissions from crop fields, especially from rice plantations. As reviewed in Chapter 1, methane accounts for 18% of the total greenhouse emissions, although a large part of these are related to livestock and agricultural activities.

The clear need to reduce greenhouse gases might generate a new challenge for the chemical industry, but nothing that the industry has not done before.

3.8.1.7 1910–1920 Steam Cracker (Ethylene, Propylene and Butadiene)

Steam crackers were invented during this period, remaining the basic of the petrochemical industry as we know it today. With the discovery and production use of steam crackers, first in Russia and then in USA, crackers started to expand around the world. Steam crackers remains still today the bases of the petrochemical industry, being ethylene its largest monomer with more that 140 million of ethylene capacity around the world.

3.8.1.8 1920–1930 – Styrene Cracking (Ethyl-benzene and Styrene) Cracking

First styrene steam crackers invented. A two step process started by the alkylation of benzene with ethylene to produce ethyl benzene and followed by dehydrogenation of the ethyl benzene to produce styrene monomer.

3.8.1.9 Polyamide Nylon (DuPont)

During the period between World War I and II, and especially after World War II (WWII) and after Japan stopped the sale of silk to the USA, silk prices skyrocketed. The US government and some chemical companies started to look into ways to obtain synthetic fibers. With that goal in mind in 1928 DuPont hired Wallace Carothers to work in polymers, and major discoveries are attributable to him, like the discovery of neoprene, the first synthetic rubber, polyester, and nylon in 1935. Nylon is a thermoplastic often referred to as a polyamide (PA) with its major target application being to replace the Asian silk.

Nylon is used in multiple applications from its original use in toothbrushes in 1938 to household brushes, ropes, tennis racquet strings, musical strings, umbrellas, shower curtains, or undergarments, including the famous nylon stockings for women in 1940s. However, it was its war applications that created its initial major demand, as nylon was used for parachutes, flack vests, some military tires, and as a rope. Indeed during 1942 to 1944 all production of nylon in the US was allocated to military purposes. After that period nylon has been used, and still is used today in multiple applications, being one of the most commonly used thermoplastics.

3.8.1.10 1930s – Synthetic Rubber

The rubber and synthetic rubber history can be structured around three milestones. In 1839, Charles Goodyear in the USA invented the process to vulcanize natural rubber. In 1910, Sergei Lebedev was the first to obtain synthetic rubber based on poly-butadiene. In the late 1930s the world's first industrial synthetic rubber plants, for emulsion styrene butadiene rubber, started operating, including those in East Germany for which Buna rubber was named. These were operated by I.G. Farben, the German chemical conglomerate.

The drop in natural rubber production in Brazil during World War I (1914–1918), triggered the need for lower-cost products with steadier supplies in order to manufacture tires. The strategic importance of natural rubber, especially during the war period, put enormous pressure on the chemical industry to develop sustainable and independent alternatives to natural rubber.

That situation became very clear when the USA lost access to the rubber from South East Asia. Between 1939 and 1941 the US launched the US Program – A two year program that put together, government, academics, scientific laboratories, and the young petrochemical industry into the mission to get enough synthetic rubber for the USA and its allies. That was the origin of the Governmental Rubber-Styrene (GR-S), program that increased the US synthetic rubber capacity from an annual output of 231 tons of synthetic in 1941 to 70 000 tons a month in 1945. To further illustrate how critical was synthetic rubber during the war periods let us share this short event. Japan attacked Pearl Harbor on 7 December 1941, to destroy the US troops and avoid their potential interference in Japan's planned invasion of Malaysia and the Dutch East Indies, currently Indonesia, and their ultimate goal to obtain access to their vast reserves of oil and natural rubber. One day after the Pearl Harbor attack, the USA declared war on Japan, and four days later the use of rubber in any product that was not essential to the war drive was banned. Further, the speed limit on US highways was reduced from 45 to 35 miles an hour, in order to reduce wear and tear on tires countrywide. Rubber chips were sold for a penny or more per pound weight at over 400 000 depots all over the country. Even President Franklin Roosevelt's pet dog "Fala" saw his rubber toys melted. This was the largest recycling campaign ever recorded in history, ensuring the success of the Allies through to 1942.

The US industrial sector had never been called upon to shoulder such a massive task, achieving so much so quickly. The engineers were given just two years to reach this target. If the synthetic rubber program failed, the capacity of the USA to fight the war would be blunted. This US drive was to help spread synthetic rubber throughout the world's markets, even in Brazil, as it strove to consolidate its industrial position during the post-War years.

Sources:

- IIRSP – A brief history of Rubber
  http://www.iisrp.com/WebPolymers/00Rubber_Intro.pdf
- American Chemical Society: United States Synthetic Rubber Program, 1939-1945
  http://acswebcontent.acs.org/landmarks/landmarks/rbb/index.html

3.8.1.11 1950s – Plastics Demand Explodes

On the back of the massive economic growth in the ADV economies and population growth globally, the whole chemical industry, but specifically the petrochemical and plastics sector, have grown very rapidly over the last 60 years, defining what some have called the "golden period of plastics and petrochemicals". In 1970 the per capita expenditure on chemicals was $46. In 2010 that amount increased tenfold, with a staggering $447 of chemical consumption per capita. On a percentile basis the role of the chemical industry in the global GDP was slightly reduced from the 6 and 7% achieved during the 1970s, 1980s, and 1990s, but the growth on a per capita basis has been spectacular.

By plastics here we are referring mainly to those related to the petrochemical industry. Please see below a short list with some of the most important plastics in chronological order, including the best available information about their inventors and major applications.

List of Major Plastics and Products Applications

- **1839 – Polystyrene (PS)**, originally invented by a German Eduard Simon who did not really know what he had discovered. In 1922, Hermann Staudinger, defined PS as a polymer, while defining its major characteristics and some of its potential uses. In 1953, Hermann Staudinger won the Nobel Prize for Chemistry for his research.
- **1926 – Polyvinyl chloride (PVC)**, invented by Waldo Lonsbury Semon while working for the B.F. Goodrich Company in the USA.
- **1930 – Polystyrene (PS)**, invented in 1839, but it was not until 1930 when BASF developed a way to manufacture PS. It can be used in many applications including packaging foams, yoghurt containers, food containers, tableware, home insulation panels, disposable cups, plates, and cutlery, CDs, cassettes and plastic boxes.
- **1933 – Polyethylene (PE)**, discovered by Reginald Gibson and Eric Fawcett at the Imperial Chemical Industries (ICI) in England. First commercial production started in 1939 in the UK and since then has become the "king" of the plastics world. Polyethylene can be found in millions of day to day and common plastics applications, from supermarket bags, to plastic bottles or elastics bands.
- **1933 – Low Density Polyethylene (LDPE)**, the first and original form of polyethylene, discovered by Imperial Chemical Industries (ICI).
- **1933 – Polyvinylidene Chloride (PVDC – Saran)**, invented by Ralph Wiley while working at Dow Chemical in the USA. It has been mainly used for food packaging. **PVC** applications range from pipes, frames, flooring, and shower curtains.
- **1935 – Polyamides (PA – Nylon)**, invented by Wallace Carothers at DuPont in the USA. In 1935, DuPont patented the new fiber known as Nylon, being finally introduced to the world in 1938. There are many applications of nylon ranging from fibers, to textiles, toothbrush bristles, ropes, and cleaning brushes.
- **1937 – Polyurethane (PU)**, invented and patented by Otto Bayer and coworkers, while working for Bayer in Germany. PU is also one of the larger plastics in volume. PU can be found in most cushion foams in mattresses, also in thermal insulation, furniture padding, coatings, and all short of foams within cars.
- **1941 – Polyethylene terephthalate (PET), i**nvented by Rex Whinfield and James Dickson, working at a small English company called "Calico Printer's Association" in Manchester. PET has been used traditionally in bottles, now specifically for carbonated drinks, as well as a plastic film.

- **1945 – Polyester (PES)**, invented by British scientists John Winfield and James Dickson in 1941 in England. After WWII, DuPont bought the right to make polyester and by 1950 started commercial production. Fibers and textiles are its major applications.
- **1948 – ABS – Acrylonitrile Butadiene Styrene Copolymer (ABS)**, was patented in 1948 and commercialized by the Borg–Warner Corporation in 1954. ABS can be found in almost all electronic equipment cases, from televisions to computers, printers, mouses, keyboards on any other similar device. ABS is normally defined as the "Prince" of electronics.
- **1951 – High Density Polyethylene (HDPE)**, was invented by Paul Hogan and Robert Banks while working at Phillips Petroleum. HDPE also has multiple applications among the most common are detergent bottles or certain resilient containers and pipes.
- **1951 – Polypropylene (PP)**, invented by Paul Hogan and Robert Banks while working at Phillips Petroleum. PP can be found in several applications, including caps, yogurt containers, appliances, car fenders, and car dashboards.
- **1954 – Foamed Polystyrene (PS – Styrofoam)** – invented by scientist, Ray McIntire at The Dow Chemical Company in the USA and first introduced in the United States in 1954.
- **1958 – Polycarbonate (PC)** – in a sort of ironic and unbelievable coincidence, its seems that two scientists from two different companies and continents developed the same material one week apart. They were Dr. Hermann Schnell at Bayer in Germany and Dr. Daniel Fox at GE Polycarbonate, in 1955. Commercial production did not start until 1957/58. PC not only has the unusual nature of two inventors, but also also has the honor of being the first plastic to go to the moon, on the helmets used for the Apollo expeditions. Additionally, it can be found in multiple applications including compact discs, eyeglasses, security windows, helmets, traffic lights, lenses, car interiors, and parts of mobile phones.
- **1978 – Linear Low Density Polyethylene – (LLDPE)** with first production in 1980 is mainly used for outdoor plastic furniture and food packaging.
- **1990 – Polyhydroxybutyrate (PHB)** is a polymer that belongs to the class of polyesters first characterized in 1925 by Indian microbiologist Abhilash Singh. Imperial Chemical Industries commercialized this first bio-plastic under the name BIOPOL in 1990.
- **2000 – Methalocene PE**, first commercial metallocene catalyzed polyolefin introduced. The fact that plastics are all around us and are tremendously correlated with our higher living standards, explains the tremendous growth in the plastics and petrochemical industry during the last 60 years.

That massive growth in plastics demand and its huge growth potential was immortalized in the famous 1967 film "The Graduate" with Dustin Hoffman. Dustin Hoffman as a recent graduate searching for a job got the following advice from a senior father's colleague on his search for a job: "just one word: PLASTICS! There is a great future in plastics!"

Indeed, that statement became the social realization of decades of massive growth for that industry. However, that might not be the end. The expected massive economic growth in the BRIC and REST economies will open the door for a much larger plastics demand, demand that might lead to another period of strong growth globally.

3.8.1.12 1960s Internationalization

In the light of that growth, chemical companies started to go beyond their national borders. It is not by chance that the largest chemicals companies in the world, like Dow Chemical or BASF started to expand globally. Dow Chemical for instance created Dow Chemical International in 1951, opening its European headquarters in Horgen, Zurich, Switzerland in 1952.

In 1958 BASF started to build facilities outside Germany, opening new markets and establishing joint ventures all around the world. Since then, both companies have achieved more than 40% of their sales outside their country of origin. That trend has never been reversed since then and the industry remains as one of the most global ones.

3.8.1.13 2010–2050 – The Chemical Industry Leads the Revolution against Climate Change

The arrival and confirmation of climate change as one of the major megatrends for the next decades, and therefore the clear need to reduce CO_2 emissions across all industries, will drive the chemical industry to the core of the solution. During the thrilling history of the chemical industry, the industry has been transitioning through different episodes, shifting its focus from innovation during its early days into mass production during the last 40 years, as global demand for chemicals exploded globally.

Climate change might bring the opportunity to the chemical industry to transition again into a new phase, moving back to its innovation days while keeping focused on the massive expected economic growth for chemical products. During the next decades the industry will be called to play a critical role in the fight against climate change, resources and energy scarcity; which will open the door to a new period of innovation and growth.

3.9 Conclusion

The history of the industry is certainly the history of human progress and innovation. The chemical industry has been "used" during its history to enable many other industries, from the automotive industry to housing, not to mention the food industry, electronics, or the appliances; but more importantly it has enabled the current levels of quality of life our societies enjoy today. Thanks to the chemical industry our society has reached the highest level of wealth and living standards in human history, and it is in these foundations where his future will be written.

The chemical industry has been "used" by governments and nations as the ultimate source of competitive advantage, solving the most difficult problems when solutions were most needed. Companies like, DuPont, BASF, The Dow Chemical Company, and Pharmaceuticals like Bayer, Pfizer, Merck, Johnson & Johnson, or Novartis, just to mention some, have entered into the history of human progress solving some of the most challenging problems governments, citizens, and society has posed. Without their inventions and products the current living standards would simply not be possible.

However, the technology in the industry is several decades old, for almost 50 years there have been no major breakthroughs or new products. The spur of innovation our industry had during the first half of the century, turned into a second half focused on the optimization and maximization of these technologies.

During the history of the chemical industry we have seen on many occasions how governments and society have been tapping into the wonders of the industry to solve some of their major problems. In the light of the upcoming trends and overwhelmingly complex issues that climate change or energy and resources scarcity might bring, the chemical industry may soon be called into service again.

How long will it take for governments, industries, and citizens to start asking the chemical industry to get into action and start solving some of the major problems that threaten our world? They did it in the past, why not this time? The industry will soon be called into action, addressing some of the negative aspects of the upcoming megatrends, enabling our industries and societies to reach the next level of technology and innovation, and allowing the wealthiest and healthiest society in human history.

3.10
Summary – Industry Major Features and Upcoming Megatrends

After reviewing the current status of the industry and the most recent history of the chemical industry we will devote the following pages to presenting the major identified features of the industry, as well as a short summary of the major megatrends that will shape our world.

These features and the selected megatrends and explanations will serve as the basis for the analysis of the future megatrends in the chemical industry (Chapter 4) and for the final projections of the chemical industry into 2050 (Chapter 5).

In terms of the major features of the industry, six major features have been identified, not necessarily in this order: the relevance of the industry, its feedstock and energy, its products, its emissions, and its structure and the awareness of society about the industry.

For each of these features we have observed additional sub-features or characteristics and when possible we will go deeper into them (Table 3.4).

For instance, for the relevance of the industry, we have observed three major ways to validate that feature. First, for its economic relevance in the society and world economy, with almost 7% of the world GDP related to the broad chemical

3 The Chemical Industry in 2010

Table 3.4 Major features and sub-features of the chemical industry.

MAJOR FEATURES OF THE INDUSTRY	
F1	ECONOMIC
	TECHNOLOGICAL
	PROFITABILITY
	RELEVANCE OF THE SECTOR
F2	NAPHTHA & GAS OIL
	GAS BASED - Ethane, Propane, Butene
	RENEWABLE
	FEEDSTOCKS
F3	**PETROCHEMICALS**
	Ethylene
	Propylene
	Butadiene
	BASIC INORGANICS
	Industrial Gases + Others Inorganics
	Fertilizers
	POLYMERS
	Plastics
	S. Rubber
	Fibers
	SPECIALTY CHEMICALS
	Dyes & pigments
	Crop Protection
	Paints & inks
	Auxiliaries of industry
	CONSUMER CHEMICALS
	PHARMACEUTICAL
	PRODUCTS
F4	**GREEHOUSE EMISIONS = C02e**
F5	FRAGMENTED
	TOP PRODUCERS - ASIA + ADVANCED
	TOP SALES - ASIA + ADVANCED
	INDUSTRY STRUCTURE
F6	REGULATION
	SOCIAL AWARENESS
	SOCIAL AWARENESS

MAJOR FEATURES	
F1	RELEVANCE
F2	FEEDSTOCKS & ENERGY
F3	PRODUCTS
F4	GREEHOUSE EMISIONS
F5	INDUSTRY STRUCTURE
F6	SOCIAL AWARENESS

industry. Secondly, for the technological relevance of the industry we have observed a very high percentage of R&D spending on sales versus other industries, a fact that constitutes a major feature of this industry, and more specifically of the pharmaceutical industry. Finally, the clear and marked difference in the profitability of its different segments constitutes another clear feature of this industry.

Along these lines, the major inputs of the industry, like its different feedstocks (naphtha, ethane, butane, propane, etc.) and energies (oil, gas, electricity, heat, etc.) as well as the different outputs of the industry, all its different products from pharmaceuticals to chemicals, and from petrochemicals, to polymers, or specialty chemicals, consumer products or inorganic; or even the amount of greenhouse emissions have been identified as distinguishing features of the industry.

Finally, aspects like the unique structure of the industry, its size, its major producers and consumers, in combination with the awareness and relationship of the industry to society, governments, and other industries appeared to us as some of the major characteristics of the industry.

For that purpose we have summarized in Table 3.5 the status and major aspects of the different features of the industry. Assuming that these constitute the key pillars of the industry today, we will use them to build the projections of the industry by 2050.

3.11
Major Features of the Chemical Industry

With large differences already among the pharmaceutical and chemical industry, we have divided the features of the industry into three major areas. First, we consider a generic feature for the broad industry, that is to say chemicals and pharmaceuticals; secondly we gather the major characteristics for the chemicals industry, and thirdly we do the same for the pharmaceutical industry.

In some specific areas both industries can share some common features, like social awareness of regulation, but in many others, the industries can be very different, like levels of profitability, R&D expenditure or energy, and labor intensity.

For simplification on many occasions we will refer to the broad chemical industry, that includes pharmaceuticals, but in reality we should not neglect the structural differences between both industries.

Finally, on building the future projections of the industry and in Chapters 4 and 5, we will start analyzing the potential impact of the upcoming megatrends on each of the industry features identified above. For that purpose, and as a brief reminder we have summarized in Section 3.11.1 the major expected megatrends for our world in the coming decades. In the next chapter we will make use of all these trends in shaping the chemical industry by 2050, so please take a careful look at the following table and the major assumptions we will use for these upcoming trends and the industry.

Table 3.5 Major features of the chemical industry and current status.

MAJOR FEATURES OF THE CHEMICAL INDUSTRY

	MAJOR FEATURES		INDUSTRY	CHEMICALS	PHARMACEUTICALS
F1	ECONOMIC		Very relevant	between 4% to 6% of the World GDP	between 1% to 2% of the World GDP
	TECHNOLOGICAL		Very relevant	% R&D expenses on sales fluctuates between 1% to 10%	% R&D expenses on sales fluctuates between 10% to 25%
	PROFITABILITY		High, but depending on segments	Average % EBITDA on sales around 9%, but fluctuates by segments between 3% to 25%	Average % EBITDA on sales around 17% but fluctuates by segments between 10% to 30%
	RELEVANCE			Very relevant industry for society, human progress and the world economy	
F2	FEEDSTOCKS & ENERGY			Industry sucessful and committed to reduced energy & feedstocks demand. Oil & Gas remain as major feedstocks, with "Shale Gas" opening a new "gas" era	
F3	PRODUCTS			Thousands of products across very different segments, applications, markets and uses.	
F4	GREEHOUSE EMISSIONS			Industry has achieved great success reducing emissions during the last decades, while being equally successful in enabling other industries to reduce emissions too. On the light of climate change, the chemical industry will lead on efforts in this area	
F5	FRAGMENTED		Fragmented	Highly fragmented. top 10 companies only 20% of market	Regular level of concentration. top 10 companies around 50% of market
	TOP PRODUCERS		Moving towards Asia and BRIC	Moving fast from the Advanced to the BRIC and REST economies. China is already largest chemical producer	92% located in Advanced Economies
	TOP CONSUMERS		Moving towards Asia and BRIC	47% on the Advanced Economies but moving fast to the BRICs?	5% on the Advanced Economies
	INDUSTRY STRUCTURE			Although with clear differences between Chemicals and Pharmaceuticals, the industry remains fragmented, with thousands of products and companies around the world. The Industry remains in transition from the Advanced to the BRIC and Emerging Geographies	
F6	REGULATION			Industries highly regulated, with regulation transitioning from local into national and global	
	SOCIAL AWARENESS			Society remains highly aware of the industry, specially on the negative aspects, but not so much on the positive aspects	
	SOCIAL AWARENESS			High and increasing, specially in on the light of the major upcoming mega trends.	

Source: ICIS Top 100 Chemical Companies – September 2011 – ICIS Chemical Business Magazine P. 28. IMAP – Pharmaceutical & Biotech Industry Global Report 2010.

3.11.1
Summary: Global Major Megatrends

MAJOR GLOBAL MEGATRENDS		KEY ASSUMPTIONS BY 2050
SOCIAL	POPULATION	The world population will grow from 6.8 to 9.1 B by 2050. Most of the growth will come in the "REST" countries where population will increase by 61% (1.8 billion) during the next decades until 2050. BRICs will increase by 13% (380 million) and ADVANCE countries by 8% (79 million).
	DEMOGRAPHICS	The world demographics will change globally with very different patterns across the different areas. Older and slow growing populations in the advanced economies will contrast with younger and faster growing populations in the REST and BRIC economies. The world life expectancy will increase from 65 to 75 years
	URBANIZATION	By 2050 the population living in cities will increase from 47% to 70% globally. India and China are expected to double its their urban populations from 27% to 55% and 36% to 72% respectively. The REST group will also increase from 43% to 68%.
ECONOMICS	ECONOMICS	By 2050 the World Economy is expected to triple from its 2010 levels, having the potential to reach a staggering $US 280 trillion GDP (2009 Dollars based), hosting of the wealthiest societies in human history. The Average world per capita GDP will have the potential to tippled from the $US 9,000 in 2010 up to $US 30,000 by 2050, pulling out of poverty millions of people around the world, while creating a super wealthy middle class with more than 50% of the world population enjoying levels of wealth never seen before.
POLITICAL	GOVERNANCE	In 2007, of the Top 150 world economies, 100 were companies. Since companies can grow faster than countries, in 2050 they could become ever more relevant & bigger. Governments will increase social awareness, regulation and control on all companies and any industry to limit their size and relevance. Companies will need to increase the awareness too.
	SOCIAL MEDIA	In 2004 FACEBOOK had 2,000 users. In 2011 they had more than 800 million users. More than 2,5 billion people where connected via social media. Social media is changing the way people communicate among themselves as well as with the different social stakeholders including Governments and Companies.
	GOVERNMENT	As the world and its issues become more global and companies bigger, National Governments and International Organizations will become more and more relevant in many ways, economic, social, regulation etc. International Organizations will need to reflect the new balance of power, increasing the representation of the BRIC and REST economies. In other cases new organizations will need to be created, organizations that better reflect the new international order and capture also the voice of large companies, large social stakeholders and society.
ENERGY	RRENEWABLE? SHALE GAS?	The clear need to address both climate change and a spectacular increase in energy demand, will accelerate the search for and use of renewable energies in the economy and in the chemical industry. Shale Gas will accelerate the use of gas in the overall economy but also in the chemical industry. Nuclear energy will be necessary but might be confronted with clear opposition after Fukushima.
CLIMATE CHANGE	GREEN HOUSE EMISSIONS	The clear need to address climate change will impose severe reduction of greenhouse emissions around the world. According to the IEA, the world will need to reduce by more than 50% its current CO_2 emissions from 33 Gigatons to 14 Gigatons of C02 Emission by 2050. In other words, the world will need to live with less than 4,000 grams of CO_2 per capita and day, and that is more than 3 times less than our world used in 2010 and more than 7 times less than our world is expected to used by 2050.
WILD CARDS	RESOURCE SCARCITY	The huge increases in world population with more than 2.3 billion more people globally by 2050 could trigger a vast increase in demand and prices for all sorts of basic resources, from food, water, raw materials, energy etc. Demand will increase significantly across all industries and the technical challenges, especially when adding climate change and energy scarcity will be paramount. The REST economies with the bulk of that increase and 1.8 billion additional people to feed during the next 40 years will be under tremendous pressure. The BRIC with an additional 390 million people will also have the potential to suffer this increase in the demand tension, while the advanced economies with slow growing populations and older population will be more "relaxed". This potential trend has been placed as a "wild card" as we don't want to think that scarcity will actually occurs but there is a clear risk that it could happen.
	PANDEMICS	In a truly global and hyper connected world, illness might travel around the world much faster. That, in combination with climate change, that might provoke severe changes in weather patterns, and perhaps diseases, could create the room for global pandemics to develop. Pandemics that could actually alter the way we live and work, temporarily stopping human progress.
	WAR - MASSIVE TERRORISM	In a more global and connected world, with global and strong issues to be tackled, not only would terrorism be more difficult to be prevent, but the tension among countries and within countries could increase despite higher levels of income. As a wild card we expect that a new massive war, the 3rd world war, or massive and destructive terrorism attacks will remain very unlikely; but for and illustration lets analyze the potential impact of these on the industry, so we can better understand the overall industry features, its major strengths and weaknesses, even in the most weird scenarios.

Bibliography

1. BASF Historical Milestones *http://www.basf.com/group/corporate/en/about-basf/history/index*
2. Dow Chemical History *http://www.dow.com/sustainability/commitments/history.htm* (December 2012).
3. DuPont and the Government *http://www2.dupont.com/Government/en_US/gsa_contracts/Government_Projects.html*
4. IIRSP - A Brief History of Rubber *http://www.iisrp.com/WebPolymers/00Rubber_Intro.pdf*
5. American Chemical Society: United States Synthetic Rubber Program, 1939–1945 *http://acswebcontent.acs.org/landmarks/landmarks/rbb/index.html*
6. The History of Plastics - Timeline of Plastics. By Mary Bellis, About.com Guide *http://inventors.about.com/od/pstartinventions/a/plastics.htm* (eds) 1990.
7. Arora, A., Landau, R., and Rosenberg, N. *Chemicals and Long Term Economic Growth – Insights from the Chemical Industry*, Wiley-Interscience.
8. CEFIC Facts & Figures 2011 – *http://www.cefic.org/Facts-and-Figures/*
9. Styrene Monomer History – CEFIC – *http://www.styrenemonomer.org/*
10. ICIS TOP 100 Companies in 2011 *http://www.icis.com/contact/request-free-top-100-chemical-companies-2011/*
11. IMAP Healthcare Report – Pharmaceutical and Biotechnology Industry Report *http://www.imap.com/imap/media/resources/IMAP_PharmaReport_8_272B8752E0FB3.pdf*

4
Impact Assessment of the Global Megatrends on the Chemical Industry

> *The best thing about the future is that it comes one day at a time.*
> Abraham Lincoln

4.1
Introduction

In this chapter we will review how the upcoming megatrends will influence the chemical industry by 2050. During last chapter we summarized the major megatrends, features, and segments of the industry; now we will start to analyze the possible impact of each of these megatrends on each of the industry's features. The conclusions from this assessment will be used later on to Chapter 5 to prepare the final projections for the chemical industry by 2050.

In order to do this we will collect all results in a simple matrix with the features down the left side of the matrix and the selected megatrends across the top (see Figure 4.1) Additionally this analysis will be conducted for each of the four geographical areas selected for this book, the World, the ADV economies, the BRIC economies and the REST. At the same time, the matrix will calculate for each of the megatrends and features of the chemical industry the total average, so we can see how a megatrend will impact each of the features in an absolute way, and for each of the different areas.

Answer	Value	Color
Very strong impact	3 points	🟥
Strong impact	2 points	🟧
Small impact	1 point	🟩

To answer these questions we will not look at the direction of the impact, meaning positive or negative, but at its strength. For that purpose we have ranked each of the potential answers using the following methodology and code above. So, if we expect a "very strong impact" of a particular megatrend on one feature we will assign 3 points and a red box. If we expect a "strong impact," we will assign 2

The Future of the Chemical Industry by 2050, First Edition. Rafael Cayuela Valencia.
© 2013 Wiley-VCH Verlag GmbH & Co. KGaA. Published 2013 by Wiley-VCH Verlag GmbH & Co. KGaA.

4 Impact Assessment of the Global Megatrends on the Chemical Industry

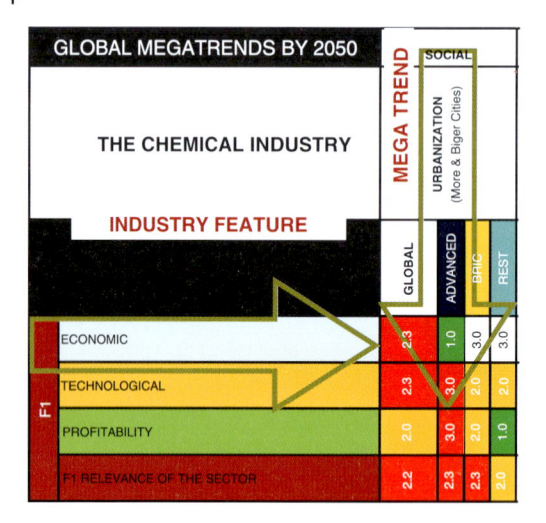

Figure 4.1 Diagram of the matrix used to collect the results of the impact of megatrends on the chemical industry.

points and an orange box, and if we expect a "small impact", we will assign 1 point and a green box.

As you might have already noticed there is an undeniably subjective component in the results of this analysis as it is actually the author who has been answering the questions. We will try to remain as factual, rigorous, transparent and generic as we can, providing the reader with the model so he or she can actually answer the questions in their own way, and, therefore, have the option to explore their own projections for the chemical industry.

Let us illustrate now how the matrix will work using one real example. For instance, if we consider the potential impact of urbanization on the first identified feature of the industry, the "relevance" of the industry (economic, technological, and profitability), we could estimate the impact of this megatrend into the industry. So as we start now to analyze the impact of these megatrends on this feature, we will ask ourselves the following question for each of the three areas, while the model will calculate the average for the world economy.

Question 1: Would the expected urbanization megatrend impact the economic relevance of the industry in this Area? being these the potential answers: very strong impact (3 points), strong impact (2 points) and small impact (1 point).

Assumption: *By 2050 the ADV economies will increase their urbanization ratio by 10% from 75% in 2000 to 85% in 2050. The BRIC economies will increase by 100% especially China and India, the REST group by 60% from 47% in 2000 to 68% in 2050. Based on these assumptions we will provide the following answers and scores.*

- **ADV economies: small impact – 1 point (green).** The major impact will be in the BRIC and REST economies. The ADV economies will see their ratio increase by

10% but not by the huge amounts that the BRIC and REST economies will see. Therefore we will expect a certain impact as there is an increase but that would be a small impact, assigning 1 point to that question (green).
- **BRIC: very strong impact – 3 points (red)**. As the BRIC are expected to double their urban population we will deem that as a very strong impact, therefore we will answer with 3 points (red) and a very strong impact.
- **REST: very strong impact – 3 points (red)**. Although the increase here will be less than that for the BRIC economies, a growth of 60% appears to us like a very strong growth. Therefore to this question we will answer with 3 points (red) and a very strong impact.
- **World**: the model will automatically calculate the average of the three areas, assigning the respective color. Based on this model the global impact of urbanization into the economic relevance of the industry would be very strong, with a calculated ratio of 2, 3 points over a potential maximum score of 3 points.

Question 2: Would the expected urbanization megatrend impact the profitability relevance of the industry in this area?

Assumption: *By 2050, the ADV economies are expected to have a GDP per capita of US$ 78 000; the BRIC economies US$ 36 000, and the REST group US$ 17 000. Based on that we will expect many more technological products and higher margins in the ADV economies compared to the BRICs and the REST. For that reason we will answer this question as follows:*

- **ADV Economies: very strong impact – 3 points (red)**. Advanced Economies, with an average GDP per capita of US$ 78000 (2009 dollar) in 2050 will be able to look for high quality products, valuing performance, and providing room for higher demand and margins for the industry.
- **BRIC: strong impact – 2 points (orange)**. The BRICs with an expected average GDP per capita equal that in Europe in 2012 will also be able to focus on products with high quality and margins, but to a lesser extent than in the ADV group.
- **REST: small impact – 1 point (green)**. The REST economies with a much smaller GDP per capita will also see the profitability of the chemical industry improve, but that improvement might come more from the volume side, with an expected urbanization growth of 68%, than from the margin side. Therefore the profitability of the industry will certainly increase, but less than in the ADV and BRIC groups.
- **World**: the model will automatically calculate the average of the three areas, assigning the respective color and value. In this case 2.3 points is the world average impact of this megatrend into the profitability of the industry, denoting a very strong impact of this megatrend into this feature of the industry.

Through this simple and detailed process we will start answering one by one all the potential combinations, while addressing more than 1000 questions and potential combinations. Then, for each of the potential combinations, we will start ranking the final results obtained through the process, with the intention to assess which of the upcoming megatrends will have the highest impact on the chemical industry by 2050, and which of the major features of the chemical industry will be most affected by the upcoming megatrends. The ultimate purpose is to start analyzing how these megatrends will define the future of the chemical industry and how the chemical industry could look by 2050.

Therefore in the next pages we will proceed to go through in a certain detail all these combinations and analyze how the chemical industry will look by 2050 for each of three major and predefined economic areas. Due to the large number of combinations analyzed in this matrix, this chapter will only focus on the major conclusions and results useful to us for extracting the major projections and trends for the future of the industry. For logical reasons we will not be able to review all combinations[1].

The complete matrix with the results from the analysis is shown in Figure 4.2. We will review the results of the analysis and its major conclusions, but before doing that, let us explain how to read this matrix and the different extracts, and views from it that we will use for our analysis and conclusions. For that purpose we have segmented the matrix and its results into three major areas, and we will measure and present the results in various ways: numerically, by ranks and graphically.

- **SET I** – this overview will help us to understand and rank which of the different megatrends will have a major impact on the chemical industry globally and for each of the areas (Figures 4.2 and 4.3). In this chapter we will review the results at the global and area level, providing some ranks and charts. For a further level of analysis the matrix will also be available in the Appendix.
- **SET II** – this set of answers will allows us to focus on and assess the potential impact of the different megatrends on each of the major features of the chemical industry. This set will rank these impacts and present the results numerically and graphically at the global and area level. This analysis will help us to understand which features of the industry are expected to be most affected by the upcoming megatrends, and in which geographical areas, helping us later to present the outlook for the industry by 2050 (Figures 4.2, 4.3 and 4.4).
- **SET III** – finally this view will allow us to go one level deeper, helping us to review for each feature of the chemical industry and area, which megatrend will have the highest potential impact (Figure 4.2).

On the analysis and results of these three set we will base our projection for the future of the chemical industry. Obviously the upcoming results as well as the following projections for the chemical industry have to be placed in the right

---

1) For those readers interested in details and individual combinations, a copy of the matrix in its real size will be provided online in an Excel format, so every reader can go through this process and validate their own view of the future of the chemical industry, and the potential role of the selected megatrends in it.

4.1 Introduction | 205

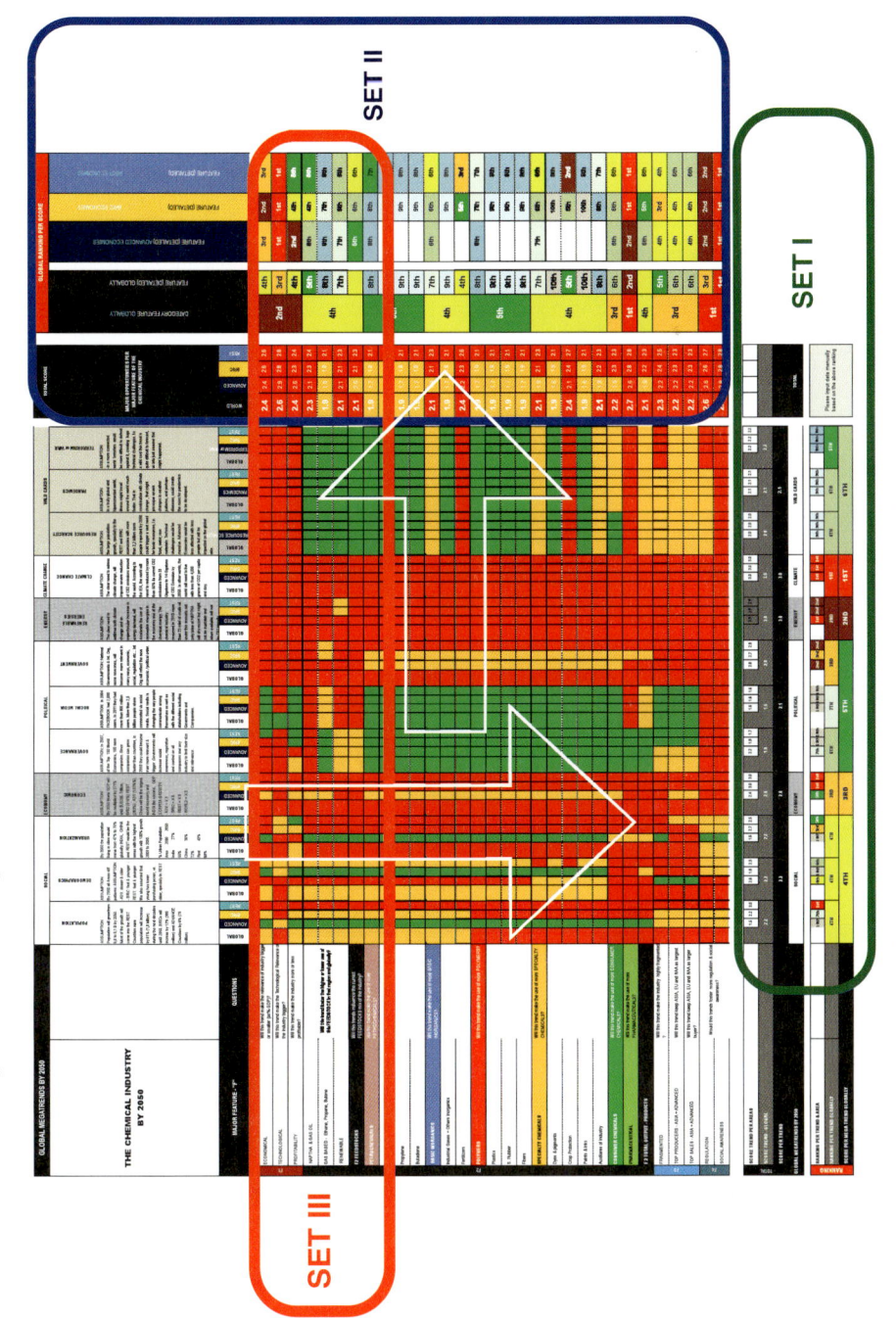

Figure 4.2 Matrix showing the impact of the megatrends on the chemical industry by 2050.

206 | *4 Impact Assessment of the Global Megatrends on the Chemical Industry*

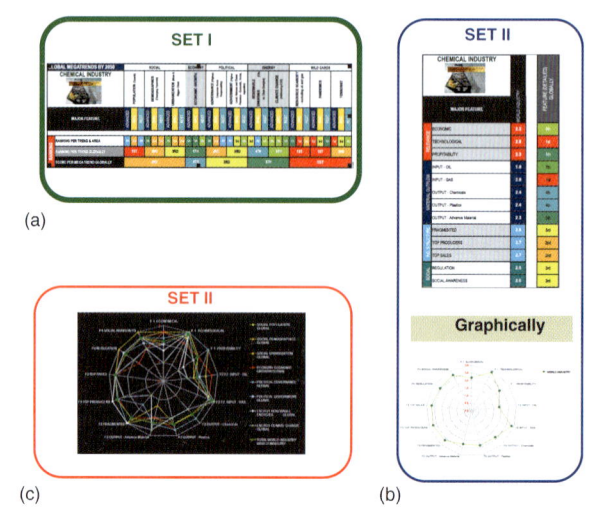

Source: Own Elaboration

(a)

(b)

(c)

THE FUTURE OF THE CHEMICAL INDUSTRY

Figure 4.3 (a) Most influential megatrends for the chemical industry globally. (b) Most impacted features of the chemical industry. (c) Most influential megatrends for each of the geographies.

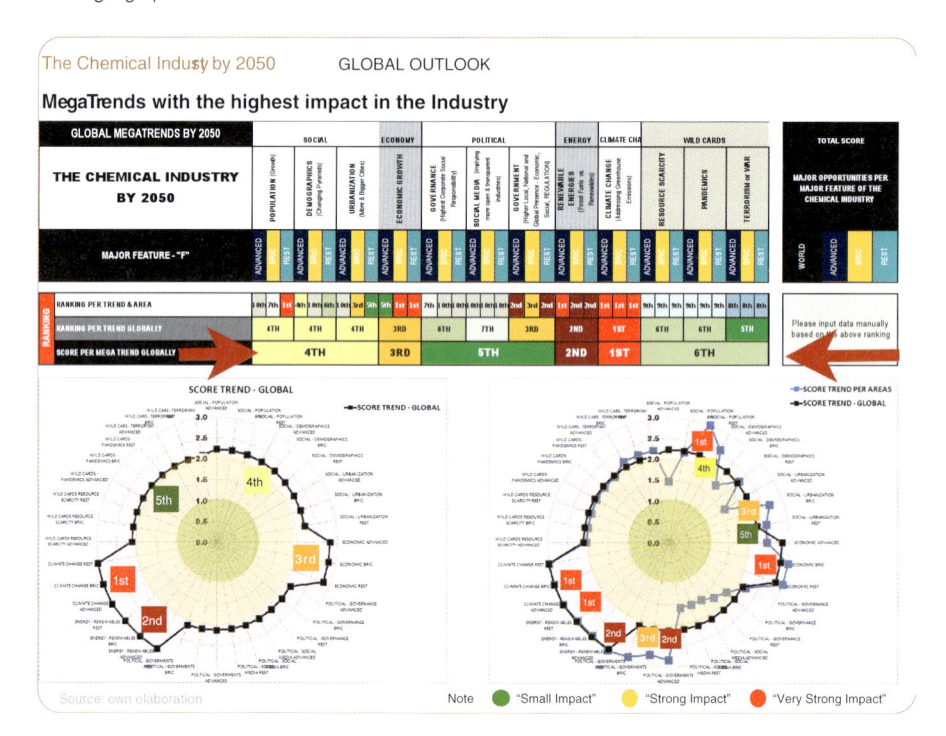

Figure 4.4 Megatrends with the highest impact on the chemical industry.

context. They are largely dependent on the choices we made in terms of megatrends and "features" for the chemical industry. Although we have been trying to be the most factual, realistic, rigorous, and conservative in our analysis, we understand that different readers might have different opinions, that is why we will provide an electronic version of the model, so every reader can build up his own model and obtain his own set of results and vision of the future of the chemical industry.

As stated in the preface, this book does not intend to come with a final solution, but more with a framework and a set of possibilities. We envision this book as a starting point for intellectual debate not as the end of it.

4.2
Megatrends with the Highest Impact into the Chemical Industry (Global & Area Level)

When looking at the global megatrends with the highest potential impact on the future of the chemical industry (Figure 4.4), climate change appears in our model as the one with the highest potential. In a much larger and wealthier world, the enormous challenge the "sustainable scenario" entails; reducing greenhouse emissions and CO_2 emissions from 12 000 g per capita per day in 2010 to 4000 g per capita per day, while still growing, will present the highest impact on the chemical industry.

The chemical industry will be one of the most affected by climate change, being doubly affected from its emissions side as well as its products. The unique capacity of the chemical industry to reduce greenhouse emissions in other industries will put tremendous pressure on the industry to maximize its production, while inventing new products and solutions that can support other industries and society in reducing their greenhouse emissions. The chemical industry with 4–5% of the world CO_2 emissions in 2010, will not only struggle cutting its own emissions while looking for new products and solutions to enable reductions in other industries and society, but also will need to grow massively to enable the expected growth in population and in the economy.

In second place, and clearly linked to climate change, appear the energy megatrends. The massive economic growth our world is expected to have will put tremendous pressure on some of the existing energies, especially crude oil. However, it is not the potential scarcity of energy, after all natural gas will be plentiful for the chemical industry, but the need to address climate change what will foster a gradual change in the world energy mix, moving from fossil fuels to energy sources with reduced greenhouse emissions. For the chemical industry, with almost 100% of its feedstocks based on fossil fuels and more than 12% of the world crude oil consumption on a daily basis, the world need to address climate change will have a tremendous impact on the feedstock side of the industry.

Third in the ranking is the economic megatrend. The expected enormous economic growth for the next decades, moving the world economy from US$ 75 trillion (2009 dollars) in 2010 to US$ 280 trillion (2009 dollars) by 2050 will create a massive growth opportunity for the chemical industry as we will see in more detail in Chapter 5. The chemical industry could have the potential to quadruple its size to more than US$ 18.7 trillion.

In fourth position appeared the social megatrends. The expected increase in population from 6.8 billion in 2010 to 9.1 billion in 2050, in life expectancy from 67 years in 2010 to 75 in 2010, and the world urbanization rates; in combination with a strong change in demographics will create significant challenges and large opportunities for the chemical industry. As we will see in the next pages when analyzing the potential impact of each megatrend on each of the different areas, features, and segments of the chemical industry; business segments like pharmaceuticals, consumer products, crop protection, or auxiliaries for the industry are expected to have significant growth on the back of some of the social megatrends.

Finally, the selected wild cards appear to be in last position. That should not be a surprise to us, indeed perhaps the surprise is why we considered these in our long term analysis. The reality is wild cards in general are events that can have a huge impact in the short term, but not much in the long term, especially over a period of 40 years. However, wild cards, as weird as they can be, are excellent indicators of the elasticity that some of the major features of the industry can have toward extreme events. In that sense, the selected wild cards will help us to think of some of the most extreme scenarios, scenarios that might never occur but when if they do could have huge impacts in the short term. Additionally, and when considering the vast range of wild cards that we could have chosen, these seems to us quite realistic, and to some extent, and unfortunately, even quite likely. Therefore they have appeared fifth in the list of megatrends with the lowest impact, however, we should not underestimate them and the industry should always be prepared for any eventuality.

4.3
Megatrends with the Highest Impact in the Industry (Area Level) – (Figure 4.4)

Now if we go one level down from the global to the area level, we can review which of the selected areas of the chemical industry is expected to be most affected.

Several combinations of megatrends in different areas will have the highest impact on the chemical industry. Climate change globally will have the highest potential impact on the chemical industry, followed by population growth in the REST economies, not in vain will they experience the highest population growth of all areas, 68% in the next 40 years; and economic growth in the BRIC and REST groups, as they are also expected to have the largest economic growth in the next decades with 518 and 283%, respectively. Finally, we can find also in first position the move to renewable energy in the ADV economies due to climate change, the fact that the ADV economies are expected to have the population with the highest per capita ratio of CO_2 emissions and the highest income on a per capita basis, will somehow force governments and citizens to lead on this front.

In second position we can find the move to renewable energies in the BRIC and REST economies, mainly due to their need to address climate change. Also in second position we find the impact of governments on the chemical industry in the ADV economies. The enormous size that chemical companies could reach by 2050, where some of the leaders of the industry like BASF, Pfizer, or Dow Chemical could have sales much greater than US$ 200 billion, will bring paramount attention from governments and society. In normal conditions these huge companies are already subject to significant attention and public and government scrutiny, however, the fact that the broad chemical industry is going to be at the core of the solution of many of the major challenges of our society during the next century will increase significantly the level of exposure and attention from governments around the world. The chemical industry will be called into action, and governments and society will look carefully at the wonders of the most sophisticated chemical industry in the world, and that is currently located in the ADV economies.

In third position we can find two major combinations: urbanization growth and the role of government in the BRIC economies. The urbanization ratio in the BRIC economies will double during the next 40 years, bringing hundreds of millions of people into cities, especially in China and India. Second the role of governments in the BRIC economies is expected to increase significantly, as economic and population growth require higher levels of control and coordination. The massive growth the BRIC economies are expected to have will also serve to sustain and even increase the high growth already experienced by the chemical industry in the BRIC economies.

In fourth position, and with a tremendous impact on the ADV economies and their chemical industry will be the change in demographics. The expected advances in the quality of life and life expectancy for citizens in the ADV economies, will generate significant opportunities for the chemical industry, and more specifically for the pharmaceutical industry.

Fifth in the ranking we find urbanization in the REST economies and economic growth in the ADV economies. The urbanization ratio in the REST economies is expected to grow from 45% in 2010 to 68% in 2050, bringing almost 2 billion people into cities. The vast economic growth the world is expected to have will also have a clear impact on the ADV economies. With the USA leading that growth, the ADV economies will almost triple their GDP during the next 40 years, from US$ 35 trillion (2009 dollars) to US$ 82 trillion (2009 dollars) by 2050. This impressive economic growth will serve to develop one of the wealthiest societies in human history, with a staggering average GDP per capita of US$ 30 675 by 2050.

In sixth and seventh position we can find the change in demographics in the REST group and the population growth in the BRIC economies. The spectacular population growth the REST group will face during the next decades, a 68% increase, will serve to create one of the youngest populations in the world. This vast number of young people will serve to stimulate demand on a quantitative basis, especially for commodity products. In parallel the increase in population in the BRIC economies, although much more modest at 13%, will bring an additional 400 million people into the world.

Finally, and in the eighth and ninth positions we can find the different wild card megatrends for each of the different areas. As previously explained megatrends can certainly have a significant impact on the chemical industry but that tends to be more short term oriented than structural. A large pandemic, massive terrorism attacks, or war can certainly radically change the supply and demand for certain chemicals and pharmaceuticals but as said, these impacts tend to be more temporal rather than structural.

4.4
Megatrends with the Highest Impact into the Different Features of the Industry

In terms of the megatrends with the highest impact on the global features and segments of the chemical industry; and the features with highest impact; it appears that the technological relevance of the industry and the level of greenhouse emissions are the segments most affected by the upcoming megatrends (Figure 4.5).

According to our analysis all of the upcoming megatrends, including social media, although with only a small impact, are expected to have a strong impact on the technological relevance of the industry. Considering all the challenges the world and the industry are expected to face it does seem strange to think the technological relevance of the industry will be highly impacted during the next decades.

In second position with almost all megatrends impacting very strongly, including social media, and with just two scores (population and urbanization) with a strong impact instead of very strong, naphtha and gas oil as feedstocks appear as very strongly impacted. In a highly regulated and connected world the pressure to address climate change and reduce fossil fuel consumption and greenhouse emissions will put enormous pressure on the industry to replace its naphtha and gas oil demand for lighter and more sustainable feedstocks.

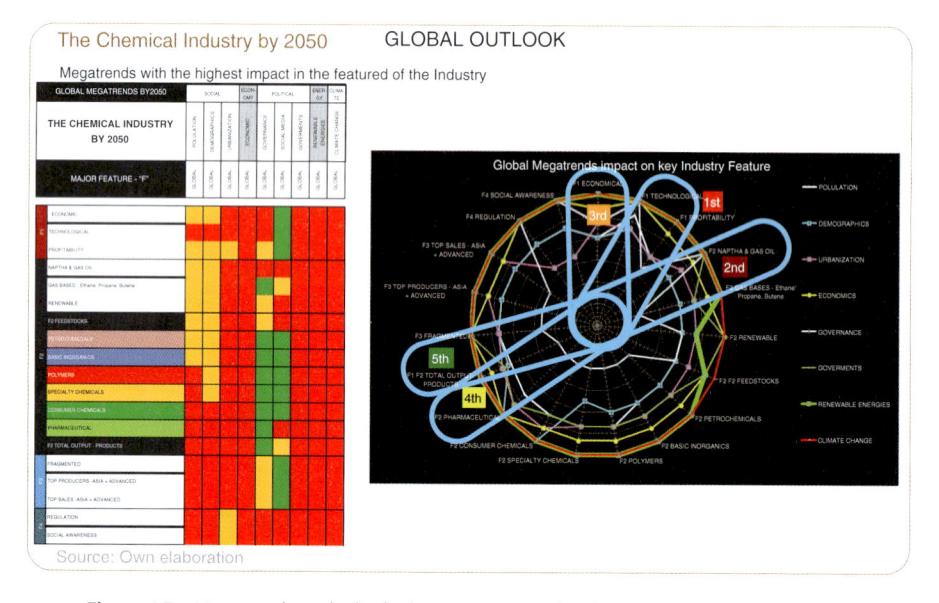

Figure 4.5 Megatrends with the highest impact on the features of the chemical industry.

A combination of events will drive this change including the fact that naphtha and gas oil as feedstocks can double their emissions per metric ton of ethylene when compared with ethane and that fact in a world eager to reduce emissions will make this transition simply a must. Secondly, and if even when neglecting for a second the impact of climate change, the fact that currently 50% of the industry feedstock is based on naphtha and gas – equivalent to 12–15% of the world crude oil daily demand – and under certain scenarios either these ratios will not be sustainable nor even affordable, this feature will be subject to large changes in the next decades. Finally, the recent developments in shale gas globally, doubling the global reserves of natural gas, in combination with the previous two points might trigger a gradual but massive change in the feedstocks of the chemical industry, a change where natural gas and other non-fossil fuel feedstocks are expected to play a much larger role.

Third in this ranking is the economic relevance of the industry. Here also almost all the upcoming megatrends scored at the highest possible level, with the exception of the population growth in the ADV and BRIC economies that scored with a strong impact and social media with a small impact. This result highlights the fact that most of the coming megatrends will have a positive impact on the economic relevance of the industry, making it bigger and more profitable. Something that appears quite logical to us, especially when considering the fact that the industry will be forced to triple its capacity while moving from a mass production business model to a more innovative and technological one.

In fourth position, the pharmaceutical industry is expected to be largely impacted by the upcoming megatrends. In a world with an older and wealthier population, where the world average life expectancy will increase to 77 years and the average world GDP per capita is expected to triple US$ 30 675 (2009 dollar) by 2050, the pharmaceutical industry is expected to experience not only significant growth but also significant changes. In 2010 almost two-thirds of the world pharmaceutical sales were made in the ADV economies, by 2050, despite the fact that the pharmaceutical industry might double its sales in the ADV economies, those sales will only represent one-third of the world total sales. Different areas with different demographics and different levels of wealth may request different levels of treatment and medicines. Ultimately, the convergence of several technologies and industries might also change the way pharmaceutical companies deliver solutions, moving from problem solving in ad hoc situations to a detection mode with constant and dynamic interaction. The convergence of pharmaceutical with IT companies and medical devices might move the industry into the next phase. A phase where human bodies are treated like a "machine" and real time information is available for prevention, treatment and cure, elevating life expectancy to levels never seen before.

Finally, the fifth feature to be highlighted in this section, is the obvious growth all segments of the chemical industry are expected to have. The projected increases in population and wealth will drive so much growth and changes in all aspects of human life, that the chemical industry, as a building block for many industries and a key enabler of living standards, will certainly be affected. Different segments of the chemical industry in different geographies will be more or less affected by the

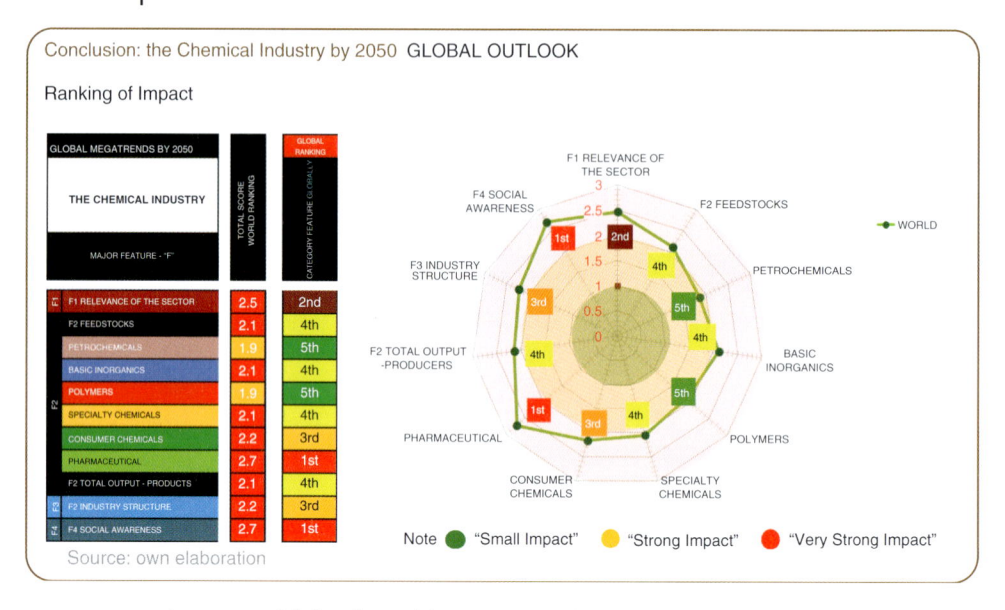

Figure 4.6 Global ranking of the impacts on the chemical industry.

upcoming megatrends but the underlying theme is a net increase in world demand, while reducing and enabling other industries to reduce greenhouse emissions and energy demand.

4.5
Major Results for the Chemical Industry Globally

At a global level (Figures 4.6 and 4.7) the segments and features of the chemical industry most affected by the upcoming megatrends are the social awareness of society with regard to the industry, its level of emissions, and the pharmaceutical sector.

The capacity of the chemical industry to enhance living standards, extend human life, and reduce greenhouse emissions and energy consumption, in combination with its expected massive economic growth, could explain why the awareness of society could increase significantly. In the same ranking position, the expected increase in life expectancy, in a world with much older and wealthier populations will trigger a spectacular increase in demand for medicines and pharmaceutical products.

Secondly, and quite related to the two previous results, the relevance of the industry will be positively and largely impacted as well. In a world with a clear need for many and better chemical products and solutions, the economic, technical, and profitability relevance of the industry is expected to increase further during the next decades.

In third position, the increases in population and GDP will have the potential to change the structure of the industry as well as to massively increasing the demand for consumer chemicals around the world. During the next decades

4.5 Major Results for the Chemical Industry Globally | 213

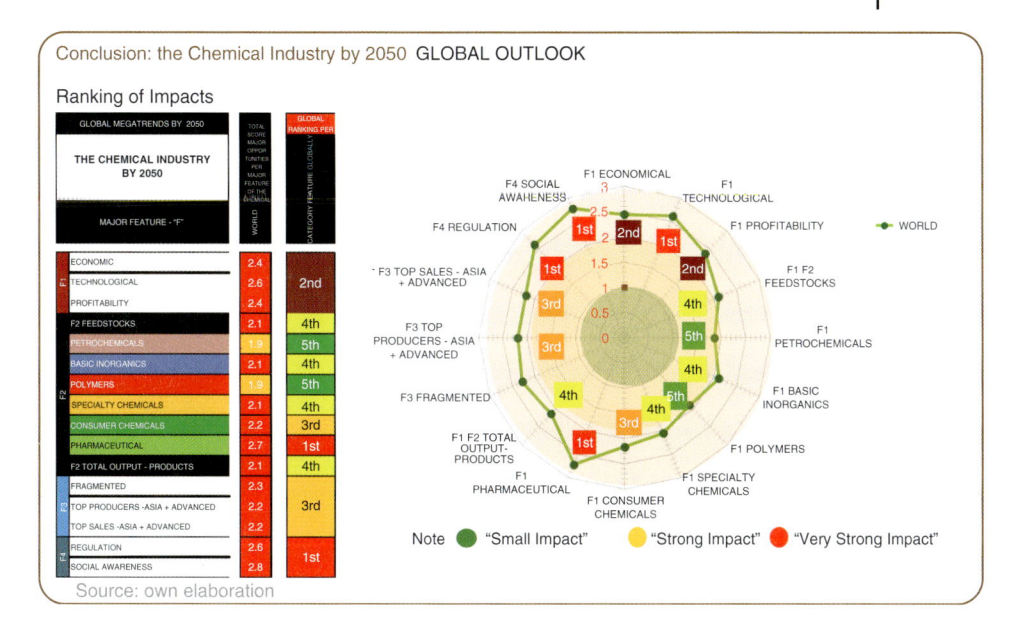

Figure 4.7 More detailed view of the global ranking of the impacts on the chemical industry.

millions of people will leave poverty forever, enabling the creation of a huge middle class with more than 50% of the world population. That expected increase in wealth for the most needy will trigger massive demand for all basic plastics and chemicals.

With regard to the structure of the industry, we will review this aspect in more detail in the next pages, but for the rest of the chapter we will start reviewing the sub-segments of the first, third, and fourth features on the industry, instead of the impact the feature as a whole. In other words as of now we will review how the upcoming megatrends will impact the different sub-segments of these three major features: the relevance of the industry, the structure of the industry, and the awareness of society with regard to the industry. At this stage we will not review the impact of the upcoming megatrends on each of the product segments of the chemical industry, as much further levels of granularity and analysis will be required, and for simplicity we will leave this analysis for later updates of this chapter. For instance, if we take the polymers group, we will not assess the impact of each of the megatrends on the different plastics, synthetic rubber, or man-made fibers.

Finally, the expected economic growth during the next forty years will drive a strong demand for all chemicals in general; from most basic chemicals to specialty chemicals and pharmaceuticals. However, climate change might be the megatrend that will create the most challenging issues and greatest opportunities for the chemical industry.

In this more detailed view where we expand further some of the major features of the industry, we can also observe how some of the sub-segments will be impacted during the next decades.

In that sense the technological relevance of the chemical industry is expected to increase greatly, becoming one the areas of the industry most affected during the next decades. As the industry tries to resolve its own challenges, but more importantly it starts to focus on how to solve the major problems the upcoming megatrends will bring to the world, the industry will again excel in innovation, product development, and new technologies. This technology supremacy will not be unnoticed by society, especially in a society increasingly concerned about the upcoming issues our world might face and more dependent on new solutions. Under these premises we expect the technological relevance of the industry to increase significantly during the next decades.

Additionally the combination of several converging trends in the industry and society, will certainly serve to increase the levels of regulation in the industry. The expected increase in importance of the industry, with much larger size and much larger players, in combination with a much bigger exposure to society at a moment in history where the wonder of the chemical industry will be at the peak of its demand, may provoke a large increase in regulation. If we add the factor that climate change will force governments to enact thousands of regulations around the world in order to curb greenhouse emissions across all industries, then we believe this feature of the chemical industry is also poised to face a large impact during the next decades.

With respect to the other incorporated new features of the industry, the massive growth the industry is expected to face, especially in the BRIC and REST economies, in combination with the massive challenges the industry will face in addressing climate change and the changing pattern in energy and feedstocks supply will serve to preserve the current high levels of fragmentation in the industry. Different areas of the world will be confronted with different dynamics here, and we will reflect further about this in the next pages and in Chapter 5.

4.6
Major Results for the Chemical Industry in the ADV Economies

For the chemical industry located in the ADV economies, three features appear to be the most affected by the upcoming megatrends: greenhouse emissions, technological relevance and social awareness (Figure 4.8). These seem to be interrelated, as the level of the industry emissions, in a context of global climate change concerns, will foster the technological relevance of the chemical industry at the same time as the awareness of society with regard to the chemical industry.

By 2050 the wealthiest, healthiest, and most technologically developed society in human history will be mainly located in the ADV economies. This super-developed society will pose significant challenges and opportunities for the chemical industry. More and better products with more efficient technologies, less energy demand, and less greenhouse emissions will be demanded. The chemical industry in the

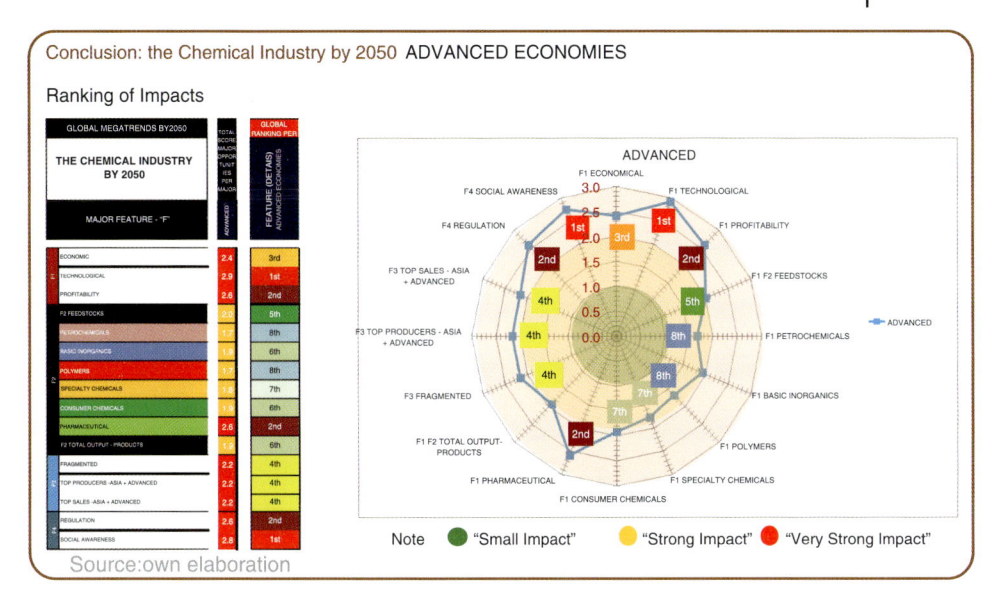

Figure 4.8 Ranking of impacts on the chemical industry for the ADV economies.

ADV economies will be requested to go beyond its own limits, doing much more with much less.

The governments and societies of the ADV economies will call the chemical industry to action, tapping into its technology relevance and innovation. This super wealthy society will be eager for solutions, solutions that will enable all the positive aspects of the upcoming growth and its associated benefits, while mitigating all the negative aspects of that growth, like energy security and supply, high energy prices, and greenhouse emissions.

The ADV group, with an estimated average GDP per capita of US$ 78,000 (2009 dollar), with USA at the front of the group with US$ 97,088 (2009 dollar) of GDP per capita will be confronted with a situation where the amount of growth in the economy and the chemical industry will necessitate an increase in qualitative standards in its products and efficiency in its production.

In this super wealthy, healthy, and developed society with all basic needs covered, and excluding for one second the sustainability concerns, its citizens will focus on raising their living standards to new levels. The expected developments in computational technology and the convergence of several sciences will extend human life to levels never seen before. The pharmaceutical industry, in a society becoming ever older, wealthier and super-educated will suffer a tremendous transformation, moving from resolution to prevention. The convergence of technologies will enable products, devices, and items that will be difficult to allocate to a particular industry. Technologies and industries will tend to converge gradually.

The need to reduce greenhouse emissions will certainly foster regulation and drastic measures on the feedstock and energy demand of the industry, however, it is on its output side where the transformation will be equally radical. The inability of any industry alone to achieve a tenfold increase in carbon productivity will force industries across the world to collaborate and work together across the whole value chain. As we saw in Chapter 1 in the business case for the European Tire Labeling, even when governments set the right regulation, stakeholders are committed, incentives are fair and defined, and the chemical industry provide the right set of innovations and products, this level of carbon productivity might not be attainable in the medium term.

The chemical industry, as a major building block connecting thousands of industries, will need to learn how to work together. As collaboration across the value chain and industries in combination with digital convergence will enable the next technological revolution, the "sustainable" one. The chemical industry as an industry of industries could lead other industries in this revolution, as it has significant experience in collaboration across the value chain already, however, decades of incremental improvements will force the chemical industry to re-ignite innovation while learning how to collaborate with other industries with a much more transformational approach.

In terms of the chemical industry all products are expected to have a significant increase. In 2010, the ADV economies per capita on average around US$ 1463 annually on chemicals and US$ 678 on pharmaceuticals. By 2050, they could spend on average under the BAU (business as usual) scenario up to US$ 2938 and 1579, respectively.

4.7
Major Results for the Chemical Industry in the BRIC Economies

An industry exposed to a growing society with expected ninefold increases in GDP per capita during the next decades and with additional 380 million people, is certainly an industry in growth, massive growth. In 2010, the average GDP of the BRIC economies was around US$ 6000 (2009 dollar). By 2050, the expected GDP per capita in the BRIC is US$ 36 000 (2009 dollar).

India and China will lead that growth in GDP. India's average GDP per capita is expected to grow from US$ 3000 in 2010 (2009 dollar) to US$ 25 761 (2009 dollar) by 2050. China's average GDP per capita is expected to grow from US$ 7000 in 2010 (2009 dollar) to US$ 45 428 (2009 dollar) by 2050. With this sort of expected growth for the area, demand for the chemical industry will be outstanding.

Under this assumption and the global and imperative need to address climate change and energy scarcity, the chemical industry will be in the challenging situation to be able to enable all this growth, while innovating new products and solutions to reduce climate change globally, and addressing its own emissions.

In that sense and according to our analysis all major features and segments of the chemical industry in the BRIC economies will be strongly affected with almost all segments expected to have strong growth. Among the highly impacted segments and features of the industry it appears to us that the awareness of the society with

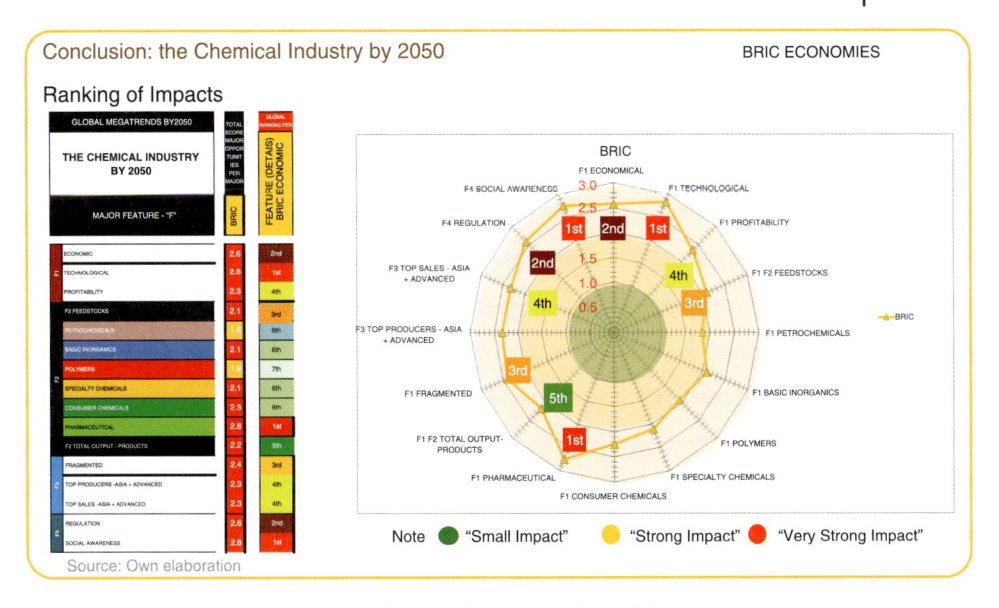

Figure 4.9 Ranking of impacts on the chemical industry for the BRIC economies.

regard to the chemical industry, its technological relevance and the pharmaceutical industry are expected to be the most impacted (Figure 4.9).

Considering the fact that by 2050 the BRIC economies on average could have the same average GDP per capita as the ADV economies in 2010, and that in 2010 the ADV economies had a per capita consumption of US$ 1463 in chemicals and $678 in pharmaceuticals while the BRIC had an average per capita consumption of US$ 344 in chemicals and just US$ 30 in pharmaceuticals, we can easily envision the huge potential growth in pharmaceuticals and chemicals in the decades to come. That formidable growth will certainly serve to enhance the awareness of society about the chemical industry. Awareness that will be complemented by its high technological components and the expectation that the chemical industry will be at the forefront of that growth and the fight against climate change.

From an industry perspective the expected growth in the chemical industry of the BRIC can drive some interesting dynamics. On the one hand it will lead to an increase in the number of chemical companies, increasing the level of fragmentation of the industry. On the other hand, it will also serve to increase the size of the chemical companies in the BRIC economies.

In 2010, only 2 of the top 10 world chemical companies (excluding pharmaceuticals) were outside the ADV economies (Sinopec and SABIC) and only one was from the BRIC countries. In the case of the top 10 world pharmaceutical companies the situation is even worse, as there was not even one representative from the BRIC economies. In that sense the expected economic growth in the BRIC economies will serve to increase the size of the local companies, companies that will be able to grow organically and inorganically in other areas. In terms of markers and production countries, China will continue being the largest producer and consumer of chemical products, and on the back of that growth we could expect many more BRIC companies and especially Chinese ones starting to top the international rankings for production and annual sales.

The need for the chemical companies to grow fast on both a quantitative and qualitative or technological basis, supported by emerging markets with high profits and valuations, will foster a global spur of mergers and acquisitions across the world, where this time the BRIC economies might be in command. In that sense it is specially remarkable that among the top 50 pharmaceutical companies in 2010 and those representing more than 80% of the world pharmaceutical sales, none were from the BRIC economies. We wonder how China and India, as the two economies expected to become the largest economies in the world by 2050, can ensure their technological supremacy and living standards without a solid chemical and pharmaceutical industry.

From a feedstocks perspective, the challenges and opportunities for the industry will be equally important. The global need to reduce greenhouse emissions and energy consumption, in combination with the fact that the BRIC, especially China and Brazil, will have vast reserves of shale gas, and the BRIC, especially India and China, are still very reliable on coal as an energy source will trigger significant changes in the feedstocks of the industry. In 2010, China had 70% of its total energy demand based on coal and India 50%. Considering the facts that China is already the largest emitter of CO_2 emissions in the world, despite being one of the lowest on a per capita basis, and that China's industries, including the chemical industry, are expected to keep growing significantly during the next decades, while China has the largest reserves of shale gas in the world with almost 1275 tcf, sufficient to ensure its current energy demand for centuries, we expect China to transition gradually to gas as a major source of energy for the country and also for the chemical industry. Brazil and India with considerable reserves of shale gas too, especially Brazil, and their high dependence on coal for India and oil for Brazil, might also undergo a similar and gradual transition to gas. Russia as one of the largest producers of gas too might not have a major change from this perspective.

The clear need to do more, much more, with much less, would get its maximum expression in the chemical industry of the BRIC economies. The BRIC are expected to have large increases in production in almost all industries. Their chemical industry will be the largest in the world, and the pressure to support the reduction in emissions and remain globally competitive will be intense. The chemical industry in the BRIC economies will be the one with the highest challenges and opportunities of all the different areas.

4.8
Major Results for the Chemical Industry in the REST Economies

As discussed in the introduction, the complexity and large heterogeneous base of the REST industry, make our observations and analysis very generic. With the underlying assumption that the REST group will experience a large increase in population, with more than 2 billion people by 2050, and the average GDP of the

4.8 Major Results for the Chemical Industry in the REST Economies

group is expected to treble from approximately US$ 7000 (2009 dollar) in 2010 US$ 17000 (2009 dollar) by 2050. The chemical industry will also be poised for spectacular growth in commodity products in the REST group and around the world.

Certainly this is a segment that later editions of this book should explore in more detail. As discussed when considering the economic megatrends, several members of this group will be among the largest economies in the world by 2050. Countries like Mexico, Indonesia, Turkey, Korea, or Nigeria are poised to be among the top 10 world economies by 2050.

In that sense, all segments of the chemical industry will experience massive growth and indeed all of them have been recognized with the highest possible ranking that for our analysis was a figure between 2 and 3. Under these premises and the global assumptions for the chemical industry, we expect similar impacts to those expected at the global level (Figure 4.10).

Massive growth in industry will need to be accompanied by reductions in energy demand and greenhouse emissions, as for the rest of the world. The technological relevance of the industry will be tested and the awareness of the chemical industry will be highlighted again. The economic relevance of the chemical industry and its profitability will also be very high, and among the highest, on the back of strong growth in some of these countries.

The low levels of average chemical and pharmaceutical consumption on a per capita basis in the REST group with US$ 233 and 44, respectively in 2010 compared with US$ 1433 and 678 in the ADV economies, highlights a fantastic opportunity for the next decades, and even for the whole century. Opportunity that will spur

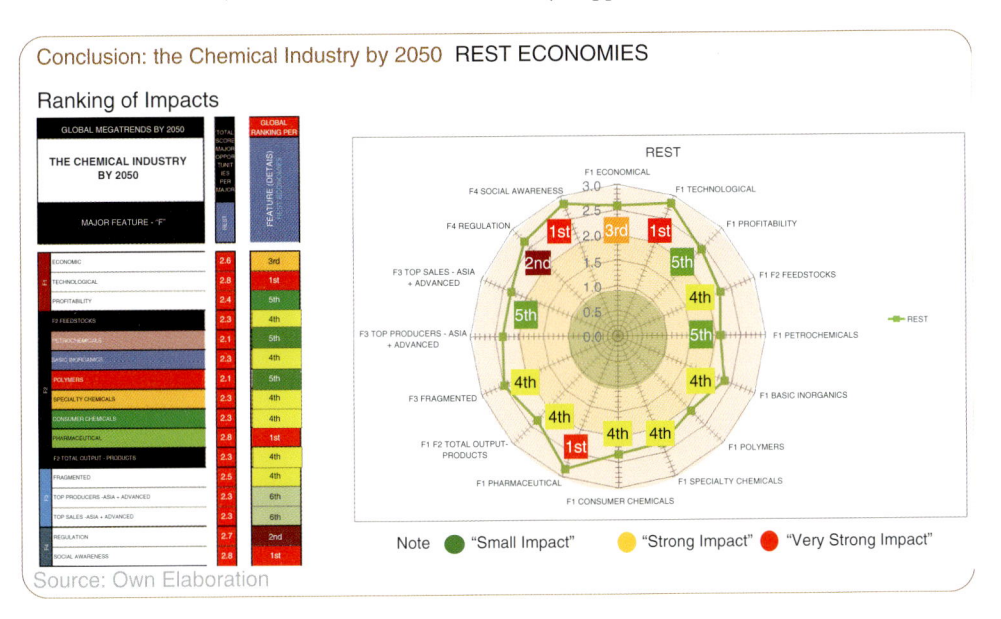

Figure 4.10 Ranking of impacts on the chemical industry for the REST economies.

a massive growth, especially on the commodity side rather than the speciality products, although it is difficult to speak in general terms due to the large variety within this group. Further analysis should be expected in further editions; analysis that will include the opportunities for the industry in places like Middle East, Korea, Indonesia, Malaysia, Thailand, Mexico, Argentina or Africa in general. We should remember that the REST economies will not only add almost 2000 billion people to the world population but will also have one of the youngest populations in the world and that will certainly influence its potential demand.

5
The Chemical Industry by 2050

Imagination is more important than knowledge.

Albert Einstein

5.1
Introduction

In this chapter we will present the major conclusions from our analysis, providing a comprehensive overview of how the chemical industry could look by 2050. However, before doing that we would like to remind our readers about the different complexities and challenges of our analysis, its context and the potential relevance and validity of our conclusions and recommendations.

First, we should be aware that all our conclusions and analysis have been based on a set of assumptions about the major characteristics of the industry and the major trends impacting our world and the chemical industry. Please keep these in mind when reviewing the conclusions and future outlook for the industry as a different set of assumptions might result in different conclusions.

For that purpose all assumptions and their interpretation, as well as most of the models, have been made available in the book so readers can have a solid understanding of the basis for our conclusions and projections. We understand that during the next decades some of the major assumptions can change, and even today different people might have a different set of assumptions and understanding of those, so let us be sure we put our conclusions for the industry into the right framework and understanding.

Secondly, we would also like to acknowledge the tremendous challenge behind our attempt to provide a generic conclusion for the whole chemical industry. As we have already reviewed in the previous chapters, the chemical industry, as an industry of industries, comprises many large and different segments and companies, most of them with different characteristics and dynamics. Starting from the obvious differences between chemicals and pharmaceuticals, there are also differences among the different sector and subsectors within these two major segments. Within chemicals, petrochemicals and polymers have very different dynamics to specialty chemicals, basic inorganic, or consumer products. Also,

The Future of the Chemical Industry by 2050, First Edition. Rafael Cayuela Valencia.
© 2013 Wiley-VCH Verlag GmbH & Co. KGaA. Published 2013 by Wiley-VCH Verlag GmbH & Co. KGaA.

even within one particular sector different subsectors might have very different dynamics. The outlook for plastics, synthetic rubber, and man-made fibers within the polymers group; or for fertilizers and industrial gases within basic inorganics may be completely different. This is a complication that becomes exponential when we start analyzing all these dynamics for the different products at the different area and country levels. Thus we will just reflect on the outlook for chemicals in general and pharmaceuticals, while occasionally we may devote some analysis and comments to some specific sectors or even subsectors within the chemicals group. Unfortunately this chapter will not review yet the outlook of the industry at the sub-segment level, and only in further updates we will start looking into these.

The additional fact that different geographic areas and countries are also expected to have different dynamics will further complicate our conclusions and the numbers of variables that we will have to consider. We will provide an overview for the industry at the global level and also for each of three major identified areas, the ADV, BRIC and REST economies. Occasionally, and for some of the major economies of the first two groups, we might provide some extra analysis or specific overviews for the chemical industry.

At this time, and despite being aware of the tremendous expected value and growth within the REST group, we will not be able to drill down into some of the most promising economies of this group. This group is not only very heterogeneous and complex, including countries so diverse as Switzerland, Iceland, Angola, Norway, Tanzania, Chile, and so on, but also includes some of the fastest growing economies in the world during the next decades. Countries like Mexico, Indonesia, Turkey, Nigeria, and Vietnam are expected to be among the top economies in the world by 2050. Therefore, a more detailed and comprehensive analysis of the major REST economies should be part of following updates.

In order to start presenting how the chemical industry will look by 2050, we would like to start by sharing the approach and process used in the following pages. To make the projections and conclusions for the industry as rigorous, intuitive, and detailed as possible, we will start reviewing and presenting the projected status of each of the major features of the chemical industry as presented during previous chapters.

In Chapter 3 we identified and presented six major features with their respective sub-features for the chemical industry (see Figure 5.1).

- **Feature 1 : Industry Relevance**
    - Economic
    - Technological
    - Profitability
- **Feature 2 : Feedstock – Inputs**
- **Feature 3 : Products – Outputs**
    - Pharmaceuticals
    - Chemicals
        - Petrochemicals
        - Polymers

5.1 Introduction | 223

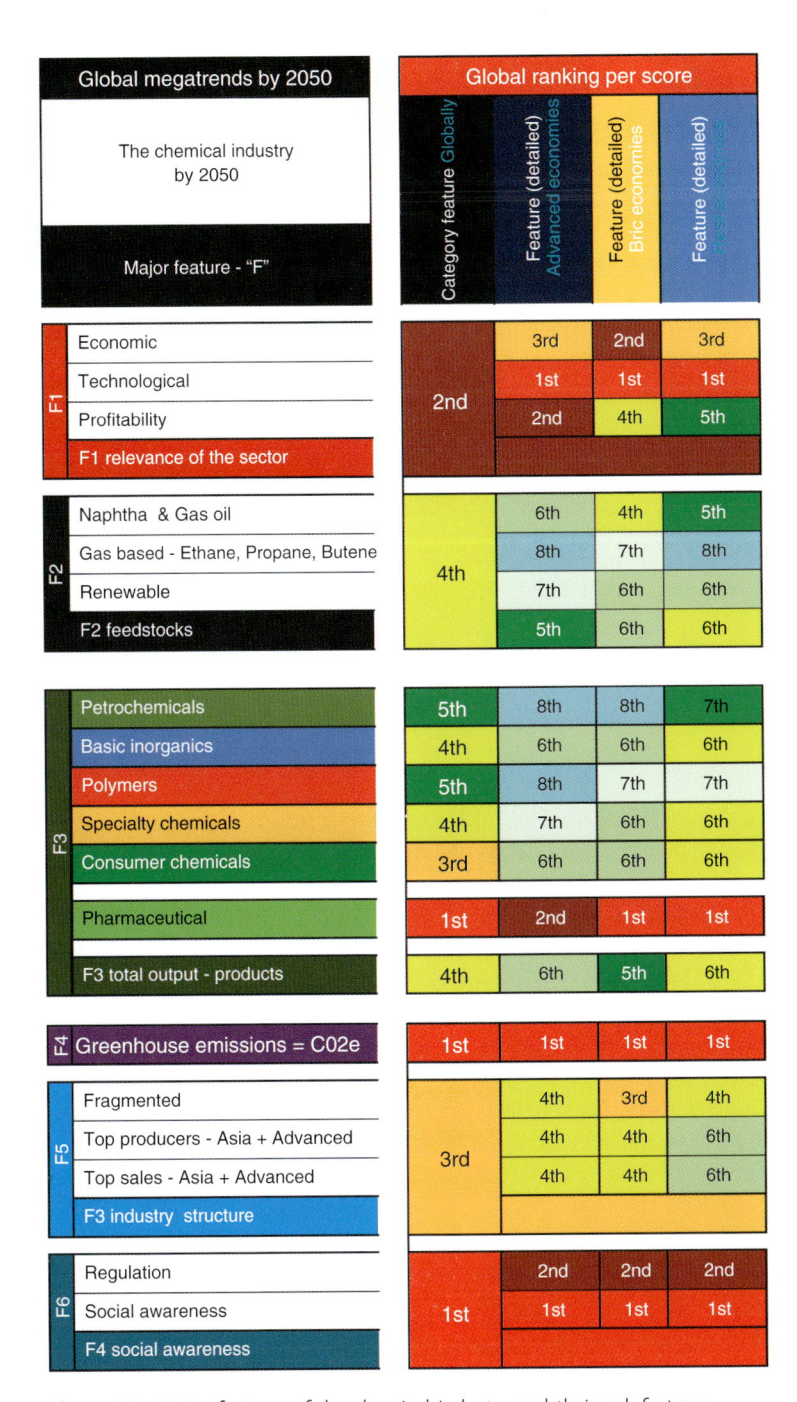

Global megatrends by 2050	Global ranking per score				
The chemical industry by 2050 Major feature - "F"	Category feature Globally	Feature (detailed) Advanced economies	Feature (detailed) Bric economies	Feature (detailed)

F1					
	Economic	2nd	3rd	2nd	3rd
	Technological		1st	1st	1st
	Profitability		2nd	4th	5th
	F1 relevance of the sector				

F2					
	Naphtha & Gas oil	4th	6th	4th	5th
	Gas based - Ethane, Propane, Butene		8th	7th	8th
	Renewable		7th	6th	6th
	F2 feedstocks		5th	6th	6th

F3					
	Petrochemicals	5th	8th	8th	7th
	Basic inorganics	4th	6th	6th	6th
	Polymers	5th	8th	7th	7th
	Specialty chemicals	4th	7th	6th	6th
	Consumer chemicals	3rd	6th	6th	6th
	Pharmaceutical	1st	2nd	1st	1st
	F3 total output - products	4th	6th	5th	6th

F4	Greenhouse emissions = C02e	1st	1st	1st	1st

F5					
	Fragmented	3rd	4th	3rd	4th
	Top producers - Asia + Advanced		4th	4th	6th
	Top sales - Asia + Advanced		4th	4th	6th
	F3 industry structure				

F6					
	Regulation	1st	2nd	2nd	2nd
	Social awareness		1st	1st	1st
	F4 social awareness				

Figure 5.1 Major features of the chemical industry and their sub-features.

- Specialty chemicals
- Basic inorganic
- Consumer products
- **Feature 4 : Emissions (CO_2) – Outputs**
- **Feature 5 : Industry Structure**
- **Feature 6 : Social Awareness**.

For each of the different features several sub-features or characteristics were identified and reviewed. For instance, for the first feature of the industry, its relevance, we identified three additional sub-features: its economic relevance, its technological relevance, and its profitability.

For some other features we went one level further down, and also identified the features of the sub-features. So, for instance, for the third feature of the industry, its products or outputs, we could drill down to the different segments of each of the different products. So for petrochemicals we could review the outlook for C2 ethylene, C3 propylene, C4 butadiene, and C5–C9 pyrolysis of gasoline and hydrogen/methane fuel gas. In this study we will usually go down from the feature level to the sub-feature level, but we may not always be able to go to levels below this.

In Chapter 4 we identified and ranked the sectors, features, and areas most affected by the upcoming megatrends. In this chapter we will present the potential outlook by 2050 for each of these features and some of the sub-features. We will start by using the conclusions from Chapter 4, while complementing these with further analyses and conclusions. We will provide, as far as possible, detailed qualitative and quantitative conclusions for each of the features of the industry.

For the quantitative conclusions, we will use the two major scenarios used in this book – the BAU and the sustainable. In some cases, as we will observe later, we may not be able to use both scenarios due to lack of data, especially for the sustainable scenario. On other occasions, we may add a modified or adjusted version of the BAU scenario where we will introduce some small adjustments to make the scenario more realistic and conservative. Additionally, and for the sake of transparency, for each scenario the key assumptions and rationale behind these will be explained and provided.

Considering all these factors, plus the logical complexity behind providing analysis and outlooks for such a long period of time and on a global basis, we would like to encourage our readers to consider the upcoming projections as broadly and directionally as possible. In this book, and specifically this chapter, we have tried to be as factual, rigorous, and conservative as possible, however, we are aware that some of the upcoming results might challenge our understanding and views of the industry. We acknowledge that fact and are confident our readers will consider the upcoming conclusions as a starting point for learning, discussion, and potential action, rather than a conclusive and final solution.

Let us proceed now to present the status of the chemical industry by 2050, by reviewing the projections for each of the selected features of the chemical industry.

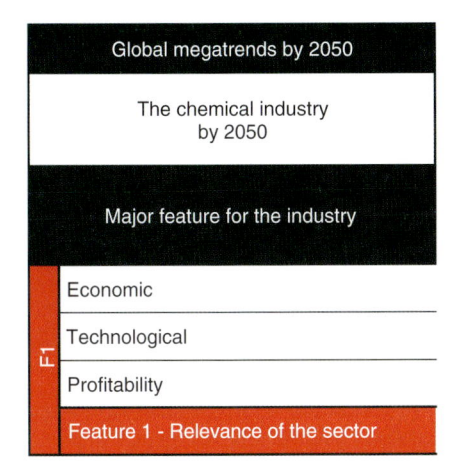

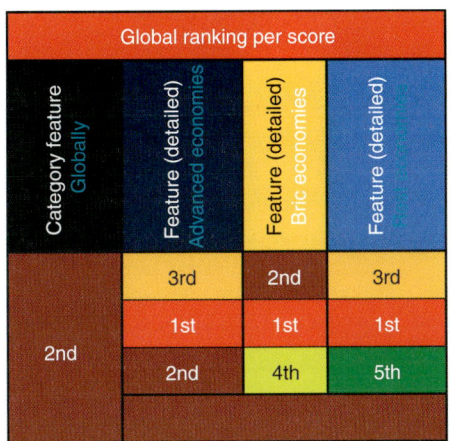

5.2
Feature 1: The Relevance of the Chemical Industry

For this first major feature of the chemical industry, three major aspects were identified and presented (see Figure 5.2). The economic relevance of the industry, as a large economic sector of the economy, with billions in revenue and millions of people and families depending on this industry, is one of the major characteristics of the industry. Its technological relevance, as an industry grounded on the pillars of innovation and technology and with high levels of R&D expenditure, is traditionally another key feature of the industry. Finally, the profitability of the industry, with two differentiated and clear dynamics and trends between the pharmaceutical and chemical sides, is certainly the last major characteristic of the industry, and one of the most important in defining its outlook potential and long term sustainability. While basic chemicals and petrochemicals companies, despite massive growth during the last decades, have been enduring reducing margins on high energy cost and commodity growth models; specialty chemicals and pharmaceutical companies have been enjoying growing markets and high margins, stimulating further growth and higher market valuations, and price to earnings ratios (P/E).

According to our analysis on the potential impact of the upcoming megatrends on the industry and its different features, described in Chapter 4, the relevance of the industry appeared to us as an area highly affected in the future. Ranking second among all the other industry features. According to our analysis the relevance of the industry is poised to increase, the technological and economic aspects being the most impacted. Different areas will have different dynamics, but certainly the relevance of the industry is one of the areas subject to increase during the next decades.

The world is poised for another long period of massive economic growth, similar to that experienced during the nineteenth century. This massive economic growth will serve to lift millions of people out of poverty, raising the world GDP per capita

Global megatrends by 2050		Global ranking per score		
The chemical industry by 2050	Category feature (Globally)	Feature (detailed) Advanced economies	Feature (detailed) Bric economies	Feature (detailed)
Major feature for the industry				
Economic	2nd	3rd	2nd	3rd
Technological		1st	1st	1st
Profitability		2nd	4th	5th
Feature 1 - Relevance of the sector				

Figure 5.2 Feature 1: The relevance of the chemical industry.

from an average of US$ 9219 in 2010 to US$ 30 675 by 2050, while raising living standards and world life expectancy from 65 years in 2010 to 75 years by 2050, and increasing the demand for chemicals and pharmaceuticals.

According to PriceWaterhouseCoopers the world economy could reach a staggering GDP of US$ 280 trillion (2009 dollar based) by 2050. Although this figure should be considered directionally, the reality is that, considering that in 2010 the world GDP amounted to US$ 63 trillion, the world economy could have the potential to at least triple. By achieving that, the world has the potential to host the wealthiest society in human history. As the world handles and resolves the major issues related to greenhouse emissions, possible resources and energy scarcity and climate change, it could be on the way to another long and large period of wealth creation.

The chemical industry, as a major "building block" of many industries and a key enabler of economic growth, living standards, and prosperity will also be highly and positively impacted. According to our analysis, the relevance of the industry – especially in its technological and economic aspects – is one of the features of the industry with the largest impact, right after greenhouse emissions, pharmaceuticals, and social awareness.

5.2.1
Economic Relevance

In an expected fast-growing global economy, the chemical industry is expected to increase significantly in size and economic relevance. The fact that the upcoming growth for the next decades will not only be quantitative, based on pure commodity and volume, but also qualitative, with a high need for new technological solutions that could assist the world's industries to reduce greenhouse emissions and further extend our quality of life, makes us believe the economic relevance of the industry will not only increase significantly but could also have a clear upward potential versus our own estimations.

If in 2010 the broad chemical industry, including chemicals and pharmaceuticals, accounted for US$ 3.9 trillion, 6.3% of the world economy, by 2050 the industry could reach a staggering figure of $20 trillion or 7% of the world economy. The largest differences in per capita consumption for chemicals across the different identified groups and their members, in combination with the large expected growth for the BRIC and REST economies, will trigger a massive growth in chemical demand, especially in these areas.

If in 2010 the world average per capita consumption of all chemicals remained at US$ 584, with US$ 456 for chemicals and US$ 128 for pharmaceuticals; by 2050 and under the BAU scenario that figure could increase to a staggering US$ 2054, with US$ 1631 for chemicals and $427 for pharmaceuticals.

Before entering into the different details of this segment, we would like to warn the readers about certain key methodological aspects for the upcoming conclusions. First, the upcoming projections for the chemical industry have been done under two scenarios: the BAU and a "modified or adjusted" version of this BAU scenario.

Aware of the long term unsustainability of either of the two BAU scenarios presented in this section, due to the clear need to reduce greenhouse emissions globally, but unable at this time to quantify the potential impact of a sustainable scenario on the chemical industry, we will just present here how the industry could look by 2050 if the world economy grows as projected and independently of climate change or energy concerns.

Therefore, please be aware of this important element as subsequent versions of the economic projections for the industry will need to start incorporating a sustainable scenario. We suggest that the following projections are used as a point for reflection on the potential outlook for the industry, as further refinements on our projections will be required.

On the other hand, and under the assumption that the chemical industry will be more positively than negatively impacted by climate change, as its products have the potential to enable large reductions of emissions in other industries, we could consider the upcoming scenarios as still conservative but directionally right. Under the additional assumptions that the chemical industry will soon be called into action and service for society – delivering new products and technologies able to address drastic greenhouse reductions – that statement can only imply a higher economic relevance for the industry.

For this purpose this section will present the broad chemical industry including chemicals and pharmaceuticals for the two discussed scenarios. In the first BAU scenario we will calculate how large the chemical industry could become if we preserve the current ratios of per capita demand for chemicals and pharmaceuticals in the different selected economic areas, while extrapolating those ratios to the 2050 economic projections.

In the modified/adjusted version of the BAU scenario we will modify some of the consumption ratios for pharmaceuticals and chemicals, based on certain assumptions and rationale.

In terms of pharmaceuticals, the facts that the GDP per capita across all countries and globally will be increased to levels never seen before, that life expectancy and older populations will increase in all regions, but especially in the ADV economies. As the BRIC economies are also expected to enjoy GDP per capita similar to that in Europe today, we deemed it reasonable to elevate the per capita demand for pharmaceuticals across all areas and countries.

This increase will be particularly noticeable in the BRIC economies, where these economies will not only have the highest increase in GDP per capita, but also the lowest per capita demand for pharmaceuticals. Let us illustrate that point with the fact that, in 2010, the Chinese estimated demand for pharmaceuticals was US$ 28 annually or 0.6% of its GDP per capita, while in the USA that figure was US$ 1036 annually or 2.2% of its GDP per capita.

Considering all these elements, we increased the per capita demand for pharmaceuticals in all regions. For the BRIC economies, the increase fluctuated between 0.7% in China, 0.4% in Russia, and 0.3% for India and Brazil. In the ADV economies the increase will be 0.2% for all countries, including the USA despite its already large demand. As the ADV economies in general, and the USA specifically, are expected to have some of the wealthiest societies in the world by 2050, with a GDP per capita of US$ 78 000 and US$ 97 088, respectively, we deemed it necessary to increase its demand by this small amount. For the REST economies we also incorporated a small increase of 0.1%, while we acknowledge that further increases, especially in some of its fastest growing economies, will be required in the future. Certainly this is an area that will require further analysis and evaluation as time progresses.

In terms of chemicals projections and for the adjusted/modified scenario, we also deemed it necessary to adjust the per capita demand rates for China and India. Under the assumption that China has already experienced a massive growth in chemicals during the last 30 years, we gradually reduced its projected demand in the future. In 2010 China had US$ 570 per capita sales of chemicals while India had just US$ 63. India is expected to have the largest economic growth during the next decades with an expected GDP growth of 2299%, more than double that in China with an expected growth of 884%.

As a consequence, we deemed it reasonable to reduce the ratio of per capita chemical sales to GDP in China from 13% to a more conservative but still fast growth rate of 7% while we increased the ratio for India from 4.3 to 7%. The ultimate rationale for this adjustment is the belief that China itself might have a higher level of development and chemical sales than India, especially after several decades of spectacular growth in China. Therefore, we expect the Chinese demand for chemicals gradually to decelerate during the following decades, while India will start to accelerate as China did during the last decades.

Readers may have different views with respect to the above assumptions, indeed some might argue that a decline in China's per capita chemical sales from 13% of the GDP per capita to 7% might be extreme, or that the decline should be done more gradually; and this indeed might be the case, but we have difficulty in accepting that under the first BAU scenario, China could have a per capita demand of chemicals double that in the USA by 2050 with US$ 5893, especially when the USA is still projected to have a GDP per capita twice that of China. Similarly for India, we could also understand that some readers might argue that this per capita chemical sales ratio could increase further than our 7%. Indeed some readers could argue that China has been having ratios well above 10% during the last decades as a consequence of their low starting point, so India could replicate that growth.

At this point we could indeed agree that a higher ratio should be considered in the future, however, under our premise to remain as conservative as possible we decided to go for a smaller increase. After all, this 2.7% increase in the per capita chemicals sales on GDP might seem small, but it is worth almost more than US$ 1 trillion chemicals sales for India. Additionally, theoretically the potential might be there, but governments, industries, and citizens still have a huge challenge in front of them in order to enable all that growth, we invite our readers to be conservative in our projections. Under these premises and at this time we rather preferred to set the increase for India per capita demand in chemicals to just 7% of the GDP. However, as time passes and as infrastructure, business environment, and practices start to improve, we might need to increase this further.

Under all the above factors and considerations and according to our adjusted/modified BAU scenario for 2050, the vast economic growth our world is expected to enjoy during the next decades will trigger a massive growth for the chemical industry. *The broad chemical industry, including pharmaceuticals, could move from US$ 3.9 trillion in 2010 to US$ 18.7 trillion by 2050* (see Table 5.1).

Table 5.1 Sales of chemicals and pharmaceuticals by 2050.

SCENARIOS	Sales (billion $US)			PER CAPITA SALES ($US)		
	INDUSTRY	Chemical	Pharma	INDUSTRY	Chemical	Pharma
2010	3996	3121	875	584	456	128
BAU 2050	20360	17252	3107	2226	1886	340
ADJUSTED BAU 2050	18786	14923	3904	2054	1631	427

Own elaboration based on best available data

The chemical industry under the adjusted BAU scenario is expected to grow from US$ 3.121 billion in 2010 to US$ 14.923 billion by 2050, while the pharmaceutical industry is expected to grow from US$ 875 billion in 2010 to US$ 3.907 billion by 2050. On a per capita basis the demand for broad chemicals is expected to grow massively, from US$ 584 in 2010 to US$ 2054 in 2050.

5.2.1.1 Chemicals and Pharmaceuticals by 2050 – per Capita Demand in $US

The per capita demand for chemicals is expected to increase globally from US$ 456 in 2010 to US$ 1631 by 2050, similar to the 2010 per capita chemicals demand in USA.

All regions are expected to increase their per capita demand for chemicals substantially (Figure 5.3); however, and despite all this growth the ADV economies are expected to remain with the highest per capita demand, almost $3000.

The BRIC countries will follow with US$ 2397 on average, but with large differences among its members. By 2050, China will have the highest per capita demand for chemicals within the BRICs, with an expected value of US$ 3120 annually. Brazil will follow at some distance with approximately US$ 2100, while India and Russia are expected to have US$ 1803 and 1630, respectively. In other words, by 2050 India is expected to be the second largest economy in the world, however, its per capita demand for chemicals will still be far from those of countries like China or the USA.

Similarly for the REST economies, the demand for chemicals is expected to rise significantly, however, with an expected per capita demand for chemicals of US$ 862 annually by 2050, when the global average will be at around US$ 1631, it will still have significant growth potential after 2050.

230 | 5 The Chemical Industry by 2050

(a)

	REST	Advanced	BRIC	World
2010	$238	$1463	$344	$456
2050 - BAU	$862	$2938	$3089	$1886
2050 - Adjusted	$862	$2938	$2367	$1631

(b)

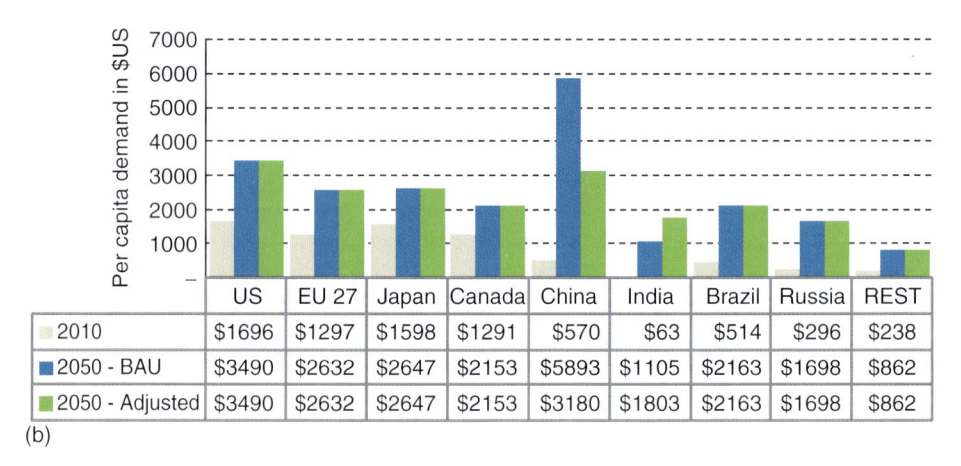

	US	EU 27	Japan	Canada	China	India	Brazil	Russia	REST
2010	$1696	$1297	$1598	$1291	$570	$63	$514	$296	$238
2050 - BAU	$3490	$2632	$2647	$2153	$5893	$1105	$2163	$1698	$862
2050 - Adjusted	$3490	$2632	$2647	$2153	$3180	$1803	$2163	$1698	$862

Figure 5.3 Chemical scenarios by 2050 – per capita demand: (a) by area, (b) by country.

The per capita demand for pharmaceuticals, under the adjusted BAU scenario is also expected to grow very fast (Figures 5.4), especially in the BRIC and REST economies, increasing the world average per capita demand from US$ 128 in 2010 to US$ 427 by 2050.

The ADV economies are expected to keep the highest per capita demand for pharmaceuticals, with almost US$ 1500 annually, an estimated average increase of 116%.

Although the BRIC economies will enjoy a gigantic increase in pharmaceutical demand, with an expected increase of 1380% by 2050, their per capita demand of US$ 444 will be only one third that of the ADV economies.

Within the BRIC, Russia is expected to have the highest per capita demand for pharmaceuticals with US 777, while China and Brazil will follow closely with around US$ 600 annually, similar to the 2010 levels in Japan. India, despite being

5.2 Feature 1: The Relevance of the Chemical Industry

(a)

	REST	ADVANCED	BRIC	WORLD
2010	$44	$678	$30	$128
2050 - BAU	$158	$1419	$262	$340
2050 - Adjusted	$167	$1579	$444	$427

(b)

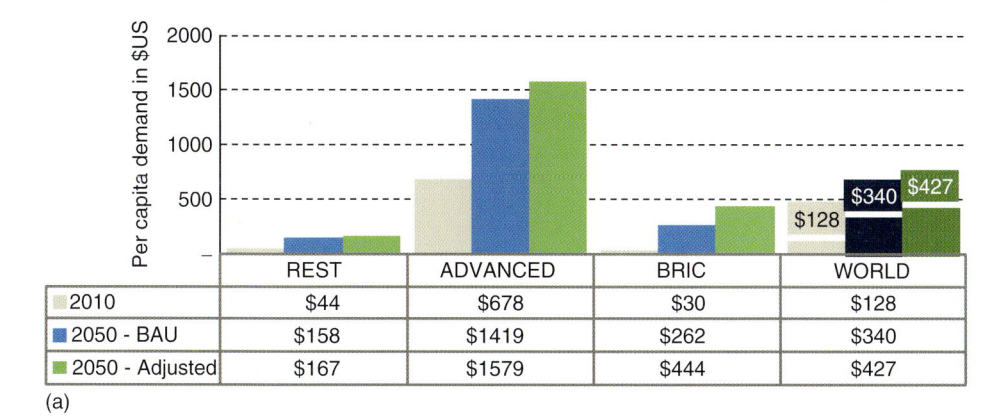

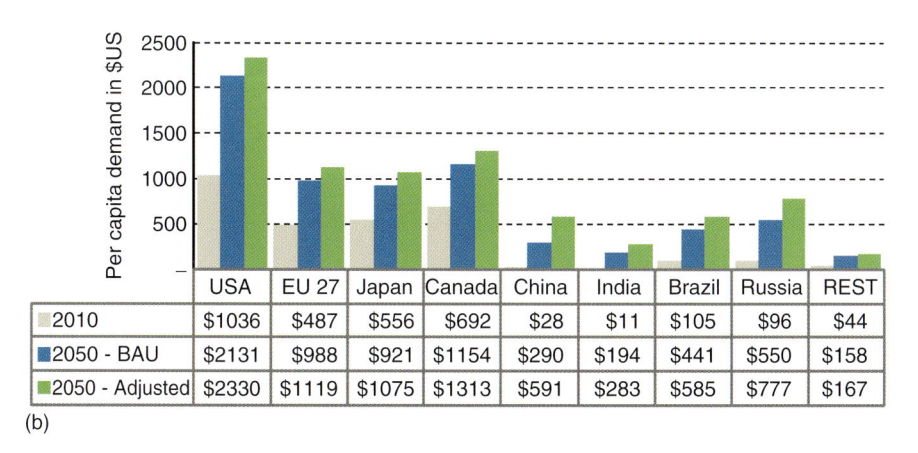

	USA	EU 27	Japan	Canada	China	India	Brazil	Russia	REST
2010	$1036	$487	$556	$692	$28	$11	$105	$96	$44
2050 - BAU	$2131	$988	$921	$1154	$290	$194	$441	$550	$158
2050 - Adjusted	$2330	$1119	$1075	$1313	$591	$283	$585	$777	$167

Figure 5.4 Pharmaceutical scenarios by 2050 – per capita demand: (a) by area, (b) by country.

expected to have the largest increase in demand for pharmaceuticals of the BRICs with a gigantic 2473% increase, will remain the lowest with just US$ 283.

Similarly the REST economies with an expected per capita demand of US$ 167 annually will remain well below the world average, highlighting further opportunities for growth in the second half of the century.

In terms of the economic relevance for the industry, this is also expected to increase see Table 5.2). If in 2010, the broad chemical industry accounted for 6.3% of the world GDP, by 2050 and under the adjusted BAU scenario that figure is expected to increase to 6.7%. Perhaps it is worth noting that under the original BAU scenario the chemical industry was poised to raise its share in the world economy even further to 7.3%.

However, this massive increase in chemical sales will not be equally distributed across the world, indeed on a percentage basis, and as logically expected based

Table 5.2 The economic relevance of the industry 2010–2050.

SCENARIOS	% ON GDP PER CAPITA		
	Total	Chem	Pharma
2010	6.3	5.0	1.4
BAU 2050	7.3	6.1	1.1
ADJUSTED BAU 2050	6.7	5.3	1.4

Based on best available data

on their lower per capita demand, the BRIC economies, followed by the REST economies are the ones that will experience the major increases.

According to our adjusted BAU scenario for 2050, the sector and area expected to have the highest growth potential will be pharmaceuticals in the BRIC economies with a huge 1595% increase, followed by chemicals also in the BRIC with a 680% increase (see Table 5.3).

The REST countries are also expected to have a huge growth until 2050, with an estimated 500% growth for both pharmaceutical and chemicals sales.

The ADV economies with much more developed economies and much larger and mature chemicals markets will enjoy the smallest increases on a percentage basis with 152% for pharmaceuticals and 117% for chemicals, however, in absolute terms the chemical industry in the ADV economies is still expected to have very large growth.

Indeed in absolute terms the broad chemical industry is expected to increase by a humongous US$ 14 834 billion, implying a growth of 370% and higher than the world GDP growth for this period (see Table 5.4).

The chemical industry, according to our adjusted BAU scenario is expected to increase by an enormous US$ 11 802 billion by 2050, accounting for 80% of the broad chemical growth.

Table 5.3 Global growth ranking in percentage terms for 2010–2050.

RANK	AREAS	INDSUTRY	% INCREASE
TOP 1	BRIC	PHARMA	1595
TOP 2	BRIC	CHEMICALS	680
TOP 3	REST	PHARMA	516
TOP 4	REST	CHEMICALS	483
TOP 5	ADVANCED	PHARMA	152
TOP 6	ADVANCED	CHEMICALS	117

Based on best available data for the Adjusted BAU Scenario for 2050

Table 5.4 Global growth ranking in $US billion for 2010–2050.

RANK	AREAS	INDUSTRY	$US BILLION	% Total	% Segment
TOP 1	BRIC	CHEMICALS	6,657	45	56
TOP 2	REST	CHEMICALS	3,477	23	29
TOP 3	ADVANCED	CHEMICALS	1,668	11	14
TOP 4	BRIC	PHARMA	1,348	9	44
TOP 5	ADVANCED	PHARMA	1,003	7	33
TOP 6	REST	PHARMA	681	5	22
TOTAL	**WORLD**		**14,834**		
GLOBAL	CHEMICALS		11,802	80	
GLOBAL	PHARMACEUTICALS		3,032	20	

Based on best available data for the Adjusted BAU Scenario for 2050

The BRIC economies with US$ 6657 billion, and on the back of huge chemical demand in India and China, will account for almost 56% of the total growth in chemical sales globally.

The REST economies will also enjoy an enormous increase in chemical demand with US$ 3477 billion by 2050, accounting for 29% of the expected increase in world chemical demand. Considering that, despite that massive increase in chemical demand, the REST economies will have a per capita demand for chemicals of just US$ 862 in 2050, when in Europe we already had US$ 1297 in 2010 and in the USA US$ 1696, it is believed that the REST group still present a significant amount of growth opportunity for the second half of this century.

Finally, the ADV economies are also expected to enjoy a huge growth during the next decades with US$ 1688 billion by 2050. As the GDP of the ADV economies is expected to double during the next four decades, while the population will increase only slightly, the per capita demand for chemicals is expected to double as well, from US$ 1463 to US$ 2938. We will review later in greater detail the country distribution of that massive growth, but some countries like the USA or Europe 27 might have the potential to double their chemical industry in the next 40 years.

The pharmaceutical industry is also expected to have enormous growth, with the potential for a fourfold increase. If the world pharmaceutical industry accounted for US$ 875 billion in 2010, by 2050 this will increase to US$ 3907 billion. In other words, each of the three regions, ADV, BRIC, and REST will have the potential to add one complete pharmaceutical industry as we know it today.

The enormous increase in wealth and GDP per capita, especially in the BRIC and REST economies, in combination with the low levels of pharmaceutical per capita demand will explain this huge increase.

In 2010, 75% of the pharmaceutical sales were in the ADV economies, 15% in the REST, and just 10% in the BRIC. Indeed, in 2010 the per capita demand for pharmaceuticals in the BRIC economies, according to our estimations, was the lowest of the three areas, with just $30 per capita annually, and with India at the bottom of the group with a mere US$ 11. The REST economies were very close to the BRIC with US$ 44 annually, while the ADV, led by USA with a record high US$ 1036, had an average demand of US$ 678 annually.

These huge disparities in global pharmaceuticals demand across countries and areas, much broader than for chemicals, in combination with the massive expected economic growth in the BRIC and REST economies will trigger spectacular growth in pharmaceuticals sales across the world. This massive growth will have the potential to change the structure of the chemical and pharmaceutical industries.

The broad chemical industry is also expected to have a massive economic growth across all sectors and geographies, the BRIC economies being the major benefactors. The pharmaceutical industry will lead that growth on a percentage basis, expecting massive growth in all regions and especially in the BRIC. This growth might have the potential to change the structure of the pharmaceutical industry, an industry traditionally dominated by the ADV economies.

On the other hand the chemical industry, much larger and maturer than the pharmaceutical industry, will grow more slowly, but the increase in absolute terms will be much greater. The BRIC economies as the larger benefactors of that growth are expected to add $7636 billion of chemicals by 2050, twice the current size of the world chemical industry. With this sort of enormous increase in mind, the world chemical industry is expected to increase by US$ 14 834 billion by 2050, increasing the economic relevance of the industry in the economy from 6.3 to 6.7%.

At the country level the growth and size of the respective markets will increase significantly too, altering the global structure of the market for both the chemical and pharmaceutical industries. The BRIC economies will lead that growth as we will see later in the feature devoted to the structure of the industry, meanwhile please see Figure 5.5 showing the largest chemical markets in the world in 2010 and 2050, and the expected growth during the next decades.

Although the potential growth for the industry remains enormous and encouraging; and the economic relevance of the industry guaranteed, we should be aware that there are still significant impediments and challenges ahead.

First, these projections have been based on a modified version of the BAU scenario for 2050, however, we are aware that greenhouse emissions, climate change, and potential energy and resources scarcity could have a significant impact on these projections. Although at this time we believe these impacts might be mainly manageable and positive for the industry, they should be monitored and clarified in later versions.

Secondly, as most of the upcoming growth for the next decades will depend on the growth potential of the BRIC, and to a large extent on the growth of China and India. Therefore we should carefully monitor the performance in these two countries.

Growth ranking in % - 2010 - 2050
Adjusted "BAU" scenario - 2050

Rank	Country	Industry	% Growth
TOP 1	India	Chemical	3816%
TOP 2	India	Pharma	3409%
TOP 3	China	Pharma	1905%
TOP 4	China	Chemical	431%
TOP 5	Russia	Pharma	609%
TOP 6	Brazil	Pharma	527%
TOP 7	Russia	Chemical	401%
TOP 8	Brazil	Chemical	372%
TOP 9	USA	Pharma	189%
TOP 10	USA	Chemical	164%
TOP 11	Canada	Pharma	140%
TOP 12	Europe 27	Pharma	132%
TOP 13	Canada	Chemical	111%
TOP 14	Europe 27	Chemical	105%
TOP 15	Japan	Pharma	61%
TOP 16	Japan	Chemical	38%

Largest world markets

Chemicals	2010		2050	
China	$ 763	China	$ 4048
Europe	$ 651	India	$ 2903
USA	$ 524	USA	$ 1385
Japan	$ 203	Europe	$ 1332
Brazil	$ 100	Brazil	$ 474
India	$ 74	Japan	$ 281
Canada	$ 44	Russia	$ 211
Russia	$ 42	Canada	$ 93

Pharma	2010		2050	
USA	$ 320	USA	$ 925
Europe	$ 244	China	$ 752
Japan	$ 71	Europe	$ 566
China	$ 37	India	$ 456
Canada	$ 24	Brazil	$ 128
Brazil	$ 20	Japan	$ 114
Russia	$ 14	Russia	$ 96
India	$ 13	Canada	$ 56

Figure 5.5 The largest chemical markets in the world in 2010 and 2050, and the expected growth during the next decades.

Indeed, China and India are expected to account for almost 50% of the world chemical growth and 40% of the pharmaceutical growth by 2050. China has shown a remarkable consistency and performance in managing to achieve its great potential. India on the contrary has been lagging behind somewhat; however it is time for India to start enabling all this growth and unleashing its full potential. The challenges are huge and the stakes are high, countries with populations of billions, huge distances, several time zones, and complex markets and regional political organizations might not be easy to be aligned and managed, even for the right purpose. However, the rewards are enormous and after all the next century might be the century of China and India.

Therefore let us suggest that we consider the above projections as a point for reflection and optimism, while remaining cautious on the upcoming developments in the sustainable growth scenarios and monitoring the developments of these two economies. Meanwhile, let us enjoy this positive outlook while future updates of this scenario will be available soon.

5.2.2
Technological Relevance

The technological relevance of the chemical industry is also expected to remain very high and with a great potential to go again to its highest historical levels.

Under the sustainable scenario the huge challenges our world will need to face in the next decades – allowing massive growth and increases in living standards and life expectancy, while reducing greenhouse emissions and energy consumption – make us believe the technological relevance of the chemical industry will increase significantly.

Through the recent history of the industry, we have seen how it has solved some of the most complicated problems of humanity, enabling significant increases in quality of life around the world. However, during the last 50 years it has gradually moved its focus from a technology and innovation business model to a more manufacturing or operational one. The enormous growth the industry experienced during the last 50 years encouraged the industry to shift gear toward mass production and, lately, to a commoditization of the industry. Graphically we could argue the chemical industry has completed its first cycle, moving from an innovation and technological focus to a mass production and optimization period. A phase where most of the major discoveries were made during the nineteenth century and the first half of the twentieth century, while during the second half of the twentieth century the industry dedicated itself to optimizing and maximizing them.

5.2.2.1   The Chemical Industry – Long Term Cycles
In a world expecting a tenfold increase in carbon productivity, and with clear energy supply concerns, the chemical industry's unique capacity to enable and empower other industries to reduce greenhouse emissions and energy demand make us

believe the industry will soon be called into action, and then its technological supremacy and innovation capabilities will be tested again. *During the next decades the industry will gradually transition from its first long super-technology cycle into its second one (Figure 5.6). During the first cycle the industry was based on high growth and mass production using fossil fuel chemistry but in the new cycle although the industry will again be focused on high growth it will be with massive emission and energy productivity. The industry will still be based on fossil fuels, at least for the first half of the century, but the focus will shift to the maximization of emissions and energy.*

Historically, the chemical industry has distinguished itself for high levels of innovation and technology, a feature that has been traditionally reflected in a high ratio of R&D versus sales. In 2008 the world average R&D to Sales ratio was 3.3%. The pharmaceutical industry stayed at the top of the list with 16.5% while chemicals were well below that figure at 2.9%. However, we should be aware that due to the complexity of the chemical industry that number might be misleading. As we start reviewing the different segments of the chemical industry we can find a large diversity in R&D expenditure (Figure 5.7).

Some sub-segments of the chemical industry, like fertilizers, industrial gases, synthetic rubber, paints and inks, and petrochemicals and polymers might be below the chemical industry average and the average for all industries. However, other large sectors, like consumer chemicals, specialty chemicals, and crop protection may be well above average.

There is also a remarkably large variation in this ratio between companies, with some companies having double the average for their sector. For instance, the average R&D on Sales ratio for the top 10 pharmaceutical companies in 2010, was 16.3% but the companies with the highest ratio, Roche and Merck, had a ratio of 20%, while those with the lowest value were Abbot and Johnson and Johnson with 11%. If we look at chemicals, for instance crop protection, the average for the top

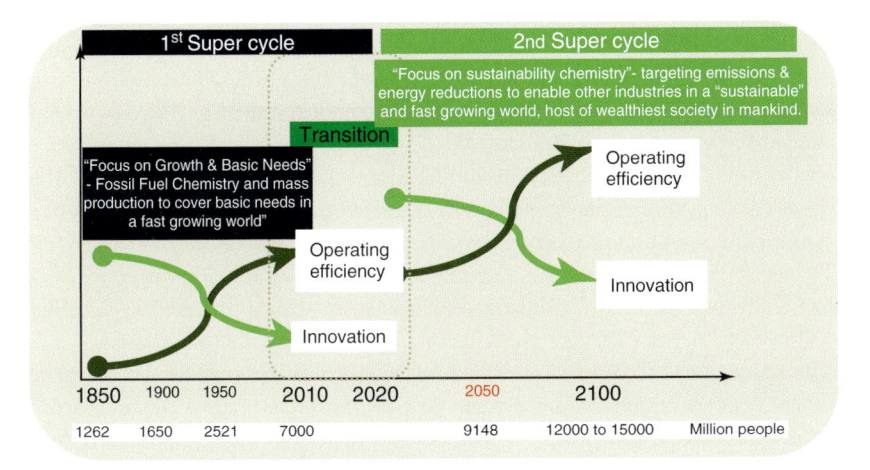

Figure 5.6 The chemical industry – long term cycles.

5 The Chemical Industry by 2050

(a)

Industry	%
Pharmaceuticals &...	16.5%
Software & Computer	9.6%
Technology hardware &...	8.6%
Health care & Equipment...	6.1%
Leisure goods	6.1%
Automobiles & Parts	4.4%
Electronics and Electrical...	4.2%
Aerospace & Defense	4.1%
All sectors	3.3%
Chemicals	2.9%
Industrial engineering	2.7%
General industries	2.3%
Household goods	2.2%
Fixed telecommunications	1.7%
Food producers	1.5%
Oil & gas exploration	0.3%

(b)

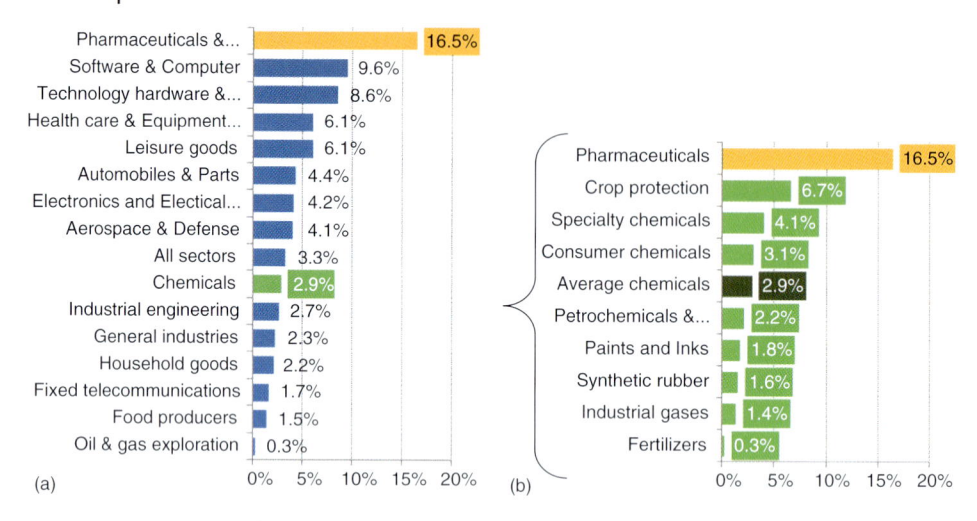

Sector	%
Pharmaceuticals	16.5%
Crop protection	6.7%
Specialty chemicals	4.1%
Consumer chemicals	3.1%
Average chemicals	2.9%
Petrochemicals &...	2.2%
Paints and Inks	1.8%
Synthetic rubber	1.6%
Industrial gases	1.4%
Fertilizers	0.3%

Figure 5.7 R&D expenditure to sales ratio: (a) for all industries, (b) for the chemical industry. (a) Source: 2009 EU Industrial R&D investment Scorecard, European Commission, JRC/DG Research. Note: Data relate to the top 1,350 companies with registered offices in the EU, Japan, the USA and the rest of the world ranked by total R&D investment. (b) Source: Analysis based on the ICIS Top 100 World Chemical companies in 2010. Note: Results based on available data for the top 100 chemical companies, excluding pharmaceutical. Assignment done by the author based on best available data.

100 chemicals companies with available data was 6.7%. However, Syngenta had a 9% ratio in 2010, while Agrium had a 4% ratio (Table 5.5).

Additionally, it seems that different areas tend to have different R&D expenditure versus sales depending on the corporate culture, technological level and level of innovation. According to CEFIC, the European and USA chemical industries have been traditionally lagging behind the Japanese chemical industry in terms of R&D expenditure (Figure 5.8). This has always been the case during the last decades, however, that tendency has increased further recently.

The chemical industry as a whole has one of the highest R&D ratio on sales of all industries, thanks in large part to the pharmaceutical industry.

The pharmaceutical industry indeed has by far the highest ratio of industry globally. The chemical industry, excluding pharmaceuticals, also has one of the largest ratios, however, here we encounter large differences among different sectors and companies within each sector, and across the industry.

However, the critical role the industry plays in enabling living standards, enhancing life expectancy and reducing greenhouse emissions, makes us believe that it will need to increase further its R&D expenditure during the next decades.

The urgent need for drastic solutions in most of the areas where the chemical industry operates, makes us believe the technological relevance of the industry is poised to increase even further. A world with higher and older populations, and with the highest life expectancy ever seen, will push the pharmaceutical industry to strive for the best. At the same time a much larger, wealthier and healthier

Table 5.5 Global R&D expenditure as a percentage of sales.

% R&D on SALES			Chemicals							
			Petrochemicals & Polymers			Specialty Chemicals			Basic Inorganic	Consumer
Based on best available data for 2010 TOP WORLD CHEMICALS	Pharma (top 10)	Chemicals (top 100)	Large Diversified Chemicals	Petrochemical & Polymers	Synthetic Rubber	Specialty Chemicals (Several)	Paints & Inks	Crop Protection	Fertilizers	Industrial Gases	Consumer Products
High End	20.0%	9%	7.0%	4.1%	3.5%	5.40%	2.3%	9.0%	0.3%	3.6%	
Avg. Group	16.5%	2.9%	3.2%	2.2%	1.6%	4.1%	1.8%	6.7%	0.3%	1.4%	3.1%
Low End	11%	0.1	0.20%	0.1%	0.1%	0.10%	0.50%	4.0%	0.1%	0.8%	
High End	Roche / Merck	Syngenta	Sumitomo/Dow	Mithubishi	n.a	Dupont/Bayer	AZKO	Syngenta	K&S	Mitsubishi	Henkel
Low End	Abbot / J&J	Several	Eastman/Mitsui	NPC	n.a	Daicel / Kurakai	Sherwim Williams	Agrium	Yara	Praxair	n.a

Source: ICIS top 100 Chemical Companies in 2010 and IMAP Pharmaceutical & Biotech Industry Global Report in 2011

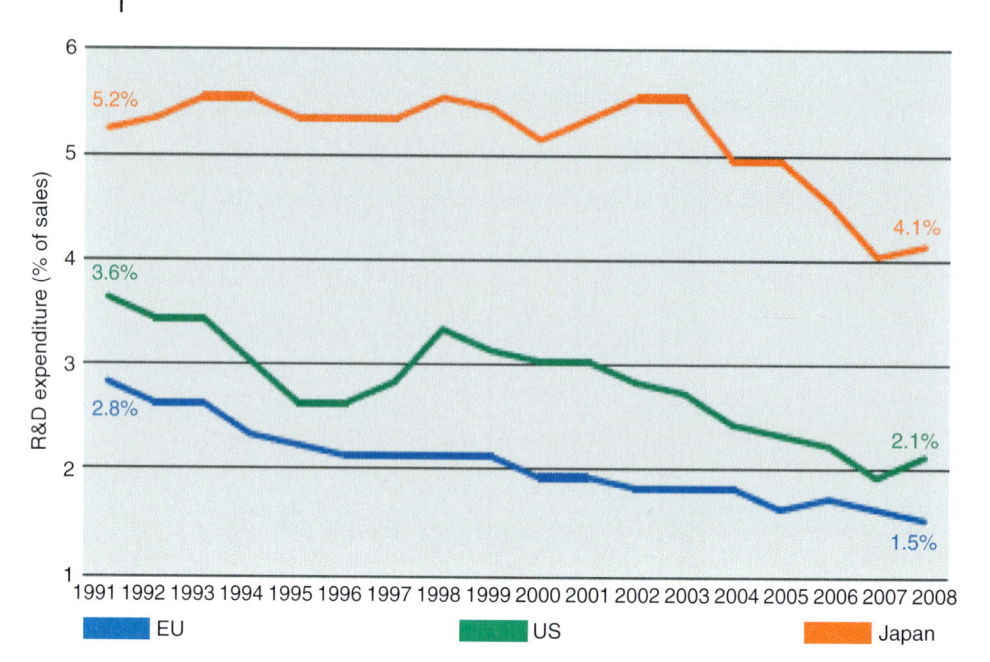

Figure 5.8 R&D expenditure as a percentage of sales for the period 1991 to 2008.

population, under the threats of climate change and energy scarcity, will push the chemical industry to enable new solutions and products that can serve to reduce emissions across the world while reducing energy consumption for the world. The chemical industry has been leading in this area for decades, now might be the time the industry will need to share its wonders again with the rest of the industries and the world. In this context we are confident the chemical industry will deliver again, but not only with higher R&D budgets, but also with different ways to innovate across the whole value chain.

The chemical industry as an industry of industries might also be in an excellent position to share its best practices and learning across the whole value chain, showing that innovation and efficiency can be achieved together. In these circumstances the technological relevance of the industry will increase furter during the next decades.

The industry and its companies will have clear opportunities to capitalize on these opportunities. Large chemical companies will have big incentives to set up consultancy businesses units, units that could share best practices and technology around the world with the smallest companies. Inside the chemical industry, consultancy and technology opportunities will be an area of high growth highlighting again the technological supremacy of the industry.

The convergence of technologies across industries will also open new areas of technological collaboration and technological progress. The collaboration across industries will increase very rapidly, and we could foresee partnerships among pharmaceutical or chemical companies not only with their consumer and end-user companies, but also with information technology companies or other remote industries.

Could we envision a world where companies like Google, Microsoft, Facebook, Baidu, Siemens, General Electric, Phillips, Toyota, Ford, GM, Volkswagen, BMW, Audi, Toyota, Nokia, Apple, Sony, McDonald, Tetrapack, Unilever, P&G, ABB, Nestlé, Boeing, Airbus, BASF, Dow, DuPont, Bayer, SABIC, Sinopec, Mitsubishi, Monsanto, Syngenta, Clariant, Henkel, Merck, Pfizer, Johnson & Johnson, Novartis, or Roche work together? Could you imagine what sort of products they could create?

Could we imagine what technology will be created when a pharmaceutical company works hand in hand with companies like Philips, GE, Google, or Apple? Could we imagine what could happen if BASF, Dow, or DuPont work together with Google, Apple, and perhaps the automotive industry? Or Syngenta, Monsanto, or Agrium work together with Yahoo, Samsung, the Weather Channel, John Deere, or Caterpillar?

The industry is in the middle of a major technological transition from its first super cycle into its second one. This transition occurs in the middle of another one; with the fast development of "shale gas". A new super cycle for an industry more focused on emissions and energy productivity, and ultimately the sustainability of the planet. With all these aspects in mind, we expect the technological relevance of the chemical industry to increase to levels of innovation never seen before.

5.2.3
Profitability

On the back of gigantic increases in demand in the coming decades in the BRIC and REST economies, the potential scarcity of energy and with most of the chemical products in high demand *the broad chemical industry is expected to upgrade its profitability and earnings profile.*

Each region and each segment of the industry, chemicals versus pharmaceuticals, or the different sub-segments within the chemical industry, will face several market dynamics but most of them will be pointing int the same direction, higher and growing demand.

Looking now at the calculated results for the top 100 chemical companies and top 10 pharmaceuticals companies in 2010, we can observe not only the large difference in profitability among the different segments of the industry but also between the upper or "specialty" side of the industry and the lower or "commodity" side. Reviewing the profitability of the selected sample we can extract the following conclusions.

First, the overall profitability for the pharmaceutical industry, measured in terms of percentage of margin over sales, appears to be higher than for the chemical industry 17% versus 9% in 2010. Secondly, within the chemical industry there are large variations in profitability across the different segments, with fertilizers, industrial gases, crop protection, and consumer products on the higher side and petrochemicals and polymers on the lower side. Finally, the difference in margin between the lower or most commodity-based side of any given segment and its most specialty side tends to be very large. For instance the delta between the lowest and the highest margin on sale can fluctuate from 28% for fertilizers, 25% for pharmaceuticals, 22% for chemicals, and 3% for paints and ink (Figure 5.9). Although these results are based on a very small sample of companies and one specific year, they might well represent the large differences in profitability across the different segments.

The expected increase in demand for the industry in the BRIC and REST economies will serve to accelerate further the demand and profitability of the more commodity or "basic" chemical products. On the other hand the so-called specialties or more "performance" oriented products will tend to enjoy higher degrees of appreciation in the ADV economies; however, the emergence of a super wealthy and large middle class with more than 50% of the world population will imply a solid and strong demand for "specialty" products elsewhere too.

Considering the large differences in profitability among the different segments of the industry, its products (commodities versus specialties) and geographical markets; and combining those with the expected difference in economic growth across the different areas and major economies, we could expect significantly different levels of profitability for the industry in accordance with the segments, products, and markets where they operate.

Reviewing now the expected growth for the two major different segments of the industry, according to the adjusted BAU scenario, we can observe how the pharmaceutical side of the industry in the BRIC economies is the area with the

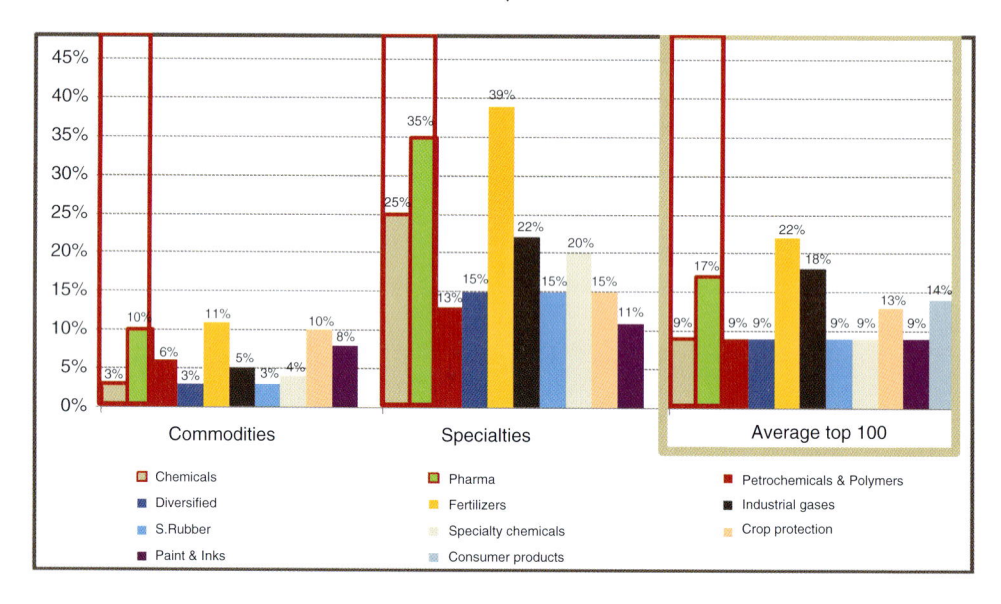

Figure 5.9 2010 margin on sales for the top chemical companies (including pharmaceuticals). Based on publicly available data for the top 100 chemical companies and the top 10 pharmaceutical companies in 2010.

highest growth potential for the industry, with an expected growth of 1595%. On the opposite side, and with the lowest growth potential, but still very high growth, we can observe how chemicals in the ADV economies are expected to grow by a more modest 117%. Table 5.6 shows a representation of the expected growth for the different segments and areas of the chemical industry.

Taking into consideration the observed correlation between industry growth, profitability, and P/E ratio (see Chapter 3 Section 1C – Industry Profitability) and the observed differences in margin between "commodity" and "specialty" products and companies; we could expect large differences in profitability and P/E ratios for the industry depending on the countries and segment where it operates.

Areas with high expected growth, like the pharmaceutical industry in the BRIC, chemicals in the BRIC, and pharmaceuticals in the REST are expected to enjoy a higher margin and ultimately higher P/E ratios. Now segmenting the chemical industry further into "Basic" or commodity products and "Performance" or specialty products, and the REST area into the NEXT fast growing countries, like Indonesia, Mexico, Turkey, Vietnam, and so on, and the REST we could map out what are the areas and segments with the highest growth, profitability, and P/E ratios for the industry.

According to our model we could expect that companies operating in the Pharmaceutical industry in the "NEXT" or BRIC groups are expected to enjoy the highest margin and market valuations in the world. Chemical companies operating in the commodity side of the industry and in the ADV economies might have the lowest margins and by default lower market valuations (Figure 5.10).

Of course each company and segment will still be subject to its own economics and its own circumstances but from a high level view, companies will be positively or negatively subject to the perception and valuation of the markets where they operate.

India, with an expected growth of 3816% in Chemicals and 3409% in Pharmaceuticals by 2050; will certainly will have the potential to enhance the profitability and the perception of the companies that operate in its market. In contrast, Japanese

Table 5.6 Expected growth for the chemical industry by 2050 (adjusted BAU scenario).

	CHEMICAL	PHARMA	
ADV	117%	152%	ADV
REST	483%	516%	REST
BRIC	680%	1,595%	BRIC
	CHEMICAL	PHARMA	

5 The Chemical Industry by 2050

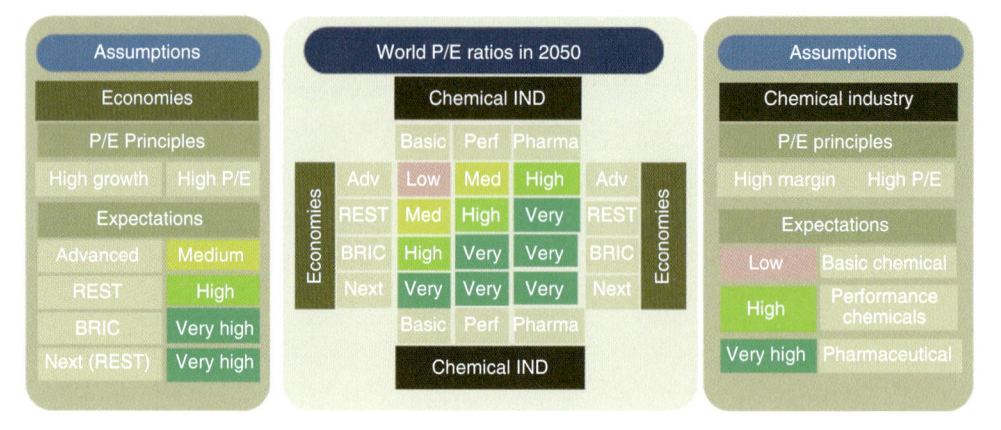

Figure 5.10 World profit/earnings (P/E) ratios in 2050.

companies relying on the growth potential of their local market might be disappointed to rely on an expected growth potential of 38% in chemicals and 61% in pharmaceuticals by 2050.

Certainly we could analyze the expected profitability of the industry in many different ways; but looking at the growth potential of the industry in the selected countries and areas might give us some good indications about the future profitability and growth potential of the companies operating in them.

During the next decades the industry is expected to have a massive increase in demand and profitability for all the segments related to the BRIC and the REST economies, and especially to those more related to the NEXT economies. On the other hand, the need for the industry to enable greenhouse emissions reductions and energy efficiency across other industries might be another enormous source of growth and profitability.

If the first aspect of growth potential is more linked to the quantitative aspect of the industry, the second is more related to its qualitative aspect. On the qualitative side, climate change and energy efficiency for the chemical industry, and higher life expectancy and higher income per capita for the pharmaceutical industry, will have the potential to elevate the technological profile of the industry to its highest levels, and with it the overall profitability of the industry.

The pharmaceutical industry will be highly and positively impacted by both the qualitative and quantitative aspects. The potential increase in growth and profitability for the industry seems remarkable. It seems to be in an extraordinary position to capitalize on the upcoming megatrends, and if the market transition from the ADV to the BRIC economies occurred as seemly and safely as for the chemical industry, the growth potential of the industry seems outstanding.

For the chemical industry, the expected growth potential of the BRIC economies, especially India and China, in combination with its capacity to enable other industries to reduce emissions and energy demand around the world might be the areas of highest growth. Some of these transitions, like that into the BRIC economies, are not completely new to the industry, after all China is already the largest chemical market in the world. However, the transition to a higher technological level in order to solve some of the most challenging issues of humanity might not be easy, after decades of focusing on maximization and optimization.

5.3
Feature 2: Inputs – Feedstocks

In this section we will review how the major feedstocks for the chemical industry, and more specifically for the petrochemical industry, could look by 2050. In Chapter 4 we divided this feature of the chemical industry into three major segments: naphtha and gas oil, gas-based feedstocks, including ethane, propane and butane, and renewable or non-fossil fuel based feedstocks (Figure 5.11).

Feedstocks for the chemical industry, and specifically for the petrochemical industry, is certainly one of the areas with the highest challenges and opportunities. In a world poised to triple its energy demand from 12 MtOE (million tons of oil equivalent) in 2010 to 35 MtOE by 2050 (under the BAU scenario) the petrochemical industry will be confronted with several dynamics simultaneously.

According to our asessment of the impact of the different megatrends on the chemical industry, please see Chapter 4 for further detail, feedstocks appeared fourth in the ranking of features of the industry most impacted.

Three major opposing dynamics will converge in the feedstock side of the petrochemical industry. First, the massive growth in chemical products will lead to an increase in demand for all sorts of feedstocks. Secondly, the recent discoveries of large reserves of shale gas around the world and the clear impact of this on the profitability and emissions of the petrochemical industry, will have the potential to change the feedstock mix of the industry. Finally, the need to address climate change and therefore reduce the world and industry greenhouse emissions might

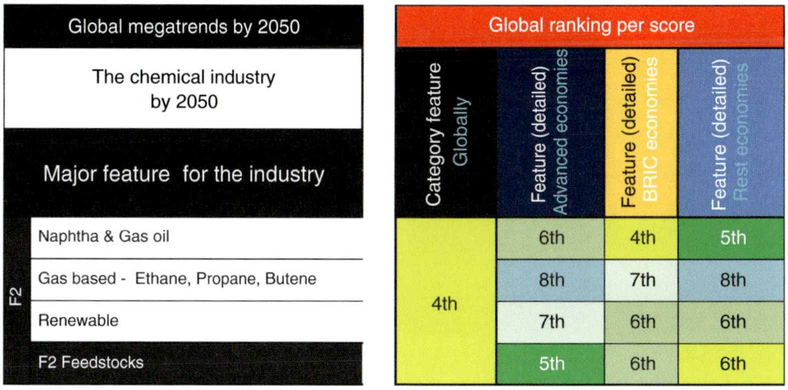

Global megatrends by 2050 – The chemical industry by 2050 – Major feature for the industry	Category feature Globally	Feature (detailed) Advanced economies	Feature (detailed) BRIC economies	Feature (detailed) Rest economies
Naphtha & Gas oil		6th	4th	5th
Gas based - Ethane, Propane, Butene	4th	8th	7th	8th
Renewable		7th	6th	6th
F2 Feedstocks		5th	6th	6th

Figure 5.11 Impact on major feedstocks.

lead to a reduction in the use of feedstocks in general, while accelerating changes in the feedstock mix of the industry toward the cleanest feedstock.

In an industry poised to quadruple during by 2050 the increase in feedstock and energy demand may be massive. According to the BAU scenario, the total feedstock demand for the petrochemical industry could increase Million metric tons (Mt) in 2010 to more than 600 million metric tons (Mt) by 2050. Naphtha and gas oil – on the back of the massive growth in China and India – could be the great benefactors with an expected growth of more than higher than 400% from 2010 levels. Ethane, propane, and butane could have growth below 400%.

At the same time, the massive discoveries of shale gas globally, and their capacity to boost and change the profitability the petrochemical industry, might serve to counterbalance the expected massive increase in naphtha demand as the major petrochemical feedstock.

Additionally, the need to reduce greenhouse emissions globally might force the petrochemical industry not only to reduce the use of feedstock in general, but also to change its current mix from naphtha and gas oil into to gas-based feedstock within the steam cracker. According to a recent estimate from SRI consulting, now part of the IHS group, if we allocate all CO_2 emissions of the steam cracker to its primary product, in this case ethylene, then ethane could generate up to 30% savings in CO_2 emissions versus naphtha and gas oil when producing 1 ton of ethylene (Figure 5.12). "These include allocation of all products by mass, by value, and by high value products, in this case – ethylene, propylene, contained butadiene, contained benzene, and hydrogen." Although this is just a first approach and alternatives will need to be considered and industry consensus will need to be built, that could serve us to illustrate what sort of impact different feedstocks could have on the steam cracker emissions, and how different feedstocks could be impacted by the need to curb emissions.

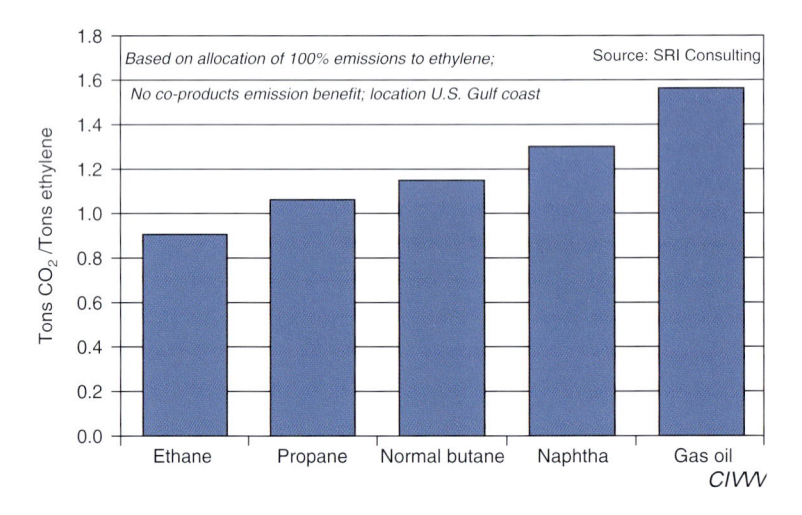

Figure 5.12 Steam cracker CO_2 emissions.

Under these premises, we will start by reviewing the basis of the current feedstock mix in the chemical industry, followed by several scenarios on the potential feedstock for the petrochemical industry by 2050. At this stage we will not elaborate in detail on the potential impact of these scenarios on greenhouse emissions of the industry, as we will review that in detail later when reviewing greenhouse emissions of the industry.

5.3.1
BAU Scenario for Feedstocks by 2050

Under the BAU scenario we have estimated that by 2050 the total amount of feedstock used by the petrochemical industry could grow by 407%, from 119 Mt in 2010 to 605 Mt by 2050. Naphtha and gas oil demand could reach 341 Mt by 2050. That implies that naphtha and gas oil alone would be almost three times more than the total feedstock demand in 2010 and almost six times more than the 2010 naphtha and gas oil demand. Ethane is also expected to increase significantly to 182 Mt, while butane, propane, and others will increase to 81 Mt (Figure 5.13).

Before proceeding to the details, we will describe the background and methodology behind these results. The BAU scenario for feedstock was based on the 2010 estimated feedstock mix of the petrochemical industry for the different economies, and the projected increased for petrochemicals demand.

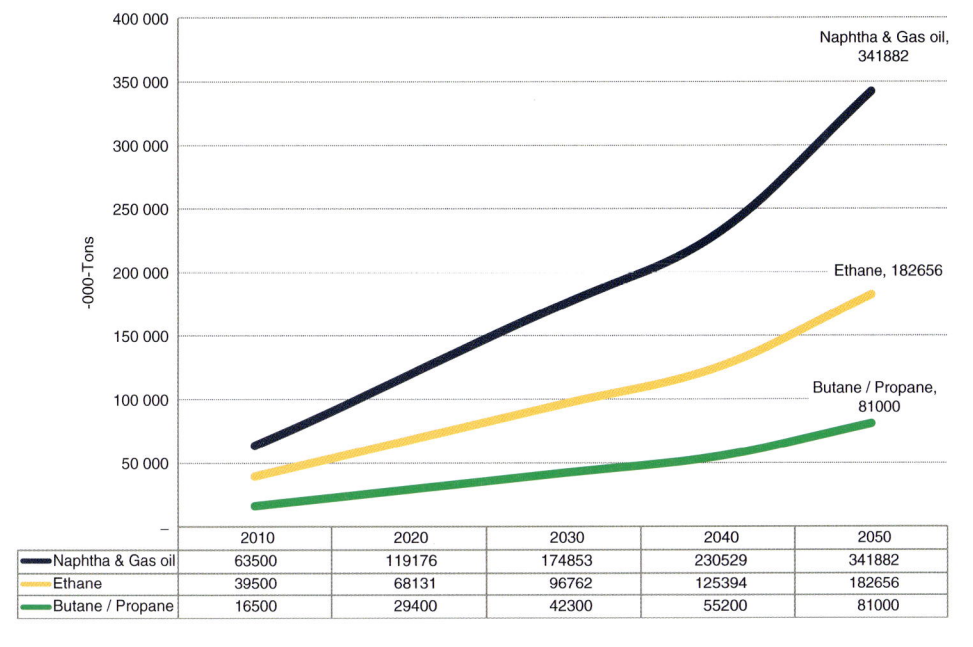

	2010	2020	2030	2040	2050
Naphtha & Gas oil	63500	119176	174853	230529	341882
Ethane	39500	68131	96762	125394	182656
Butane / Propane	16500	29400	42300	55200	81000

Figure 5.13 World feedstocks 2010–2050 based on BUA scenario.

5 The Chemical Industry by 2050

Feedstocks -000- Tons — 2050

Bau scenario	Feedstock total	Naphtha & Gas oil	Ethane	Propane & butane
USA	60795	9251	35684	15860
EU - 27	44790	35382	3477	5931
Japan	9263	8710	–	553
Canada	9175	738	7382	1055
Brazil	135278	115109	492	19677
Russia	81559	46776	20390	14393
India	16536	13701	2126	709
China	12034	9026	1755	1254
Advanced	124023	54082	46543	23398
BRIC	245406	184612	24763	36032
REST	236109	103188	111350	21570
World	605538	341882	182656	81000

BAU - Petrochemical feedstocks - 2050

Region	Naphtha & oil	Ethane	Propane / Butane / REST
World	56%	30%	13%
REST	44%	47%	9%
BRIC	75%	10%	15%
Advanced	44%	38%	19%

BAU - Petrochemical feedstocks - 2050

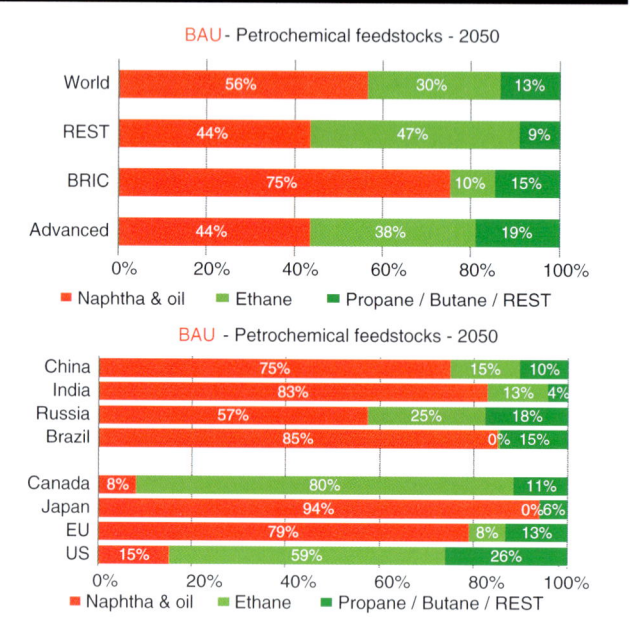

Country	Naphtha & oil	Ethane	Propane / Butane / REST
China	75%	15%	10%
India	83%	13%	4%
Russia	57%	25%	18%
Brazil	85%	0%	15%
Canada	8%	80%	11%
Japan	94%	0%	6%
EU	79%	8%	13%
US	15%	59%	26%

Figure 5.13 (continued)

For instance, in 2010 the feedstock mix for the USA petrochemical industry was 15% naphtha and gas oil, 59% ethane and 29% propane, butane, and others. So, for the BAU scenario we used this mix as our basis for the 2050 US feedstock projections, repeating this process for each of the countries.

This is a very simple approach but as previously stated the purpose of these scenarios is to provide some high level idea of where the industry is heading rather than searching for very accurate numbers.

Under this methodology it is not strange to see this massive increase in naphtha and gas oil demand globally; especially when considering that most of the growth in petrochemicals will occur in the BRIC economies, and these are heavily reliant on naphtha and gas oil.

5.3 Feature 2: Inputs – Feedstocks

Indeed, according to this scenario the BRICs will have more than 75% of the petrochemical feedstock based on naphtha and gas oil, while the REST and the ADV conomies will remain at 44%.

Within the BRICs, Brazil with 85% and India with 83% will be leading in terms of naphtha consumption. Within the ADV economies, Japan with 94% and Europe with 79% have the highest ratio of naphtha and gas oil in their feedstock mix.

In contrast, Canada with just 8% and the USA, after the massive and recent discoveries of shale gas, with 15%, appear to be countries with the lowest percentage of naphtha and gas oil in their feedstock mix.

The REST group, heavily influenced by the large ethylene capacity of the Middle East, all based on ethane, also is an area with very low level of naphtha and gas oil demand.

Therefore on this BAU scenario the total feedstock demand will increase very dramatically, naphtha and gas oil being the largest benefactors. Indeed under this BAU scenario, the percentage of naphtha and gas oil usage over the total feedstock globally will increase from 53% in 2010 to 56% by 2050. Geographically, the BRIC and the REST with 40% of the global feedstock demand each appear to be the areas will highest growth during the next decades. In the ADV economies as they lose their dominance in the production of petrochemicals, their feedstock demand will decline. In 2010, 47% of the feedstock world demand was in the ADV economies, but by 2050 under the BAU scenario it will be only 20%.

However, and as previously stated, the BAU scenario does not seem be very realistic for two fundamental reasons: the large discoveries of shale gas and the growing pressure to address climate change by reducing emissions.

The large discoveries of shale gas around the world, especially in some of the BRIC economies, Brazil, India, and China, with some of the largest shale gas reserves in the world, will provide big incentives to the petrochemical industry in those countries to move from liquid to gas-based feedstock. Considering the large cost advantage ethane-based crackers have over naphtha-based oness and the fact that the largest demand for petrochemicals will be located in the BRIC economies, it is difficult to envision a situation where the BRIC will not foster their shale gas exploration and use for their petrochemical industry. After all the BRIC economies will have the highest demand and large shale gas reserves, so why not take advantage of them?

Secondly, the fact that in the steam cracker, ethane-based petrochemicals can have up to 30% lower CO_2 emissions than naphtha-based ones, and in a global environment where greenhouse emissions will be carefully monitored, we could expect much larger use of ethane, butane, and propane versus naphtha or gas oil. In a world with constant reductions in CO_2 emissions, and where companies including petrochemicals could trade with CO_2 certificates, any reduction in CO_2 emissions could have a large market value.

Under these assumptions we expect shale gas to grow fast globally during the next decades, fostering the use of gas-based feedstock in the petrochemical industry. For that purpose, and illustration purposes, two additional simulations will be presented. These two simulations will give an overview of the possible amount of feedstock demand. We called these two scenarios just simulations as they are a very broad approximation and very inaccurate from a chemical point of view, although they serve our purpose. From a chemical point of view different feedstocks will trigger different ethylene and monomer yields, however, for simplicity we have assumed constant yields. This is not only an oversimplification but is chemically incorrect, however, as we are just interested in the potential ethane demand for the petrochemical industry if more shale gas is available globally, this simulation will serve our purpose.

A first simulation, called "Shale Gas I or ethane +20%", where gas feedstock increases by 20% in every country globally, with the only exceptions being Canada and the USA, as already more than 80% of their feedstock is gas-based. For Canada we did not propose any further increase on the 2010 figure of 92%, while for the USA we increased by just 7% from 85% in 2010 to 92%.

The second simulation, called "Shale Gas II," will present a more radical and extreme situation. A scenario where the petrochemical industry will make full use of the potential of shale gas reserves globally and the industry will move to the highest level of gas-based feedstock globally, with 92% gas-based and just 8% naphtha and gas oil.

At this time and due to poorer economics, we will not contemplate the massive use of other alternative feedstock for the petrochemical industry. We will discuss them, when discussing greenhouse emissions in the chemical industry, but we will still consider them minor, especially when considering that gas-based feedstock might be extremely competitive, while having the potential to achieve significant reduction in greenhouse emissions.

5.3.2
Feedstock Simulation II by 2050: "Shale Gas I – Ethane + 20% Globally"

Under this second simulation, a 20% reduction in naphtha and gas oil in favor of ethane globally could change the global feedstock mix significantly, making ethane the largest feedstock in the petrochemical industry with almost 300 Mt (Figure 5.14).

Globally, the percentage of naphtha and gas oil in the feedstock mix will decline very little, from 53% to 51%.

The massive expected growth in petrochemicals in the BRIC economies during the next decades, especially in India and China will offset the 20% reduction in liquid feedstock globally.

As discussed, this simulation on its own might not seem very valuable at this time, but it give us a dimension on what is the potential impact of a large exploitation and use of shale gas globally.

Although a 20% increase in ethane versus naphtha and gas oil might not seem very aggressive, in reality this scenario will imply massive investments for the gas and petrochemical industry.

The gas industry will need to invest billions around the world in infrastructure, extraction capacity, and transportation of shale gas. The petrochemical industry, especially in the BRIC economies, Europe and Japan will also need to invest vast resources to convert their crackers from naphtha to ethane or flexi-crackers. One might wonder whether this huge change would be worthwhile.

From a profitability perspective, and as reviewed when analyzing the expected profitability of the industry, ethane-based crackers make perfect sense and might be almost compulsory to remain competitive globally. However, we introduced this scenario and the following one, not to reflect on the profitability of the petrochemical industry but to reflect later on the emissions of the industry and the potential demand of the petrochemical industry for gas-related feedstock.

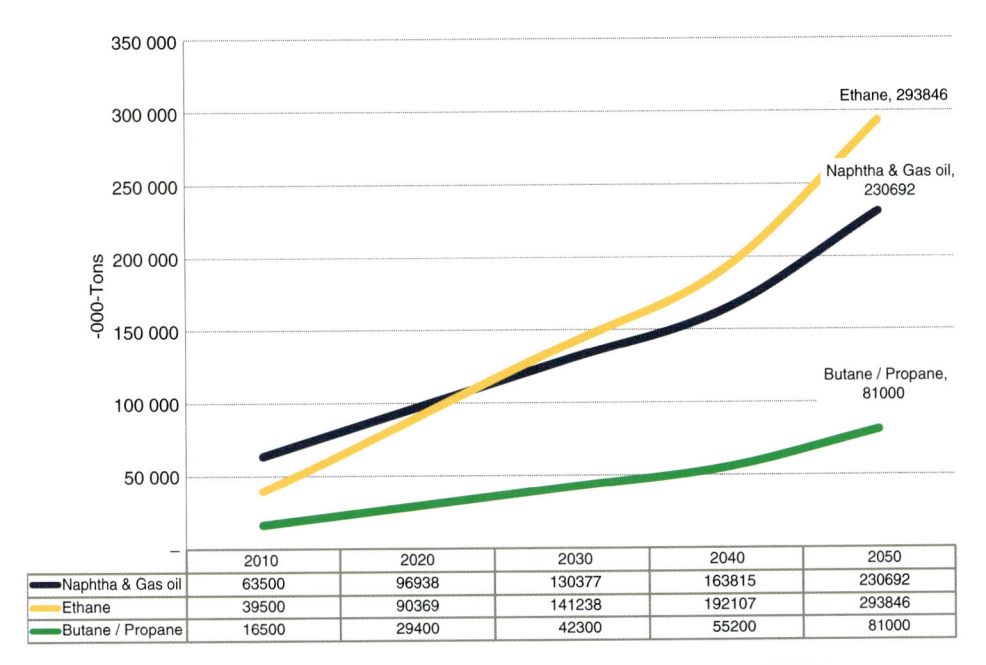

-000-Tons	2010	2020	2030	2040	2050
Naphtha & Gas oil	63500	96938	130377	163815	230692
Ethane	39500	90369	141238	192107	293846
Butane / Propane	16500	29400	42300	55200	81000

Feedstocks -000- tons 2050				
Shale gas (I) Ethane (+20%)	Feedstock total	Naphtha & Gas oil	Ethane	Propane / Butane / REST
USA	60795	4864	40072	15860
EU - 27	44790	26426	12433	5931
Japan	9263	6855	1855	553
Canada	9175	738	7382	1055
Brazil	135278	87931	27670	19677
Russia	81559	30177	36989	14393
India	16536	10417	5410	709
China	12034	6619	4162	1254
Advanced	124023	38883	61742	23398
BRIC	245406	135143	74231	36032
REST	236109	56666	157872	21570
World	605538	230692	293846	81000

Note 1: All countries, except USA and Canada, increased ethane demand by 20%
Note 2: USA & Canada, operates with just 8% naphtha

Figure 5.14 World feedstocks 2010–2050 (shale gas scenario I – 20% extra ethane).

252 | *5 The Chemical Industry by 2050*

Shale gas (I) - Petrochemical feedstocks 2050

Region	Naphtha & Oil	Ethane	Propane / Butane / REST
World	51%	36%	14%
REST	56%	35%	9%
BRIC	55%	30%	15%
Advanced	31%	48%	20%

Shale gas (I) - Petrochemical feedstocks - 2050

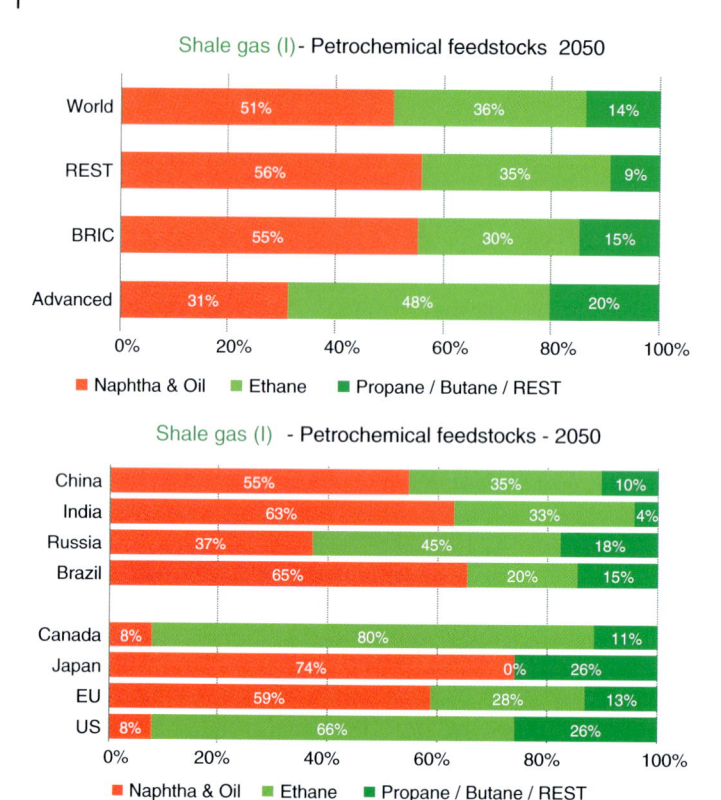

Country	Naphtha & Oil	Ethane	Propane / Butane / REST
China	55%	35%	10%
India	63%	33%	4%
Russia	37%	45%	18%
Brazil	65%	20%	15%
Canada	8%	80%	11%
Japan	74%	0%	26%
EU	59%	28%	13%
US	8%	66%	26%

Figure 5.14 *(continued)*

5.3.3
Simulation II by 2050: "Shale Gas II – Ethane at Maximum Capacity Globally" – (Unreal)

This is certainly the most radical simulation for feedstock, and indeed might be completely unattainable and unrealistic in this short period of time, especially with the expected massive growth for petrochemicals globally (Figure 5.15).

However, this scenario could be very valuable from two different perspectives. First, to have a first estimate and impression on how much ethane would be available by 2050 if the petrochemical industry decides to embrace completely shale gas and gas as feedstock, or in other words, how much naptha and gas oil will be liberated. Secondly, this scenario could serve us later to reflect on the potential benefits and cost for the industry to reduce CO_2 emissions, while reducing drastically naphtha and gas oil versus ethane. Additionally, although we will not review these critical factors, we could also elaborate on the potential impact of these massive changes in the different cracker co-products and their related industries.

So, at this time and for illustrative purposes, let us project this last scenario, while we will have some further discussions and reflection on these results when

5.3 Feature 2: Inputs – Feedstocks

	2010	2020	2030	2040	2050
Naphtha & Gas oil	63500	60489	57479	54468	48447
Ethane	39500	126382	213265	300147	473912
Butane / Propane	16500	29771	43042	56313	82856

Feedstocks -000- Tons   2050

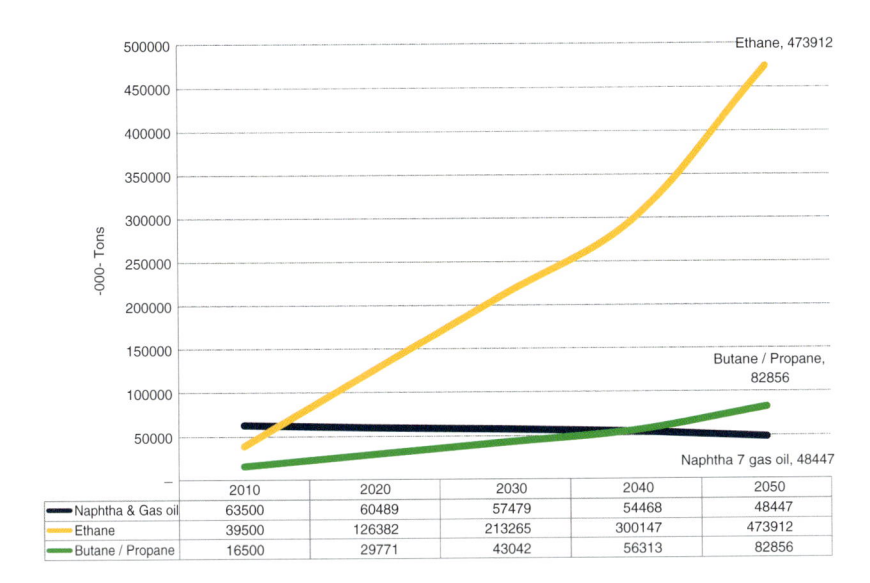

Shale has (II) Ethane at max	Feedstock total	Naphtha & Gas oil	Ethane	Propane / Butane / REST
USA	60848	4864	40125	15860
EU - 27	44898	3583	35384	5931
Japan	9263	741	6114	2408
Canada	9175	738	7382	1055
Brazil	134663	10822	104164	19677
Russia	81271	6525	60353	14393
India	16583	1323	14551	709
China	12084	963	9868	1254
Advanced	124185	9926	89005	25254
BRIC	244601	19632	188937	36032
REST	236429	18889	195970	21570
World	605215	48447	473912	82856

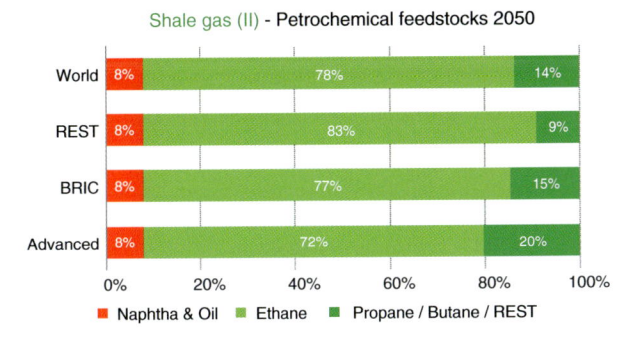

Shale gas (II) - Petrochemical feedstocks 2050

	Naphtha & Oil	Ethane	Propane / Butane / REST
World	8%	78%	14%
REST	8%	83%	9%
BRIC	8%	77%	15%
Advanced	8%	72%	20%

Figure 5.15 World feedstocks 2010–2050 (shale gas scenario II – maximum ethane).

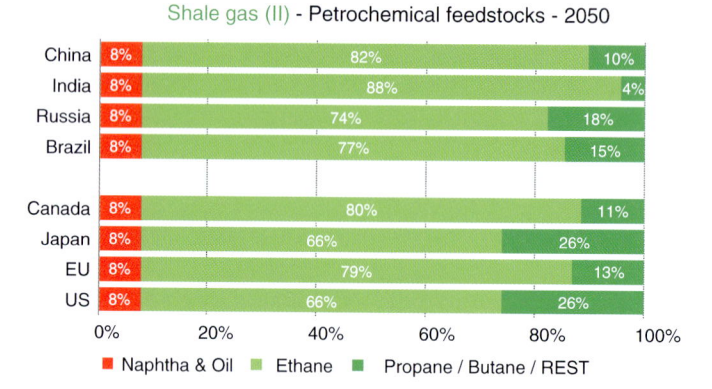

Figure 5.15 *(continued)*

analyzing the potential economics and scenarios for the greenhouse emissions of the industry.

Under this simulation, ethane will become the largest feedstock in the industry with almost 500 Mt, and propane and butane will become even bigger than naphtha and gas oil.

Indeed, despite the fact that the petrochemical industry could double by 2050, the total naphtha and gas oil demand could be just 48 Mt, less than in 2010. So large changes should be accomplished by the petrochemical industry.

This simulation although very unlikely for obvious reasons, might not be impossible from an availability perspective, or undesirable from a cost and CO_2 emissions perspective; so something to further refine and reflect further in updated new versions.

5.3.4
Conclusion and Feedstock Alternatives

The massive expected growth in chemicals will trigger an equally impressive demand for feedstock to levels never seen before. The rapid developments on shale gas in the USA and its global availability, in combination with its capacity to enhance cracker profitability, reduce emissions and the need to address climate change will accelerate the changes in the global energy and feedstock mix for the chemical and petrochemical industries.

An energy world with much more gas and less crude oil seems irrevocable. In the petrochemical world, more gas-based feedstock not only seems very plausible and logical but almost a must for petrochemical companies that want to remain competitive in the coming decades. The large benefits US crackers will enjoy in the next decades will provide large incentives around the world to accelerate the developments of shale gas and its use as a feedstock in the petrochemical industry.

According to the Chemical vision by 2020, there are several feedstock alternatives for the chemical industry but most of them remain either economically inefficient

Table 5.7 Alternative feedstocks.

RESOURCE	ACCESS	PRICE PREMIUM	CO$_2$ PRODUCED	TECHNOLOGY READINESS
COAL	HIGH easy to use and transport	LOW	MORE produces more CO$_2$ than oil but can be captured in the process	Ready and commercial
RESIDUAL PET COKE	HIGH TO MEDIUM localized at renfieries, can be shipped	LOW	MORE H$_2$ decient but can be captured in the process	Ready and commercial
OIL SHALE TAR SANDS	HIGH cannot ship without initial processing to petroleum	MIDDLE TO HIGH	MORE H$_2$ decient but can be captured in the process	Ready in short period.
BIOMASS	MEDIUM Less localized than coal, distributed, not uniform or concentrated	HIGH	NEUTRAL TO NEGATIVE (if captured)	Ready for small chemicals (fuel ethanol is subsidized)
ORGANIC WASTES (MSW)	LOW distributed not concentrated, plus health concerns?	HIGH	INSIGNIFICANT	Limited - requires more collection, community support, and political will than technology
STRANDED GAS	LOW non-transportable and difficult to access	LOW depends on the source	LOW less than coal or oil	Ready

Source: Based on the World Chemical Vision by 2020 – Chapter Alternative Feedstocks. http://www.chemicalvision2020.org/

or produce more CO_2 emissions or are not yet ready or commercially available. See Table 5.7 for a summary of the different major alternatives and their advantages and disadvantages.

Although we are confident a sort of "silver bullet" might not come during the coming decades, shale gas is certainly the closest to that, at least for the petrochemical industry, so we expect a gradual but constant introduction of it as a major energy and feedstock for the industry.

Additionally the transition to light cracking will certainly be accompanied by a transition to biofeedstock. This will take longer than that to shale gas, but its growth will be solid and continuous. Biomass or bioethanol from sugar, corn, or beet will become more and more popular. Ethylene from sugar is already available and although the economics might not be fully compelling at this time, these will need to be repositioned based on the upcoming need to reduce CO_2 emissions drastically. According to the latest information from BRASKEM, the Brazilian chemical giant, ethylene from sugar can capture or fix up to 2.5 tons of CO_2 per ton of Ethylene. Considering the global need for the chemical industry to reduce its own emissions, and under the assumption that the full cost for CO_2 emissions will soon be part of the price and cost for chemical production, biofeedstock will have a bright future in the next decades.

5.4
Feature 3: Outputs – Products

In Chapter 4 we observed that the outputs of the broad chemical industry, and more specifically its different products, will be subject to significant changes during the next decades. According to our analysis the pharmaceutical industry will be the area with the highest growth and highest impact by the upcoming megatrends, although all products will be subject to tremendous growth. For this feature we will present the outlook for the broad chemical industry by 2050, including a high level view of the outlook for its two major segments, chemicals and pharmaceuticals. Within the chemical industry we will also present a high level outlook for ethylene demand by 2050, with some mention of different growth potentials by region (Figure 5.16).

For that purpose we will use the adjusted BAU scenario as it is difficult to find enough data available on the potential impact of the sustainable scenario. *According to our adjusted BAU scenario for the chemical industry, the broad chemical industry, including pharmaceuticals, is expected to grow from US$ 3.9 trillion in 2010 to US$ 18.7 trillion by 2050.* The chemicals part is expected to grow from US$ 3121 billion in 2010 to US$ 14 923 billion by 2050, while the pharmaceutical industry is expected to grow from the US$ 875 billion in 2010 to US 3907 billion by 2050 (Table 5.1).

On a per capita basis the demand for broad chemicals is expected to grow massively from US$ 584 in 2010 to US$ 2054 in 2050. Chemicals per capita demand is expected to increase from US$ 456 in 2010 to US$ 1631 in 2050, while that for pharmaceuticals is expected to grow from US$ 128 in 2010 to US$ 427 in 2050.

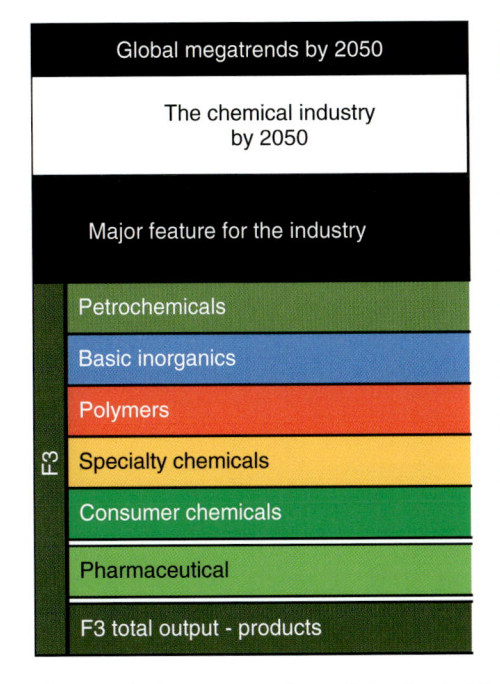

5.4 Feature 3: Outputs – Products

Major feature for the industry	Globally	Advanced economies	BRIC economies	REST economies
Petrochemicals	5th	8th	8th	7th
Basic inorganics	4th	6th	6th	6th
Polymers	5th	8th	7th	7th
Specialty chemicals	4th	7th	6th	6th
Consumer chemicals	3rd	6th	6th	6th
Pharmaceutical	1st	2nd	1st	1st
F3 total output - products	4th	6th	5th	6th

Figure 5.16 Impact on products of the chemical industry.

From a product point of view, and taking as a basis the 2010 industry ratios and prices, we have estimated the following sales for each of the industry segments by 2050 (Figure 5.17). This is certainly an oversimplification, based on a very simple set of assumptions. As previously stated, our major intent is not to be 100% accurate, which would be impossible, but to start providing a high level overview and understanding of where our industry could go.

For those readers interested in more precise and detailed scenarios much more complex analysis will be required. At this time we will just present some high level views to show the potential of the industry and its different major segments and products under this basic adjusted BAU scenario and the mentioned assumptions.

Under this linear and simple model, we can observe how large each of the different segments and products could become. Pharmaceuticals will remain as the largest with US$ 3907 billion, followed by specialty chemicals with US$ 3804 billion, petrochemicals with US$ 3607 billion, polymers with US$ 3559 billion, basic inorganic with US$ 2304 billion, and consumer products with US$ 1901 billion (Figure 5.18).

Within the chemical portfolio, plastics will remain the largest segment with US$ 3014 billion sales. In second place at around $2000 billion we could find auxiliaries for the industry, consumer products, and pyrolysis gasoline (C5–C8). In a smaller category we can find all the other products, ranging from paints and inks with expected sales of US$ 1186 billion to C4-butadiene with $188 billion (Figure 5.19).

258 | 5 The Chemical Industry by 2050

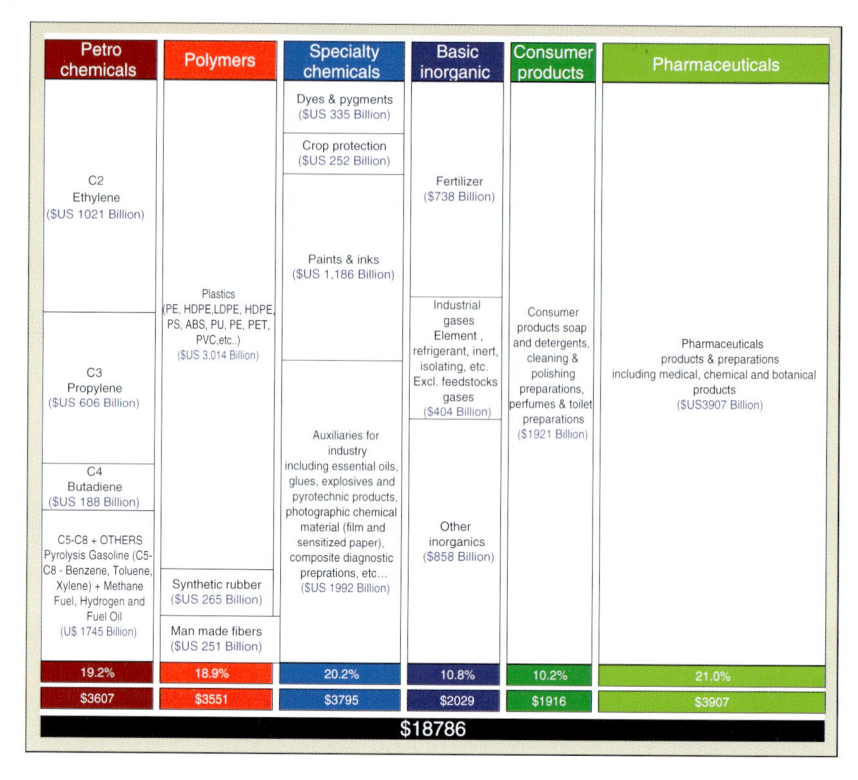

Petro chemicals	Polymers	Specialty chemicals	Basic inorganic	Consumer products	Pharmaceuticals
C2 Ethylene ($US 1021 Billion)	Plastics (PE, HDPE, LDPE, HDPE, PS, ABS, PU, PE, PET, PVC, etc..) ($US 3.014 Billion)	Dyes & pygments ($US 335 Billion)	Fertilizer ($738 Billion)	Consumer products soap and detergents, cleaning & polishing preparations, perfumes & toilet preparations ($1921 Billion)	Pharmaceuticals products & preparations including medical, chemical and botanical products ($US3907 Billion)
C3 Propylene ($US 606 Billion)		Crop protection ($US 252 Billion)	Industrial gases Element, refrigerant, inert, isolating, etc. Excl. feedstocks gases ($404 Billion)		
		Paints & inks ($US 1,186 Billion)			
C4 Butadiene ($US 188 Billion)	Synthetic rubber ($US 265 Billion)	Auxiliaries for industry including essential oils, glues, explosives and pyrotechnic products, photographic chemical material (film and sensitized paper), composite diagnostic preprations, etc... ($US 1992 Billion)	Other inorganics ($858 Billion)		
C5-C8 + OTHERS Pyrolysis Gasoline (C5-C8 - Benzene, Toluene, Xylene) + Methane Fuel, Hydrogen and Fuel Oil (U$ 1745 Billion)	Man made fibers ($US 251 Billion)				
19.2%	18.9%	20.2%	10.8%	10.2%	21.0%
$3607	$3551	$3795	$2029	$1916	$3907

$18786

Figure 5.17 Sales (billion dollars) for each of the industry segments by 2050 (adjusted BAU scenario).

Considering the simplicity of this basic model and the complexity of the industry and upcoming dynamics, we do not see much more value now in drilling further down in each of the different segments and regions. Indeed much more complex and detailed analysis will be required, so at this time we will just provide this high level view of the industry from a sales perspective.

However, and for the sake of illustration on the potential of the industry and later debate, we would like to introduce a deeper overview for perhaps one the most critical segments of the chemical industry, the ethylene market.

5.4.1
Global Ethylene Market by 2050 – BAU Scenario

For this BAU projection we have used the estimated per capita ethylene demand ratio on GDP per capita in 2010, and applied that ratio constantly to the PWC's GDP projections for 2050 for each of the selected countries, areas and globally. As with the other BAU scenarios this is a large oversimplification and a very simple and straightforward calculation, but this projection is just intended to provide us with a first flavor or indication of where the market could go, while further refinements should be introduced later.

5.4 Feature 3: Outputs – Products

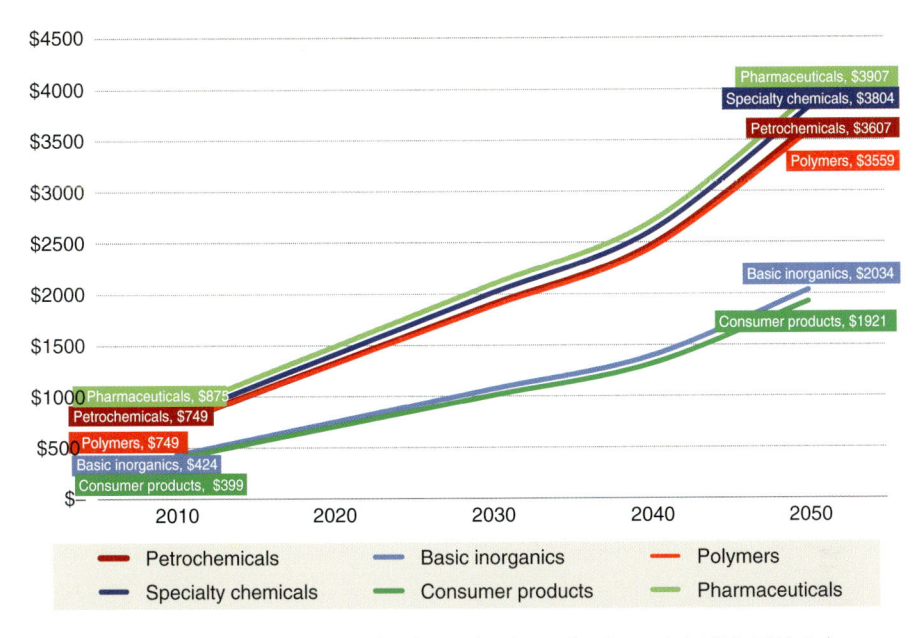

Figure 5.18 Sales (billion dollars) in the chemical industry for the period 2010–2050 (adjusted BAU scenario).

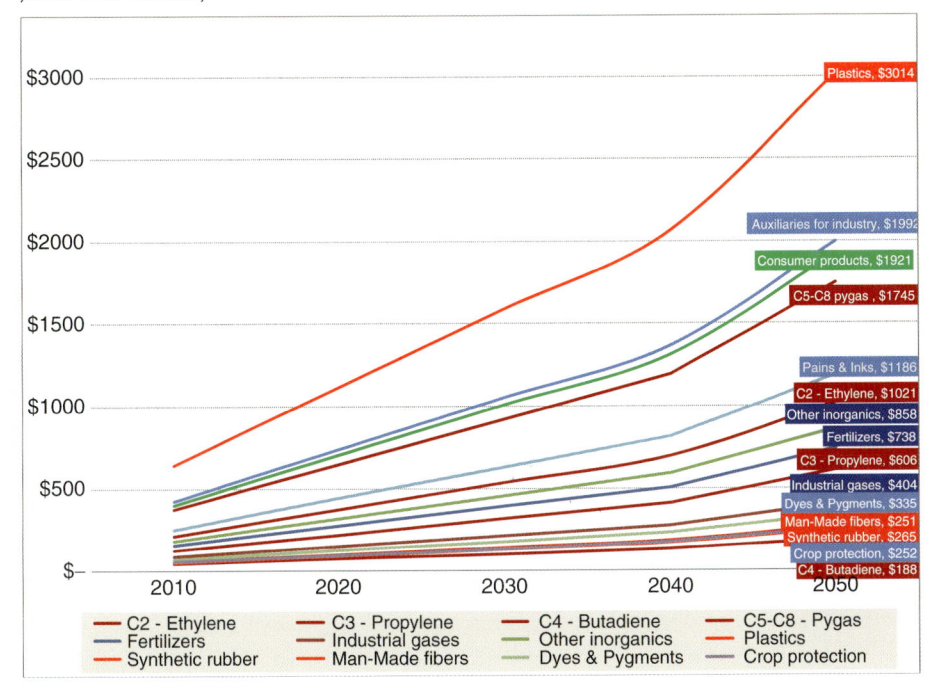

Figure 5.19 Sales (billion dollars) of products in the chemical industry for the period 2010–2050 (adjusted BAU scenario).

5 The Chemical Industry by 2050

According to this BAU Scenario, the world ethylene market could quintuple by 2050, from the estimated 120 Mt of in 2010 to 600 Mt by 2050. The massive economic and population growth expected in the BRIC and REST economies, in combination with the rise in living standards and GDP per capita in the ADV economies could support such an enormous growth in demand.

In 2010 the ADV economies had the largest ethylene demand in the world, 55 Mt annually, almost double that in the BRIC economies, 22 Mt. The REST's ethylene demand remained in the middle with 41 Mt annually. On a per capita basis the differences across the three areas remained much greater with a citizen in the ADV group consuming more than four times the amount of a citizen in the REST group and more than seven times that of a citizen in the BRICs.

In 2010 the world per capita ethylene demand was 18 kg annually. The ADV economies had an annual per capita demand of 58 kg, the REST 14 kg and the BRICs 8 kg.

According to the projections for this BAU scenario, the world ethylene demand could grow by 400% by 2050, with the global distribution shifting from the ADV to the REST and BRIC economies.

Indeed, the REST and the BRIC economies are expected to have the largest annual ethylene demand in the world by 2050 with almost 240 Mt each, implying a staggering 971% growth for the BRIC, 483% for the REST and 122% for the ADV economies (Figure 5.20).

	2010	2020	2030	2040	2050
Advanced	55900	65486	80532	100041	124012
BRIC	22500	72790	113599	168458	241073
REST	41600	69924	97906	134849	242492

Figure 5.20 Ethylene demand 2010–2050 (BAU scenario).

In terms of annual per capita demand, and due to the linearity of our very simple model, the ADV economies will continue to enjoy the highest per capita demand by 2050, with "perhaps" an unrealistically high 118 kg.

The BRIC in a more reasonable projection could reach 75 kg per capita per year (similar to the USA in 2010) while the REST will remain at 50 kg per capita per year (Figure 5.21b)

Looking now at this projection at the country level, China and India are supposed to lead the global growth in ethylene demand, although please bear in mind that at this time we are not considering the countries within the REST group. This area is critically important for ethylene, not only because it is expected to have a huge growth in demand but also because it includes the Middle East as a large ethylene producer (Figure 5.21c).

In any case and according to the BAU Scenario, China is expected to have the largest ethylene demand in the world with 117 Mt by 2050. In other words, the Chinese ethylene demand in 2050 will equal the 2010 world demand.

That figure also entails an impressive 884% growth from 2010, implying that the Chinese ethylene demand will become higher than that in the USA or Europe over the next decade.

At the same *India is expected to become the second largest ethylene market in the world with 88 Mt of demand by 2050, a gigantic 2298% growth versus 2010.* According to this scenario India could surpass Europe and the USA as the second and third largest ethylene markets by 2030–2040.

US ethylene demand is also projected to grow on the back of higher economic growth and still growing and older populations. However, and logically, this growth is expected to be less than for the BRIC or REST economies, at least on a percentage basis, but still a very solid one. According to our projection the USA ethylene demand is expected to grow by 164% by 2050, a total of 60 Mt. Europe will follow a similar pattern with a growth in demand versus 2010 of 104% and total estimated demand by 2050 of 45 Mt.

Russia and Brazil are also expected to become very large markets with 16 and 12 Mt by 2050. In contrast, Japan is expected to have a very modest growth over the next decades on the back of low economic growth and shrinking population.

Looking at these countries on a per capita basis (Figure 5.21a), we can observe how the low levels of ethylene demand in the BRIC economies, in combination with the projected economic growth for that region, could support the upcoming growth in ethylene demand. India or China with their respective 3 and 10 kg of per capita demand in 2010, are in huge contrast with the 44 kg in Europe, 74 kg in USA, or even with the extraordinarily high 129 kg in Canada. Considering the extremely low levels of per capita demand the BRIC economies had in 2010 could somehow make more palatable a projection that seems simple enormous. Looking now at the ADV economies, the high levels of ethylene demand already existing in some of its members, like the USA or Canada, make us wonder about the validity of the results from this BAU ethylene scenario. In our opinion further refinements with more realistic assumptions should be incorporated into this projection, however, at this time it serves our purpose to highlight direction and trends.

262 | *5 The Chemical Industry by 2050*

(a)

	2010	2020	2030	2040	2050
US	74	88	105	127	153
Europa	44	49	59	72	89
Japan	51	49	60	71	85
Canada	129	128	148	178	216
Brazil	18	26	37	53	76
Russia	18	49	66	84	101
India	3	12	20	33	52
China	10	32	49	70	100
World Avg	18	27	36	46	66

(b)

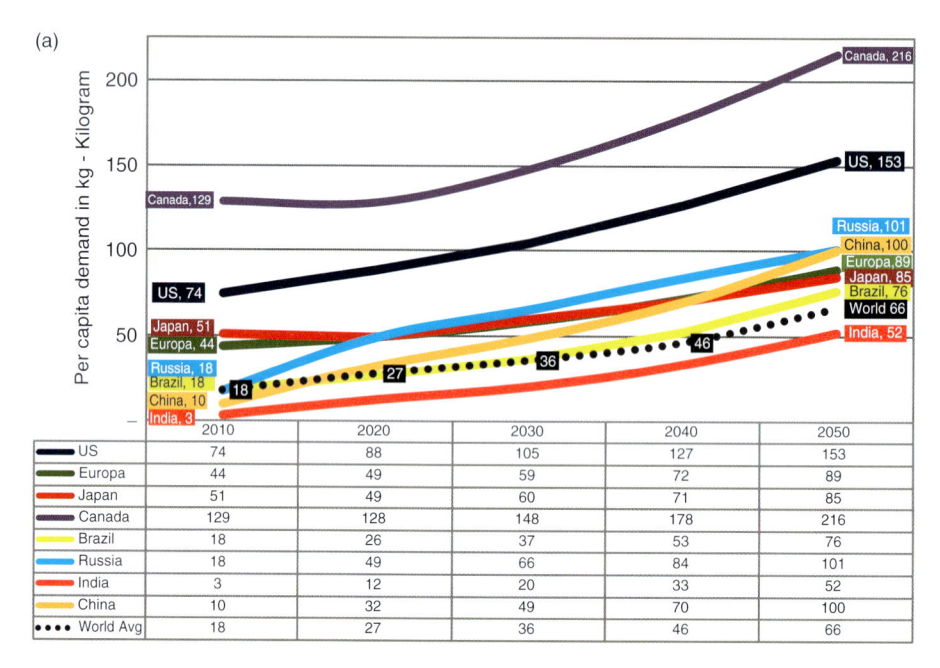

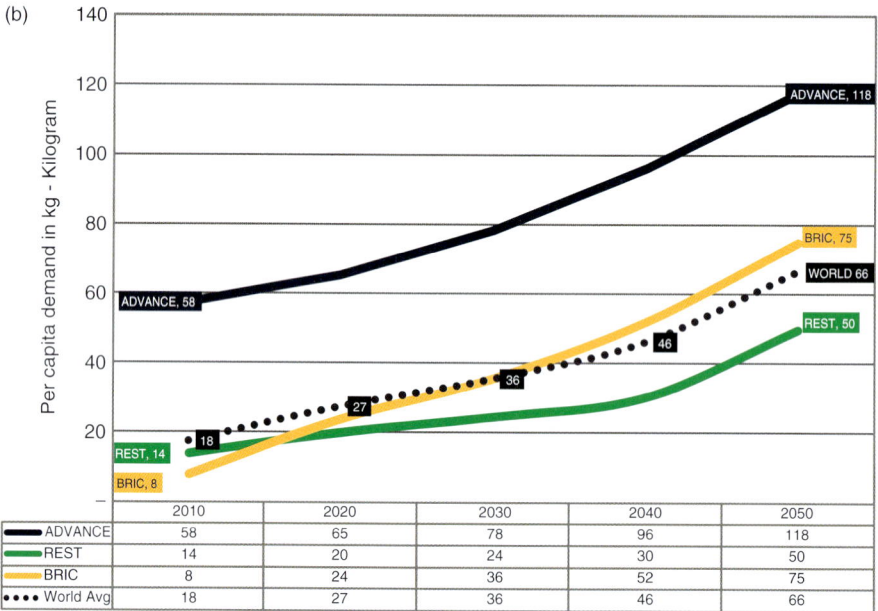

	2010	2020	2030	2040	2050
ADVANCE	58	65	78	96	118
REST	14	20	24	30	50
BRIC	8	24	36	52	75
World Avg	18	27	36	46	66

Figure 5.21 Ethylene demand 2010–2050 (BAU scenario). (a) Per capita demand by country, (b) per capita demand by area, (c) demand in thousand tons by country.

(c)

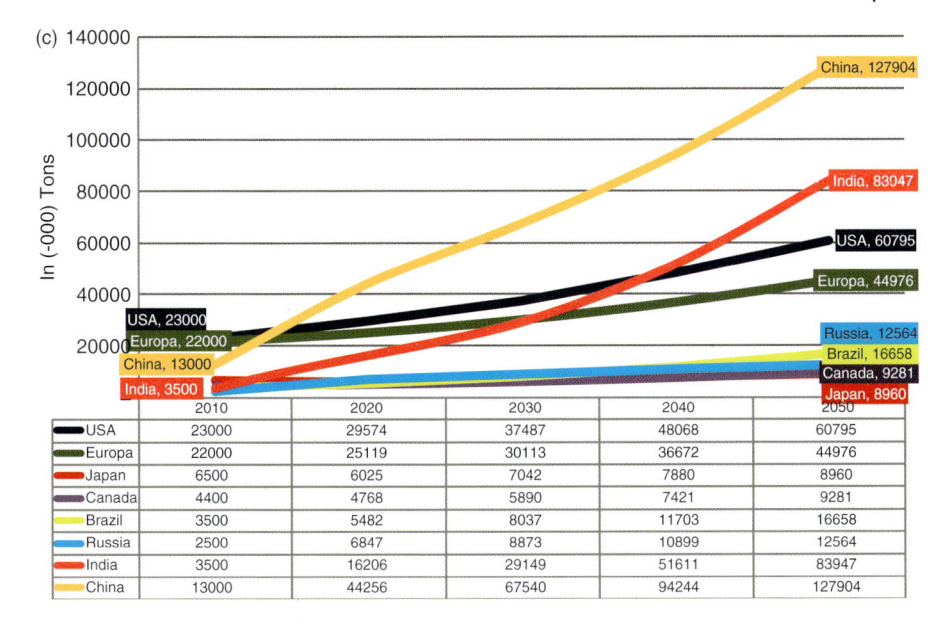

	2010	2020	2030	2040	2050
USA	23000	29574	37487	48068	60795
Europa	22000	25119	30113	36672	44976
Japan	6500	6025	7042	7880	8960
Canada	4400	4768	5890	7421	9281
Brazil	3500	5482	8037	11703	16658
Russia	2500	6847	8873	10899	12564
India	3500	16206	29149	51611	83047
China	13000	44256	67540	94244	127904

Figure 5.21 *(continued)*

To conclude this overview, we would like to provide an overview of what that could mean *from a steam cracker perspective, so we could also have an idea of what size investment the industry might need to satisfy this enormous demand.*

According to this BAU scenario, if the world needs to satisfy such a massive increase in ethylene demand then more than 500 steam crackers will be needed globally by 2050 – with an optimal average capacity of 800 000 tonnes of ethylene capacity. As previously discussed, this scenario is based on very simple and linear assumptions. For instance, this calculation not consider any cracker replacement due to longevity, when indeed many crackers in the ADV economies are more than forty years old; but in case we have a first indication based on growth, while further adjustments will be required.

According to this, China and India are expected to be the countries with room for further capacity expansion with more than 137 and 94 new steam crackers by 2050. The REST group, most probably led by the Middle East, is also expected to play a large part in capacity expansion during the next decades with more than 200 additional cracker sites (Figure 5.22).

Within the ADV economies, the USA with the expectation of growing populations, the highest GDP per capita of the group, and the highest per capita demand of ethylene equivalent could need 46 additional stream crackers by 2050.

Europe could also need another 12, while Japan might need to rationalize some of its crackers as at the current operating rates they might not be competitive. Considering the differences in size and operating rate of the current steam crackers around the world, we could expect further optimization and increasing operating rates before all these investments will be implemented. Countries and areas with

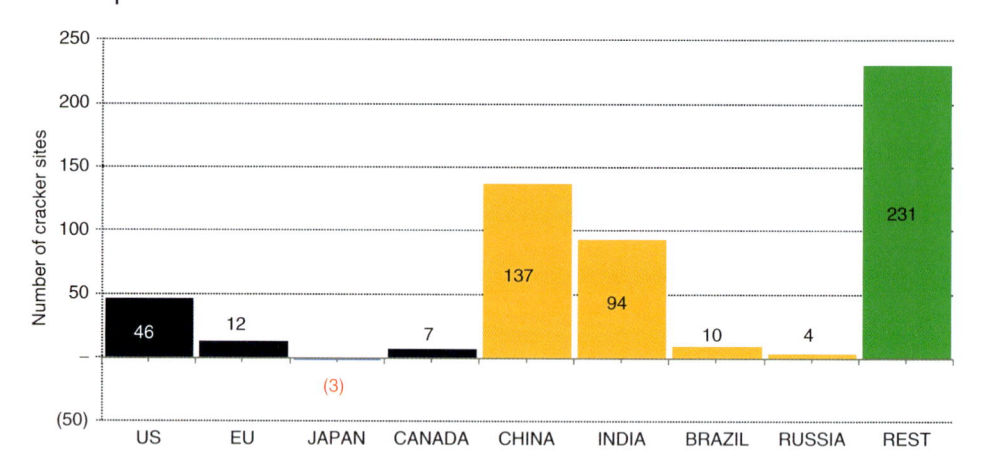

Figure 5.22 Location of additional steam crackers by 2050 (capacity 800 kt per annum).

small crackers and low operating rates will have large incentives to increase rates and gradually move into new and large flexible crackers, especially when considering the expected massive impact of shale gas on the ethylene market.

In any case, even under the more conservative and stringent scenarios not shown at this time, the world ethylene demand could certainly triple to 360 Mt by 2050, implying massive cracker expansion in all areas and almost all the considered countries.

5.5
Feature 4: Climate Change – Greenhouse Emissions – CO$_2$ Emissions

Climate change is probably the most pervasive of all the upcoming megatrends, not only to the chemical industry but to our World and society. The urgent necessity to address climate change at all levels – national, industrial, and social – and with all available tools, will create a significant challenge and opportunity for the chemical industry and our society (Figure 5.23).

This massive challenge will have the capacity to alter most of the key aspects of the chemical industry, from its feedstock to manufacturing, its products, its supply and chain, research & development (R&D), human resources (H&R), public affairs, margin, and economics; determining the whole industry success or failure.

In the review of this feature we will provide a comprehensive overview of the potential impact of climate change, and more specifically greenhouse, emissions on the chemical industry and vice versa. For that purpose several scenarios on CO$_2$ emissions for the world and the chemical industry by 2050 will be introduced. These will provide us with detailed insight on the carbon productivity required by the world, and our industry, in order to avoid climate change.

The unique capacity of the chemical industry to enable greenhouse emission reductions through its products and technologies will also serve to increase the

5.5 Feature 4: Climate Change – Greenhouse Emissions – CO₂ Emissions

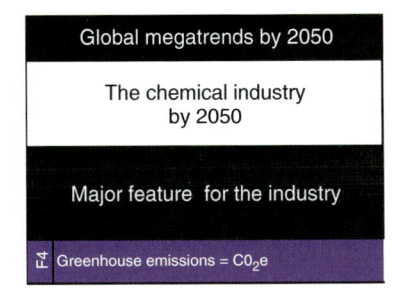

Global megatrends by 2050	Global ranking per score				
The chemical industry by 2050	Category feature Globally	Feature (detailed) Advanced economies	Feature (detailed) Bric economies	Feature (detailed) Rest economies
Major feature for the industry				
F4 Greenhouse emissions = CO₂e	1st	1st	1st	1st

Figure 5.23 Impact of climate change.

technological relevance of the industry. During this segment we will also introduce the concept of "abatement," while reviewing the latest figures on greenhouse emissions abatement allowed by the industry.

Finally, under the scenario for the industry of massive increases in chemicals demand globally, increasing greenhouse emissions, especially in the BRIC and REST economies, and the urgent need to address climate change; the industry will face tremendous pressure to reduce further its greenhouse emissions. In that context, programs like the European Emission Trading System (EU ETS) to enforce emissions reductions will become even more notorious and their experience might soon expand to other regions or become global soon. We will review the latest thoughts in this area while discussing the implication for the industry of a potential extension of these systems at a global level.

The chemical industry, as of the largest industries in the world, with a high energy demand and intensity, is one of the industries with the highest greenhouse emissions in the world. *Based on the 2007 IEA (International Energy Agency) results, the author estimates that the total CO_2 emissions of the chemical industry in 2010 could reach 4.1% of the total world CO_2 emissions, being equivalent to 1.4 Gt of CO_2.*

During the production of chemicals greenhouse emissions occur at different parts of the lifecycle, from the extraction of fossil fuels used for the industry as energy and feedstock, the production of chemicals, their transportation and others. According to the International Council of Chemical Associations – (ICCA) the total lifecycle emissions of the chemical industry could be divided as shown in Figure 5.24.

During the next decades and until 2050 the world amount of greenhouse emissions is expected to increase significantly to levels never seen before and completely unsustainable for our planet. Looking at the different scenarios for CO_2 emissions by 2050 (Figure 5.25), the world CO_2 emissions could go from 33 Gt in 2010, to an impressive 94 Gt by 2050 under the BAU scenario where regular emissions efficiencies are still taking place, up to an enormous and unrealistic 212 Gt if the world keeps releasing CO_2 in the same way as it did in 2010, and no emissions efficiencies take place.

However, and according to the IEA, if our world would like to avoid the negative aspects of climate change the world CO_2 emissions by 2050 will need to be reduced by more than half from the 2010 levels, from 33 Gt of CO_2 to just 14 GT.

266 | 5 The Chemical Industry by 2050

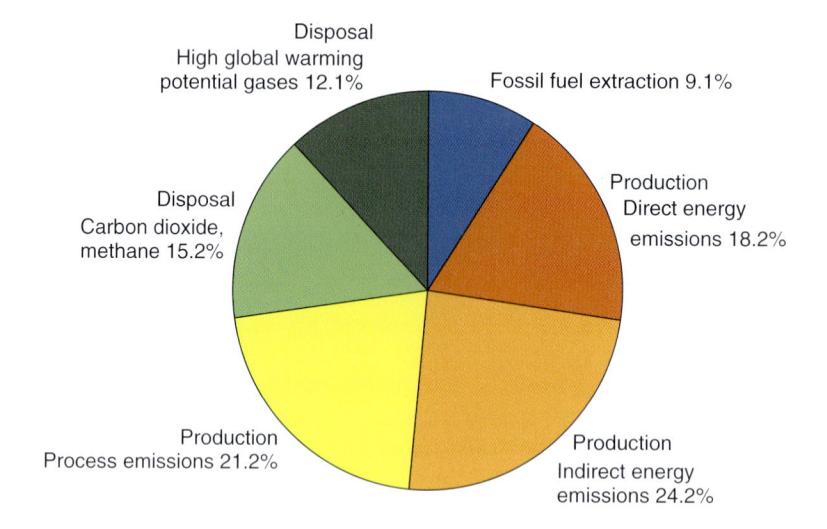

- Disposal High global warming potential gases 12.1%
- Fossil fuel extraction 9.1%
- Production Direct energy emissions 18.2%
- Production Indirect energy emissions 24.2%
- Production Process emissions 21.2%
- Disposal Carbon dioxide, methane 15.2%

Total greenhouse gas emissions of the chemical industry were 3,300 million tons of carbon dioxide equivalent in 2005. This figure does not include emissions savings enabled by products of the chemical industry.

Source: ICCA Report

Figure 5.24 Total lifecycle emissions of chemical industry products (2005). Source: ICCA – Extracted from the Publication "Turning the tide on climate change: The climate change challenge and the chemical industry."

The enormous economic growth expected for our world, in combination with the increase in population and the massive dependence on fossil fuel as a major source of energy, will be at the core of this formidable and sustainable increase in greenhouse emissions. Under a so-called "unrealistic" scenario, where no emissions efficiencies take place during the next decades and the world continues increasing with the current carbon productivity, the world CO_2 emissions could reach the staggering and completely unsustainable 212 Gt by 2050. The ADV economies could increase their CO_2 emissions from 12 to 28 Gt, the BRIC from 12 Gt to a spectacular 154 Gt, and the REST from 9 Gt in 2010 to 51 Gt. As previously stated this is a completely unrealistic scenario, only used for illustrative purposes.

Under the more realistic BAU scenario, where the different economies are able to reproduce at least the same emissions efficiencies observed during the last four decades, 1970–2010, the world CO_2 emissions will still triple from 33 to 94 Gt.

Under this BAU but still unsustainable scenario the ADV economies will be able to enable all the expected economic growth while reducing their total emissions from 12 to 11 Gt. This remarkable situation will be in clear contrast to that observed in the BRIC and REST economies.

If in 2010 the BRIC economies accounted for 37% of the total emissions under the BAU scenario the BRIC are expected to account for 55% of the total; increasing its total emissions fivefold from 12 to 52 Gt. The REST economies will face a similar

5.5 Feature 4: Climate Change – Greenhouse Emissions – CO_2 Emissions

(a)

	2010	No efficiencies – 2050	"BAU" 2050 incl. efficiencies	IEA sustainable 2050
WORLD	33	212	94	14
REST	9	51	32	7
BRIC	12	134	52	5
ADVANCED	12	28	11	2
CHEMICAL INDUSTRY	1.4	8.7	3.3	0.5

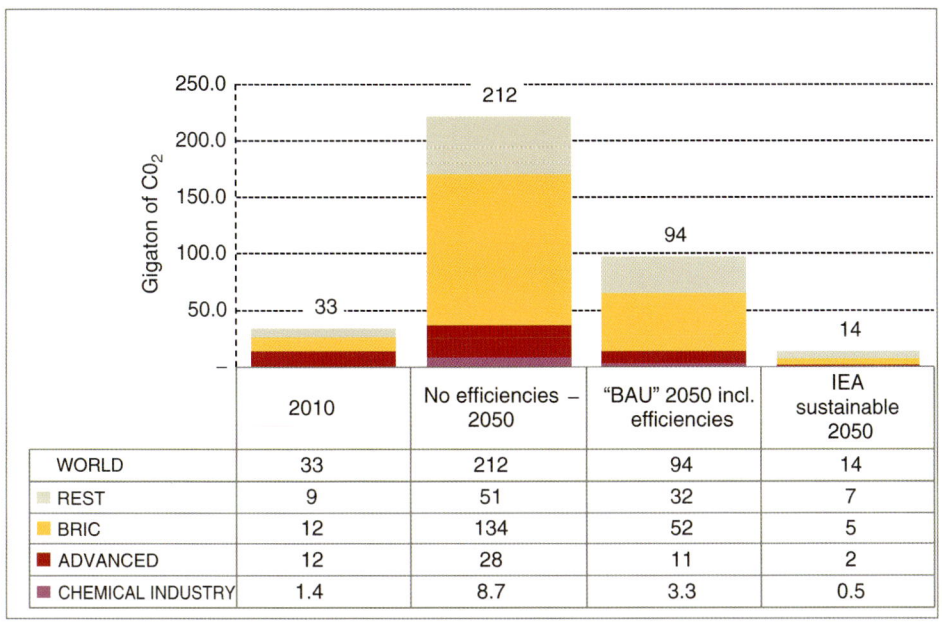

(b)

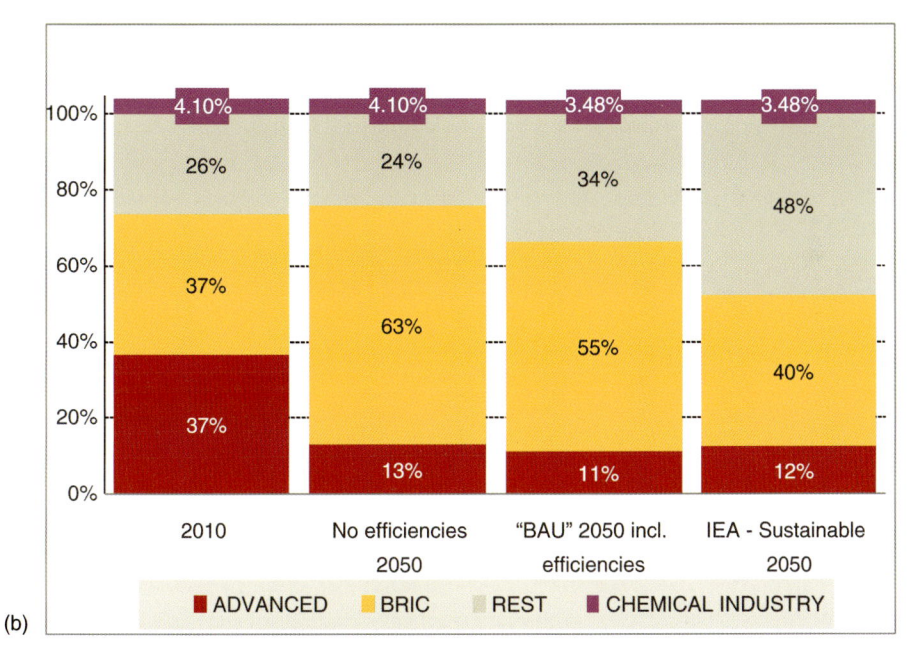

Figure 5.25 World CO_2 emissions by area: (a) in Gt, (b) on a percentage basis. Note: all these scenarios with all the assumptions, methodology, and data at the country and per capita level are available in the appendix.

scenario, tripling their total emissions from 9 to 32 Gt by 2050, and increasing their global share from 28 to 34%.

Finally under the sustainable scenario (IEA-based) the world CO_2 emissions should be reduced by half giving world annual CO_2 emissions of just 14 Gt. For this scenario we considered the 14 Gt of CO_2 emissions as our reference and distributed the emissions reduction by area in a simple mathematical, homogenous, and equitable manner.

On acknowledgment of the direct correlation between economic growth and CO_2 emissions, and the fact that still today different countries and areas remain at different stages in their economic development, if we wanted to maximize the global economic growth while reducing emissions, we assumed it fair and logical to start reducing further where the GDP per capita and the per capita emissions are higher, while allowing higher per capita emissions where economic growth still needs to occur. Ultimately, all countries should aim for the same level of emissions on GDP, below 2%, but in the short term we could not treat all countries equallly due to their different starting points.

Under these premises and to illustrate how challenging it would be to address climate change and CO_2 emissions reductions, we set different efficiency ratios for the different areas and countries. For the ADV economies CO_2 emissions will be limited to 2% of their total GDP by 2050. For the BRIC the average will be around 5%, with small differences across the group: Russia will be at 3%, China and Brazil at 4%, and India will be limited to 6%. For the REST group the average for the group remains at 8%.

According to this calculated sustainable scenario the ADV economies should reduce their CO_2 emissions from 12 to just 2 Gt by 2050, the BRIC economies from 12 to 5 Gt and the REST from 9 to 7 Gt.

In other words, the world economy under the sustainable scenario will need to increase its GDP by more than four times, while halving its total CO_2 emissions.

5.5.1
Historical and Future Scenarios on World CO_2 Emissions

To further illustrate the complexity of the challenge let us briefly review the recent performance and experience of our world during the most recent and similar time period (Figure 5.26). From 1970 to 2010 the world GDP experienced a similar growth to that expected for the next decades until 2050, quadrupling from US$ 15 to US$ 63 trillion, however, the world CO_2 emissions were not halved as we are expecting in our sustainable scenario but actually doubled from 15 to 33 Gt. *Despite clear improvements in carbon productivity during the last four decades – for each trillion of GDP the world generated 1 Gt of CO_2 in 1970 but only 0.5 Gt in 2010 – the reality is that the total emissions doubled in absolute quantities. With an expected world GDP of US$ 280 trillion (2009 dollar) by 2050, CO_2 emissions of just 14 Gt annually imply a carbon productivity almost 10 times greater than the current carbon productivity, certainly a challenge never seen before.*

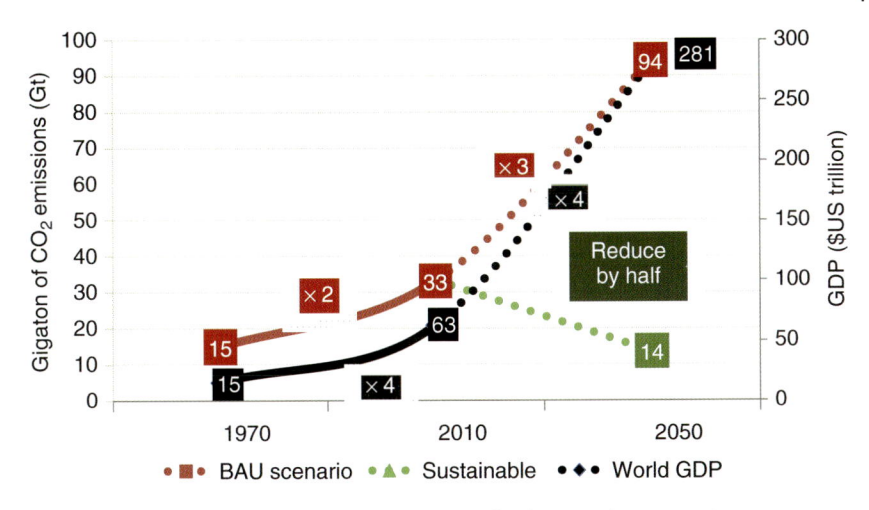

Figure 5.26 World CO_2 emissions and world GDP for the period 1970–2050.

For the chemical industry the challenge will also be enormous. On the one hand the industry will need to enable all the expected economic growth and expected improvements in living standards around the world, enabling CO_2 reductions in many other industries, investing billions in additional capacity and new technologies to support other industries' sustainable growth. On the other hand the industry will need to reduce significantly its CO_2 emissions in accordance with the sustainable scenario, while addressing major changes in demand, markets, energy, and feedstock.

From a CO_2 perspective and under these scenarios the chemical industry CO_2 emissions could increase from the estimated 1.4 Gt to 3.3 Gt under the BAU scenario and up to 8.7 Gt under the unrealistic scenario. However, and according to our calculations, if the world needs to reduce its CO_2 emissions according to the sustainable scenario, the chemical industry will need to reduce its total CO_2 emissions from 1.36 to just 0.47 Gt (Figure 5.27a).

In other words the chemical industry will need to be able to quadruple its size while reducing its CO_2 emissions by more than half.

On a per capita basis our world CO_2 emissions will need to go from the current 13 283 to just 4076 g per capita per day by 2050 (Figure 5.27b).

Living in a world of just 4000 g of CO_2 per capita per day implies severe CO_2 reductions in all aspect of our life, impacting the way we live and how all industries work. Considering that a regular car in Europe can generate around 130 g of CO_2 per km, driving just 30 km per day will be equal to the total daily allocation of a citizen by 2050.

Under this challenging scenario, the chemical industry will also need to accomplish severe reductions in CO_2 emissions. In 2010 it accounted for 545 g of CO_2 per capita per day. By 2050 it will need to reduce its emissions to just 142 g per capita per day or, in other words, by more than one third despite being expected to quadruple its size and revenue during this period.

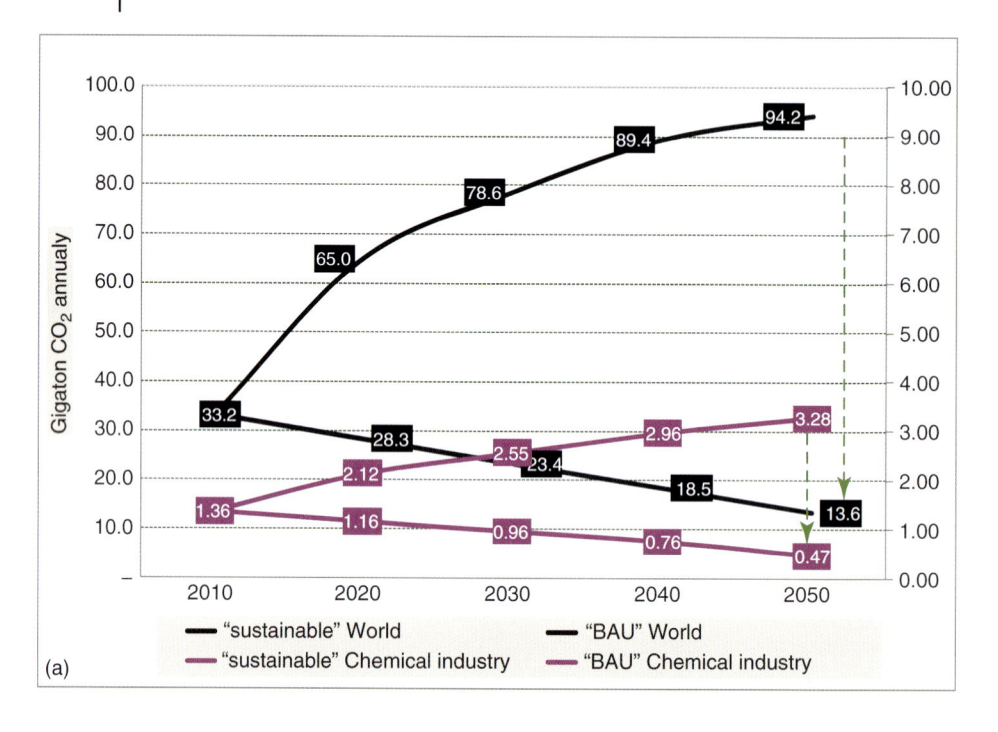

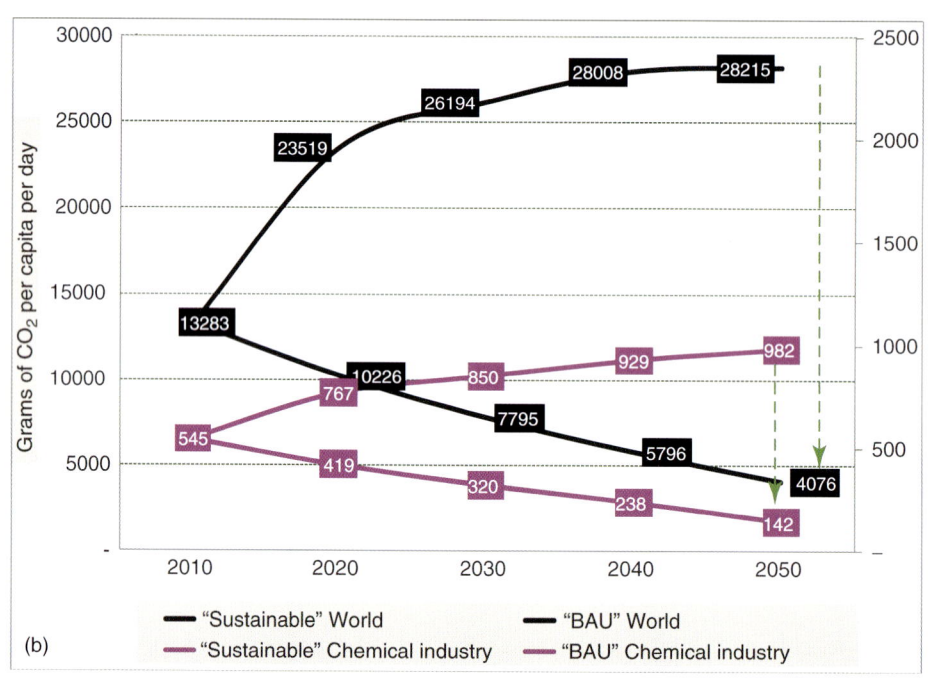

Figure 5.27 CO$_2$ emissions – scenarios for the chemical industry. (a) In Gt, (b) in g per capita per day. Source: Author elaboration based on IEA CO$_2$ Emissions projections and International Council of Chemical Association on the industry emissions.

This seems to us a remarkable task and a formidable challenge, combining massive growth and huge carbon productivity reductions in a limited period of time.

In order to further illustrate the challenge we could now try to compare it to the latest observed performance of the European chemical industry, as perhaps one of the areas with one of the most developed chemical industries in the world. During the last 20 years, from 1990 to 2010, the European chemical industry, including pharmaceuticals, was able to achieve an outstanding growth of 60% in chemical production while reducing greenhouse emissions by 50%. Please see Chapter 3 Section 2 – Feedstocks & Energy.

Assuming now that the whole chemical industry could replicate the same outstanding performance as the European chemical industry, in terms of growth and carbon productivity – and that statement might already be a very large assumption – the chemical industry might still face a major challenge. Indeed, assuming the global chemical industry were able to replicate the carbon efficiency of the European Union, while still accomplishing the expected growth by 2050, reaching projected sales of US$ 19.7 trillion, the CO_2 emissions of the industry would remain at 0.87 Gt, almost twice as high as requested under the sustainable scenario.

To reduce to just 0.47 Gt of CO_2 as requested by the sustainable scenario, even if the whole industry were able to replicate the observed and outstanding carbon productivity of the European chemical industry, seems to be a formidable and complex challenge. A challenge that will become more or less complicated depending on the areas and countries where the industry operates.

The sustainable scenario will impose carbon productivity requirements that will be tremendously challenging for all the selected areas and countries, however, among them we will still be able to observe some major differences. At this time and under the sustainable scenario we have not been able to calculate the CO_2 emissions and carbon productivity requirements for the chemical industry in each of the different areas and countries, however, by looking at the requirements of the selected countries we can start figuring out in which areas and countries the industry will have more pressure to reduce emissions. In any case, under the sustainable scenario the chemical industry as a whole will need to learn to work in an environment where it does not generate more than 0.47 Gt of CO_2 annually or 142 g per capita per day.

Looking at the carbon productivity requirements of each of the different areas (Figure 5.28); and calculating that by simply dividing the expected CO_2 emissions per capita per day under the BAU scenario versus the emissions required under the sustainable scenario; we can observe how the pressure on the BRIC economies might be the highest among all the different areas. The expected "explosion" in CO_2 emissions in the BRIC economies, partly due to the equally "explosive" economic growth, in combination with increasing populations and lower observed carbon productivity than the ADV economies could explain these higher multiples.

Under the sustainable scenario, the required world carbon productivity would need an increase of 6.9 times, implying an 86% reduction in emissions versus the BAU scenario by 2050. For the BRIC economies the expected carbon productivity need an increase of 9.6 times, implying a 90% reduction in emissions versus the BAU Scenario for 2050. For

5 The Chemical Industry by 2050

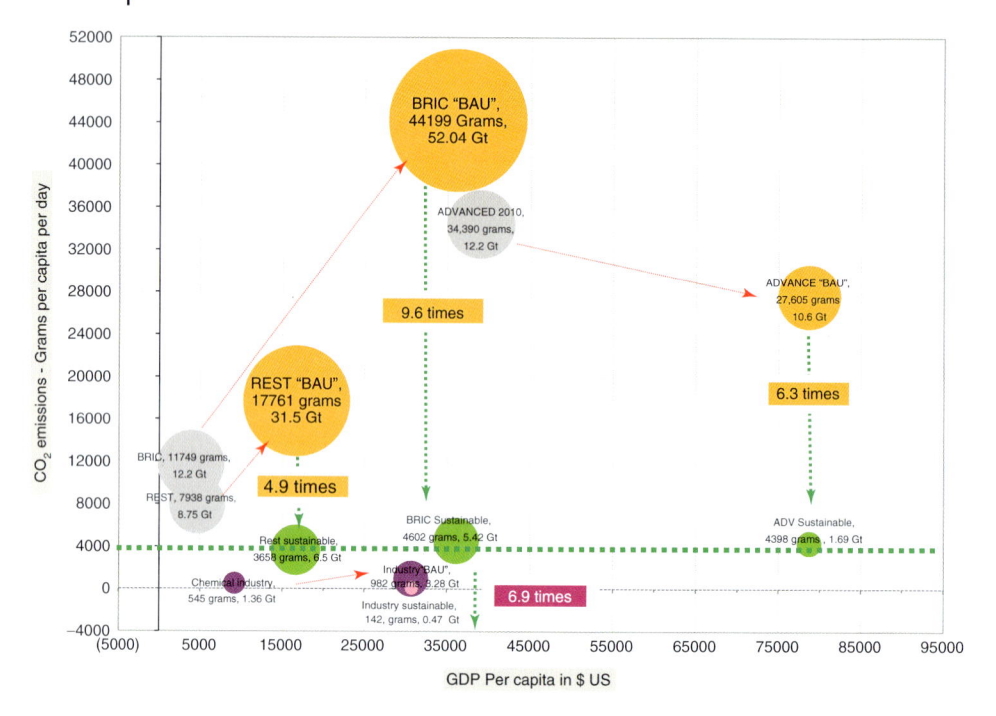

Figure 5.28 Daily CO_2 emissions per capita versus GDP per capita. Illustration of the increases in carbon productivity required by area to achieve the sustainable scenario of 4000 g per capita per day in 2050.

the ADV economies the requested carbon productivity increase will be slightly less than for the BRIC economies at 6.3 times, or 84% reduction versus the BAU scenario. Finally, for the REST economies that ratio will be the lowest of all the groups with an increase carbon productivity of just 4.9 times or 79% reduction. For the world chemical industry including pharmaceuticals, the requested carbon productivity will be the same as for the average of the world with a carbon productivity increase of 6.9 times or 86% reduction versus the projected emissions under the BAU scenario for 2050.

As we can see, even if we could expect differences in carbon productivity requirements across the different selected areas, the reality is that even for the areas with the lowest requirements, like the REST countries, a reduction in CO_2 emissions versus the "BAU" scenario of 79% seems pretty challenging. For illustration purposes let us briefly review the expected carbon productivity of the different countries selected for this analysis.

Among the different countries (Figure 5.29) we can separate them into two categories, those with requested carbon productivity "above" or "below" the requested world average carbon productivity by 2050.

In the first group, above the world average, we can find Russia at the top of the list. The expected large economic growth in combination with the high energy intensity of the country and its "limited" carbon efficiency will force Russia to accomplish one of the largest reductions in CO_2 emissions in the world. According

5.5 Feature 4: Climate Change – Greenhouse Emissions – CO_2 Emissions

(a)

(b)

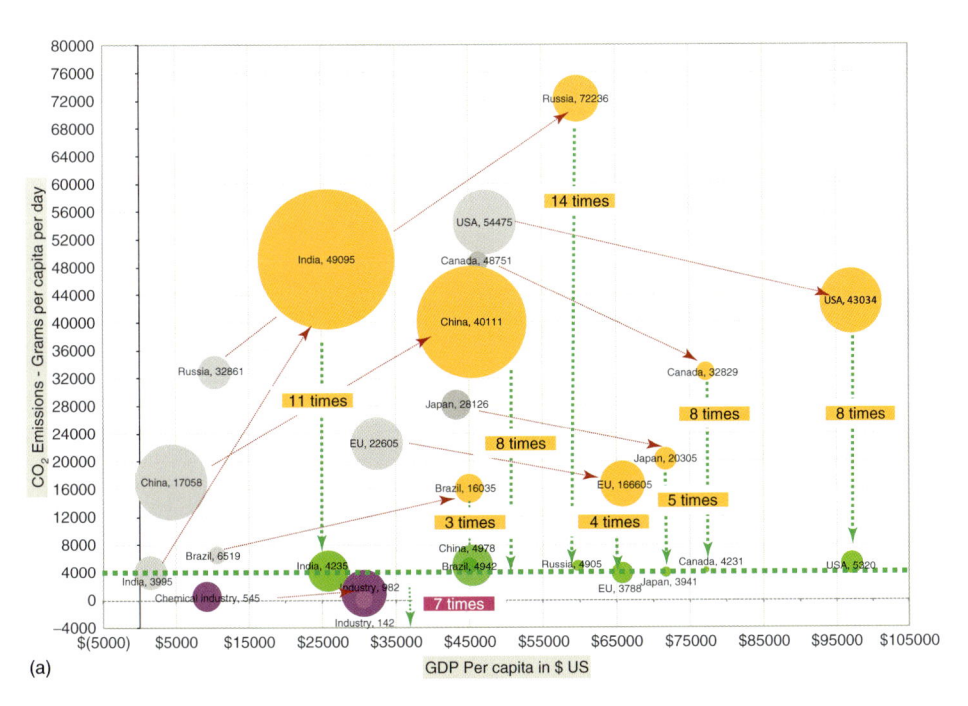

Global ranking		CO_2 emissions per capita per day in grams			
Carbon productivity 2050		"BAU"	"Sustainable"	% Reduction	Times
BRIC	RUSSIA	72236	4905	93	14.7
BRIC	INDIA	49095	4235	91	11.6
ADV	USA	43034	5320	88	8.1
BRIC	CHINA	40111	4978	88	8.1
ADV	CANADA	32899	4231	87	7.8
WORLD	TOTAL	28215	4076	86	6.9
WORLD	INDUSTRY	982	142	86	6.9
ADV	JAPAN	20305	3941	81	5.2
REST	REST	17761	3658	79	4.9
ADV	EUROPE	16605	3788	77	4.4
BRIC	BRAZIL	16035	4942	69	3.2

Figure 5.29 Daily CO_2 emissions per capita versus GDP per capita. Illustration of the increases in carbon productivity required by countries to achieve the sustainable scenario of 4000 g per capita per day in 2050. The results are also shown in tabular form.

to the BAU scenario Russia might have CO_2 emissions of 72 236 g of CO_2 per capita per day. However, under the world sustainable scenario, it will need to reduce its daily CO_2 emissions to just 4905 g of CO_2 per capita. Under these scenarios, Russia will need to accomplish a severe reduction of 93% in CO_2 emissions by 2050 versus the projected BAU scenario. This reduction will be equivalent to accomplishing a carbon productivity increase of more than 14 times, the highest for the selected economies.

In this global ranking of expected carbon productivity by 2050, India appears in second position. Despite the fact that India had one of the lowest per capita daily CO_2 emissions in the world, and by far the lowest among the selected countries, with just 3995 g in 2010, below the sustainable target of 4000 g by 2050; the fact that its economy is expected to grow by an unprecedented 2299% by 2050 and it has the lowest carbon productivity of these groups during the last four decades, could explain this paradoxical situation.

According to the BAU scenario India could increase its CO_2 emissions from 3995 to 49 095 g by 2050. However, according to the sustainable scenario, India will need to reduce its daily per capita CO_2 emissions to just 4235 g by 2050.

In other words India will need to be able to increase its GDP and economy by a staggering 2299% by 2050, its population by 37% from 1171 to 1610 million people, while increasing its daily per capita emissions from 3995 to just 4235 g. This seems a not unattainable target but certainly a formidable challenge. India presents probably the most fascinating challenge of all the selected countries; the potential of the country to raise the living standards of millions of people, elevating the average GDP per capita of the country from just US$ 1477 to 25 761 by 2050 should be enough incentive for the governments, industry, and society to accomplish such a gigantic challenge. However the potential lack of awareness, coordination and focus could undermine this great potential.

The USA, China, and Canada appear third, fourth, and fifth, respectively on this global ranking with a similar need to reduce CO_2 emissions by approximately 88% versus their respective BAU scenarios. Despite the fact that China appeared to have the highest carbon productivity during the last four decades almost double that of the USA and Canada (please see Appendix on world carbon productivity ratios 1970–2010), China had one third of the CO_2 emissions per capita per day of the USA and Canada in 2010, with just 17 058 versus the 54 475 g for the USA and 48 751 g for Canada; the fact that the Chinese economy is expected to grow by a staggering 884% by 2050, five times faster than the USA and Canada can explain why all three countries will need to accomplish similar carbon productivities.

In the lower part of the ranking Brazil and Europe appear to have the lowest need to reduce CO_2 emissions versus the BAU scenario with just 69 and 71%, respectively. These reductions, despite being in the lower ranking of this group, still imply severe improvements in carbon productivity for these economies.

Despite the fact that the Brazilian economy could more than triple by 2050, the fact that its CO_2 emissions per capita per day in 2010, 6519 g, were among the lowest in the BRICs, just above India, could explain why the expected reductions in the CO_2 emissions are expected to be the lowest among these countries.

Europe on the other hand, being one of the most efficient economies in terms of carbon productivity, could balance its expected growth during the next decades with regular reductions of CO_2. Indeed all the ADV economies under the BAU Scenario will manage to reduce their emissions despite the expected economic growth by 2050, due to their strong efficiency in reducing emissions and their "reduced" growth expectations.

Europe for instance, under the BAU scenario, is expected to reduce its per capita per day emissions from 22 065 g in 2010, to just 16 605 g by 2050, simply by replicating the same carbon productivity as in the last four decades. However, the clear need to reduce its total emissions under the sustainable scenario to a figure closer to or below 4000 g will trigger further reductions. Under the sustainable scenario, by 2050 Europe is expected to have 3788 g of emissions, the lowest of the ADV and BRIC countries. That target will imply a reduction of emissions of 71% versus the expected emissions under the BAU scenario, and 83% versus the 2010 CO_2 emissions.

Although at this time we cannot calculate the expected CO_2 emissions reductions for the chemical industry in each of the respective countries or areas, we can already foresee which countries or areas will be subject to the most stringent and severe reductions. Certainly all countries will need to accomplish severe emissions reductions although small differences can already be envisioned.

Under the sustainable scenario we expect unprecedented emission reductions for all areas, countries, and industries; reductions that will go beyond previous experience and will test the capacity of our world and the chemical industry to enable growth while reduce emissions.

In some areas like Europe the need to reduce greenhouse emissions in the chemical industry will not only be a nice aspiration but an obligation, an obligation severely monitored and subject to strong penalties. In January 2013 the EU European Trading Scheme (ETS) will be further expanded into the petrochemicals, ammonia and aluminum industries and to additional greenhouse gases beyond CO_2. Although it is very early and difficult to estimate the potential impact on the industry some major trends can be observed.

In 2005 the EU as the cornerstone of its policy to combat climate change introduced the EU ETS. The EU ETS was the first and largest mandatory "cap" and "trade" scheme for CO_2 emissions globally. The EU ETS is a multi-country, multi-period, and multi-industry program. It covers 30 countries, the EU 27 plus Norway, Iceland, and Liechtenstein, across multiple periods (2005–2008, 2008–2012, and the third period starting in January 2013), and multiple industries, like power stations, combustion plants, oil refineries, iron and steel works, and factories making cement, glass, lime brick, ceramics, pulp, paper, and board. Airlines joined in 2012 and, as of January 2013, petrochemicals, ammonia and aluminum industries, and other greenhouse gases will be added. Indeed nitrous oxide emissions from certain processes are also covered.

The EU ETS covers CO_2 emissions from more than 11 000 power plants and industrial plants in 30 countries, accounting in 2010 for almost 50% of the EU's CO_2 emissions and 40% of the total greenhouse gas emissions. The ultimate target of the EU ETS is to have by 2020 European emissions 21% lower than in 2005.

276 | *5 The Chemical Industry by 2050*

The EU ETS works on the principle of "cap" and "trade". There is a cap or limit annually on the total amount of certain greenhouse gases that can be emitted by the plants, and factories of the system. Within this cap companies receive certain emission unit allowances (EUA) which they can sell or buy from one another as needed. At the end of the year each company must surrender enough EUA to cover all its emissions, otherwise large fines will be imposed. If companies manage to reduce their emissions below their cap they can sell EUA, and vice versa. The ultimate goal of the scheme is to gradually reduce the total emissions in Europe by reducing the number of EUA available.

Looking at the recent prices for the EUA to emit 1 tonne of CO_2, this has been fluctuating between €10 and €35 in the period from 2007 to 2011 (Figure 5.30). As the number of EUA will be reduced in the future that price is estimated to go from €10 in 2011 to €50 by the end of the scheme (2020).

Although it is very difficult and early to calculate the potential impact of this program for the European petrochemical industry, without knowing at this time the number of certificates available, we could start using some of these figures to visualize later the potential impact for the chemical industry.

In this sense two additional factors should be considered: (i) the emission trading schemes are expanding very fast into other countries and regions, with a clear target to become global. (ii) The new versions will become increasingly stringent in their targets, and the expected cost of the EUAs will continue to rise.

Currently, there are several emissions trading schemes in operation globally; some are voluntary, like in Japan and Switzerland, while other remains compulsory,

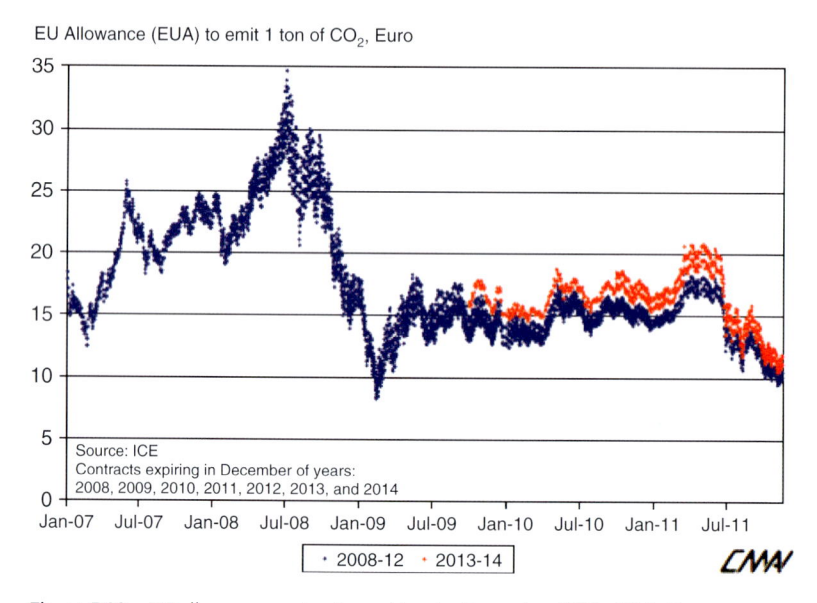

Figure 5.30 EU allowance contracts expiring in December 2008–2014. Source: HIS – CMAI – Europe and Middle East Report Olefins and Derivatives. End November 20112 – Early December 2011 – Issue Number 302.

like that in Europe. However, there is a clear trend to have more schemes in place globally, and in time these are expected to become not only more stringent but even global. Please see in Table 5.8 a summary list of some of the schemes in operation around the world.

As can be seen that list has been growing very fast since 2005 when the schemes started in Europe and Japan. Switzerland and New Zealand joined in 2008 and the UK extended the EU ETS to large non-energy-intensive organizations in 2010.

In 2012 two important new schemes began operation, in January the Western Climate Initiative (WCI) that includes California in the USA and Canada British Columbia, Manitoba, Ontario, and Quebec in Canada and in June the Australian Carbon Price Mechanism. This is certainly an important milestone as the USA, Canada, and Australia in 2012 had some of the highest CO_2 emissions per capita.

By 2015 two critical additions are expected to take place with the start of the Korean ETS and the potential start of the Chinese ETS. The incorporation of China, currently the largest generator of CO_2 emissions globally, in combination with the first scheme in the USA, even if it is only in California, is a tremendous step toward making these schemes global.

5.5.2
Summary – Global Emission Trading Systems in Operation

With the assumption and belief that the ETS will become increasingly global and stringent by 2050, we can estimate what could be the potential impact on the chemical industry.

Under the BAU scenario the CO_2 emissions of the broad chemical industry, including pharmaceuticals could almost triple from the estimated 1.36 to 3.28 Gt by 2050. However, and according to the results from the sustainable scenario, if the world wishes to avoid climate change, the total emissions of the chemical industry will need to be capped at 0.47 Gt by 2050.

Assuming the whole world will move into one ETS similar to that in Europe and the cost to emit 1 tonne of CO_2 would be equal to the 2011 value in the EU ETS of €10 or US$13; the potential cost for the chemical industry to procure all the EUA needed to close the gap between the emissions the industry could have under the BAU (3.28 Gt) and the emissions they will need to have under the sustainable scenario (0.47 Gt) could be a staggering €28 trillion or more than US$ 36 trillion. In the likely scenario where the prices for the EUA could increase even further to €30 or €50 the cost appears simply unbearable.

Let us consider a more optimistic scenario, where we assume that the whole chemical industry would be able to replicate the outstanding performance of the European chemical industry, growing by 60% while reducing its CO_2 emissions by 50% over a period of 20 years, while all the other variables remain unchanged. The potential cost to procure the EUA needed to close the gap between the expected emissions of 1.36 Gt and the sustainable scenario of 0.47 Gt could cost the industry up to €8.9 trillion or US$ 11.5 trillion. This again would increase in line with any increase in value of the EUA.

278 | 5 The Chemical Industry by 2050

Table 5.8 Summary of global emission trading systems in operation.

START	COUNTRY/AREA - EMISSION STRADING SCHEME	STATUS	TARGET
January, 2005	**EUROPEAN UNION Emissions Trading Scheme (EUETS)** Commencing in 2005, it is the world's first and largest mandatory cap and trade scheme for CO_2 emissions, covers all 27 EU member states, and three non-members (Iceland, Liechtenstein and Norway). The EU ETS covers roughly 1100 facilities in the electricity generation sector and major energy-intensive industries which are collectively responsible for roughly 50% of the EU's emissions of CO_2 and 40% of its total greenhouse gas emissions. As of January 2013, th EU ETS will be further expanded to the petrochemicals, ammonia and aluminum industries and to additional gases. Note: The EU 27 includes the following countries, Austria, Belgium, Bulgaria, Cyprus, Czech Republic, Denmark, Estonia, Finland, France, Germany, Greece, Hungary, Ireland, Italy, Latvia, Lithuania, Luxembourg, Melta, Netherlands, Poland, Portugal, Rumania, Slovakia, Slovenia, Spain, Sweden and United Kingdom.	Mandatory	by 2020, 21% emission reduction vs. 2005
January, 2005	**JAPAN Voluntary Emission Trading Scheme (JVETS)** Japan has operated a voluntary emission strading scheme since 2005 that covers CO_2 emissions from fuel consumption, electricity and heat, waste management and industrial processes from over 300 companies. Chemical companies represented 13% of the participants. Consideration of a mandatory ETS in Japan is ongoing. The JV ETS covers approximately 75% of the greenhouse emission already.	Voluntary	1st phase was 21% versus base year. 2nd phase was 19% vs. base year
January, 2008	**The SWISS Emissions Trading Scheme** The Swiss Emissions Trading Scheme run in conjunction with an exemption from the mandatory CO_2 taxes. The scheme covers CO_2 emissions from large companies or groups of companies that opt in to the scheme	Voluntary	

5.5 Feature 4: Climate Change – Greenhouse Emissions – CO$_2$ Emissions

January, 2008	**NEW ZEALAND Emissions Trading Scheme (NZETS)** NZ ETS currently covers emissions from forestry, stationary energy, industrial processes and liquid fossil fuels, which are collectively responsible for roughly 50% of New Zealand's emissions. Emissions from waste and synthetic gases are scheduled to enter the scheme in 2013, while agriculture is legislated to enter in 2015	Mandatory	
January, 2010	**UK CRC Energy Efficiency Scheme** The CRC Energy Efficiency Scheme is a "cap" and "trade" scheme applying to large non energy-intensive organizations in the public and private sectors that are not covered by the EU ETS. These organizations are responsible for around 10% of the United Kingdom's emissions	Mandatory	
January, 2012	**USA & CANADA Western Climate Initiative - WCI** The WCI is a collaboration between California and the Canadian provinces British Columbia, Manitoba, Ontario, and Quebec. Initially the WCI included 10 Western US States and Canadian Provinces. On 18 November 2011, Arizona, Montana, New Mexico, Oregon, Utah and Washington formally left WCI with no intention to implement the scheme. The cap-and trade scheme covers emissions from electricity, electricity imports, industrial combustion, and industrial process emissions. It is expected to be expanded in 2015 to include transportation fuels and residential, commercial and industrial fuels. Those expected to implement the program when it begins in January 2012 comprise approximately two-thirds of total emissions in the WCI jurisdictions. When fully implemented in 2015, this program will cover nearly 90% of the GHG emissions in WCI states and provinces.	Mandatory	Reduce regional GHG emissions to 15% below 2005 levels by 2020

continued over leaf

Table 5.8 (continued)

START	COUNTRY/AREA - EMISSIONS TRADING SCHEME	STATUS	TARGET
July, 2012	**AUSTRALIA Carbon Pricing Mechanism** Australia's carbon pricing mechanism will require around 500 of Australia's biggest polluters to pay a fixed carbon price for their emissions from 1 July 2012. The scheme, which covers around 60% of carbon pollution, will transition to a flexible price cap-and-trade emission strading scheme from 1 July 2015. Both aprice floor and price ceiling will be in place for the first three years of the flexible price period	Mandatory	
2013-2015	**CHINA Emission Trading Scheme** China is scheduled tol aunch pilot emissions trading schemes in six provinces and cities in 2013 with a v iew to develop a nation wide trading scheme by 2015	Mandatory as of 2015?	
2015	**KOREAN Emission Trading System** In 2012 Korea approved the Korea's emissions trading scheme, startingi n January 2 015. This scheme will require about 470 of Korea's largest polluters to pay for their CO_2 emissions, collectively covering roughly 60% of Korea's greenhouse gas emissions. Companies which exceed emissions limits will pay a penalty equal to three times the market value of the credits – but the penalty will be limited to a maximum 100,000 won (around $88.5) per ton of emission	Mandatory	30% from projected levelsi n 2020

Both scenarios are certainly an oversimplification and for illustration purposes. Indeed more specific analysis by industry segment and area should be conducted but, as previously stated, the purpose of this chapter is to provide some first figures to indicate the potential effort required and cost to the industry to address the clear need to reduce greenhouse emissions. We can understand that some of these figures are so huge that most readers might be wondering about the validity of the calculations, their veracity and eventually what to do with them. We should also remember that these scenarios have been based on multiple assumptions, assumptions that, although we have been trying to be as conservative and realistic as possible, can bring significant figures like those above.

In the light of these figures we can assume that the challenge to reduce climate change for the chemical industry is going to be technically and economically formidable. The chemical industry has a remarkable track record in reducing emissions while growing capacity, however, the upcoming decades present a challenge never seen before. A challenge that comes with a significant value proposition even in the most conservative scenario; please bear in mind that the above scenarios only contemplated reductions in CO_2 emissions not all the greenhouse gases, and the selected cost per ton of CO_2 was one of the lowest observed during the last years, while the expectation is that the cost could increase significantly during the next decades.

Under these scenarios we expect the industry to reinforce its strong commitment to carbon and energy productivity, reducing even further its greenhouse emissions while enabling the massive growth expected during the next decades.

On the positive side there is another critical element to be considered, and an additional reason for the industry to excel on carbon productivity during the next decades; its unique capacity to enable emissions reductions in many other industries. The need for the industry to grow is not only critical for the chemical industry and our society to enable the expected living standards, but is also vital for many other industries and society in general as its products can generate major savings of greenhouse emissions.

Although this aspect is generally disregarded by the existing ETS, the reality is that the chemical industry through its different products and technologies can enable major emissions reductions globally. Reductions that if they are valued at their market price, on CO_2 emissions reductions per tonne, could clearly offset most of the potential cost the industry could incur by producing its own products.

According to the ICCA – the total net greenhouse emissions abatement enabled by the chemical industry was 6 Gt of CO_2 equivalent in 2005.

5.5.2.1 Chemical Industry – Greenhouse Emissions Abatement in 2005

According to the ICCA report (Figure 5.31), 39% of these contributions were made through the use of chemicals in insulation applications, 26.6% through the use of fertilizers and crop protection, lighting accounted for 11.6%, packaging for 3.7%, and transportation 3.6%. The rest was distributed across multiple applications across several industries.

5 The Chemical Industry by 2050

Chemical industry applications and the net abatement (final product saving in industry emissions) they allowed in 2005

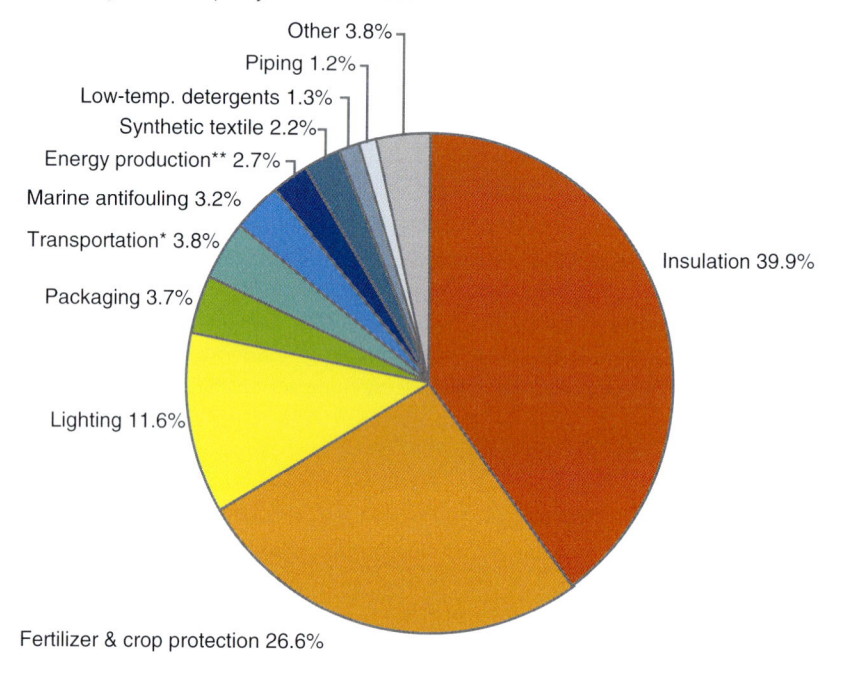

- Other 3.8%
- Piping 1.2%
- Low-temp. detergents 1.3%
- Synthetic textile 2.2%
- Energy production** 2.7%
- Marine antifouling 3.2%
- Transportation* 3.8%
- Packaging 3.7%
- Lighting 11.6%
- Fertilizer & crop protection 26.6%
- Insulation 39.9%

Figure 5.31 Chemical industry applications and the net abatement (final product savings in industry emissions) they allowed in 2005. Source: ICCA – Extracted from the publication "Turning the tide on climate change: The climate change challenge and the chemical industry".

Taking into consideration these enormous abatement figures we can understand the clear need for our world to have more and even better chemicals.

The capacity of the industry to enable large emissions reductions is something that should be maximized and valued accordingly. The fact that the industry net greenhouse emissions are actually negative, might not serve to reduce its total greenhouse emissions, or avoid further focus on additional carbon productivity, but it is something that has to be understood, explained, valued, and recognized across the industry and society in general.

Climate change is probably by far the most challenging and complex of all the upcoming megatrends. The expectation of the chemical industry to quadruple its production by 2050, while reducing its actual emissions by more than half, while supporting other industries to reduce their emissions, is not only a formidable challenge but a tremendous opportunity for the industry to excel at technology and innovation.

The need and urgency to overcome climate change is clear, the value for our society is vital and the incentives for the chemical industry paramount. Under this scenario we expect the chemical industry to play a critical and leading role in the sustainability of our planet; not only by reducing its own emissions but, even more importantly, by enabling other industries to do the same.

The chemical industry has been able to overcome some of the most challenging situations during the last centuries, enabling growth and prosperity, this time will not be different.

5.6
Feature 5: Industry Structure

As the world economy becomes larger and wealthier in general, and especially in the BRIC and REST economies, the world demand for chemicals and pharmaceuticals at both absolute and per capita levels is expected to increase massively, changing the overall structure of the industry (Figure 5.32).

The chemical industry, excluding pharmaceuticals, as the largest component of the broad chemical industry with around 75–80% of the industry, is the part of the industry with the highest changes on an absolute basis. Changes that will serve to consolidate the trends observed in the chemical industry during the last decades, trends that will confirm China as the largest world consumer of chemicals.

On the other hand, the gigantic differences in pharmaceutical per capita demand around the world with deltas up to hundreds of times – in 2010 India pharmaceutical per capita demand was more than 100 times lower than in the USA – in combination with the rapid growth of the BRIC and REST economies, make us believe *the pharmaceutical industry will undergo a historical transformation during the next decades. This transformation will serve to start the rebalancing of the pharmaceutical industry, from the ADV economies to the BRIC and REST group.*

Therefore the broad chemical industry is expected to suffer major shifts in its structure. Three major aspects will be the subjects of our evaluation in the following sections. First, the global ranking of the different markets for both the chemical and pharmaceutical industries, with a much stronger relevance of the BRIC economies. Secondly, the per capita demand for chemicals and pharmaceuticals. While increases in demand on a per capita basis will be observed in both chemicals and pharmaceuticals, for pharmaceuticals the ADV economies will keep most of the demand. Finally, the massive growth expected in the BRIC economies, will trigger a large wave of trans-national integration and mergers and acquisitions (M&A), with the BRIC at the forefront of that trend.

Figure 5.32 Impact on industry structure.

5.6.1
Markets – Largest World Markets

5.6.1.1 Chemicals

In 2010, the largest chemical market was China, with 25% of the global sales and US$ 763 billion, followed by Europe with 21% (US$ 651 billion) and the USA with 17% (US$ 524 billion). The REST group was also significantly large with 23% of the global chemicals sales and almost US$ 720 billion (Figure 5.33).

By 2050 and under the BAU modified scenario several changes are expected.

The REST group will be the largest market for chemicals with 28% of the world market, accounting for almost US$ 4.2 billion.

China will be the largest single market, excluding the REST group, with 27% of the market, US$ 4 billion.

However, Europe and the USA will see their global presence reduced in favor of India. The EU will move from being the second largest chemical market in the world in 2010 to being the fifth largest when including the REST group with just 9% of the market and US$ 1.3 billion.

The USA similarly will move from being the third largest chemical market to being the fourth including the REST, with 9% of the chemical market and US$ 1.3 billion. Per contrary India is expected to become the second largest individual market after China, excluding the REST group, with 20% of the global chemical sales, accounting for almost US$ 3 billion by 2050.

India, in both the BAU and the modified BAU scenarios, is expected to become the world's second largest chemical market in the period 2030–2040, when it will surpass the EU and the USA, while the USA is expected to overtake the EU by 2040.

Brazil is also expected to go through a major growth and transformation, becoming the fifth largest chemical market in the world by 2030, after overtaking Japan. Brazil's chemical market will grow from US$ 100 billion in 2010 to US$ 477 billion by 2050. The Russian chemical market is also expected to surpass the Canadian chemical market in this decade and is expected to grow from US$ 42 billion in 2010 to US$ 208 billion by 2050.

In terms of group areas, *the BRIC area already has the largest chemical market in the world in 2010 and will preserve it. On the other hand the chemical markets of the REST economies are expected to surpass those in the ADV economies by 2030, becoming the second largest chemical market in the world.*

5.6.1.2 Pharmaceuticals

The pharmaceutical market is also expected to undergo a huge structural shift, with an unprecedented change in focus from the ADV economies to the BRIC. As in the previous cases, massive increases in market size will be the result of a combination of large populations and increases in per capita demand, however, there will still be a large difference in per capita demand among the different countries and areas of our analysis.

In 2010, 75% of the world pharmaceuticals sales were in the ADV economies, with the USA alone accounting for more than one third of the world pharmaceutical sales and EU 27 another 28% (Figure 5.34).

5.6 Feature 5: Industry Structure | 285

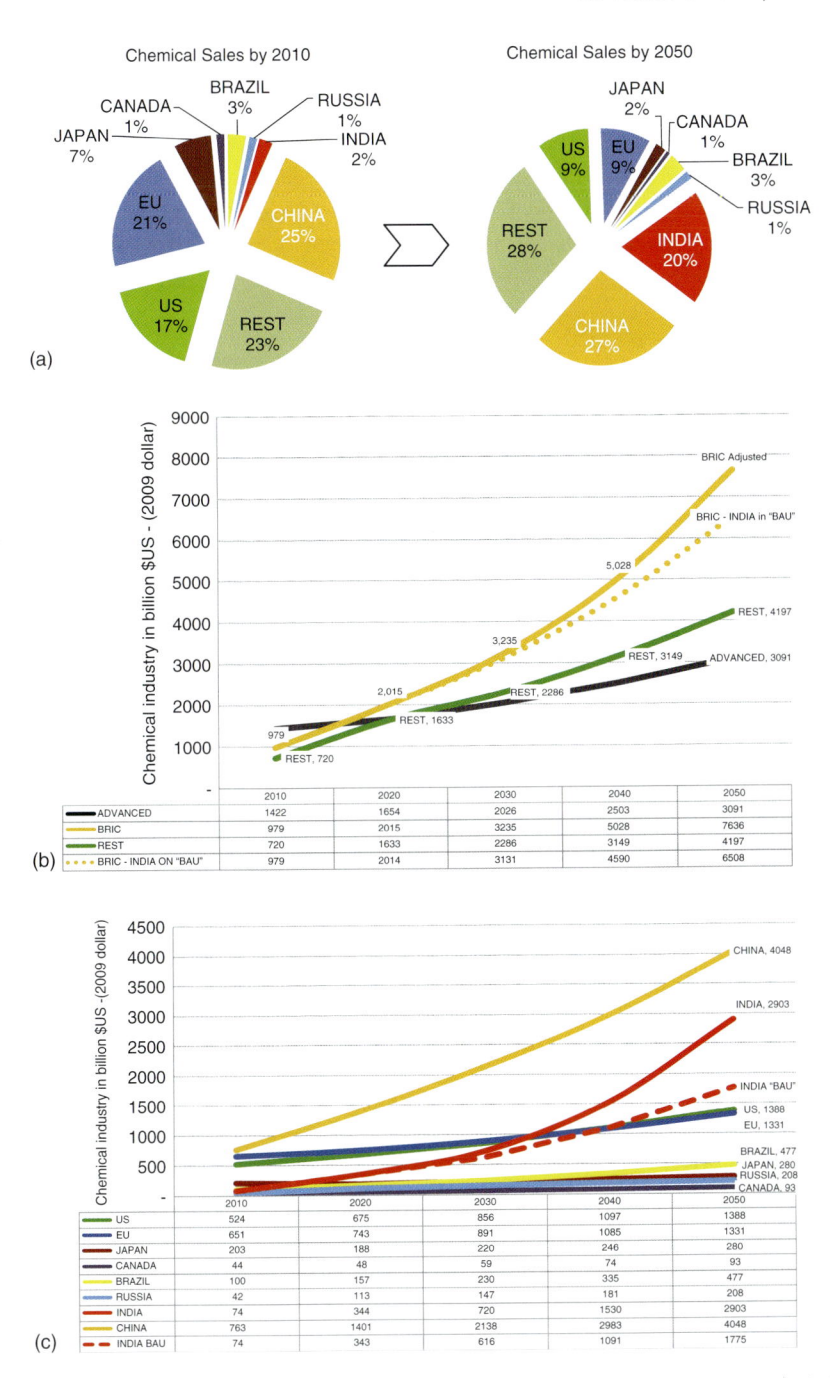

Figure 5.33 (a) Chemical sales (excluding pharmaceuticals) by country in 2010 and 2050 expressed as a percentage. (b) Chemical sales (excluding pharmaceuticals) by area in 2010–2050 (adjusted BAU scenario). (c) Chemical sales (excluding pharmaceuticals) by country in 2010–2050 (adjusted BAU scenario).

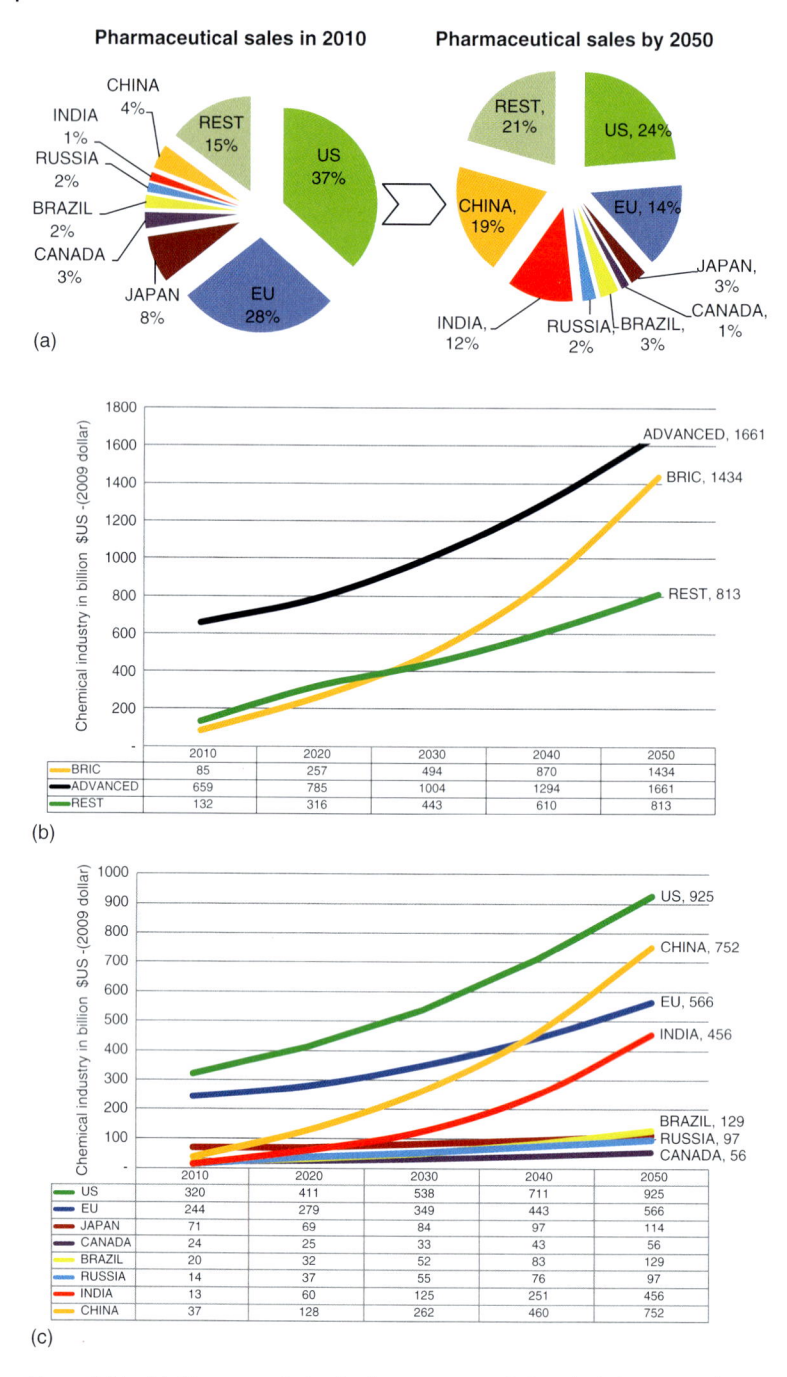

Figure 5.34 (a) Pharmaceutical sales by country in 2010 and 2050 expressed as a percentage. (b) Pharmaceutical sales by area in 2010–2050 (adjusted BAU scenario). (c) Pharmaceutical sales by country in 2010–2050 (adjusted BAU scenario).

By 2050 and under the modified BAU scenario several changes are expected, although the ADV economies will still have the world's largest pharmaceutical market.

The USA is expected to remain as the largest pharmaceutical market in the world with 24%, accounting for US$ 925 billion by 2050. In other words the USA pharmaceutical market in 2050 will be as big as the world pharmaceutical market in 2010.

China with 19% of the world demand and US$ 752 billion will appear as the second largest pharmaceutical market. Europe will follow with 14% (US$ 566 billion) and India will be fourth with 12% (US$ 456 billion).

Japan, despite increasing the total size of its market by a staggering US$ 43 billion, from US$ 71 billion in 2010 to US$ 114 billion in 2050, will lose its third position in the world pharmaceuticals markets in 2010, in favor of China and India. Despite the massive growth expected in pharmaceuticals sales in the BRIC and REST economies, and as we will review later, there will still be significant differences among the different economic groups and countries in pharmaceutical demand on a per capita basis.

In terms of areas, *the ADV will remain as the largest market for pharmaceuticals with 43% of the world total and US$ 1.6 billion by 2050. The BRIC economies will become the second largest pharmaceutical market by 2030, surpassing the REST group, and are expected to have 37% of the total pharmaceutical sales by 2050, with a US$ 1.4 billion market. The REST group, with expected sales of US$ 813 billion by 2050, will grow its market by a staggering 681% during the next four decades.*

Within the BRIC, China is expected to become the third largest pharmaceutical market, surpassing Japan in 2020 and the second largest pharmaceutical market in the world by 2040 when surpassing Europe 27. India according to our modified BAU scenario could become the fourth largest pharmaceutical market between 2020 and 2030.

5.6.2
Per Capita Demand

5.6.2.1 Chemicals

Under the modified BAU scenario, the world average per capita demand for chemicals is expected to grow dramatically, increasing from $456 in 2010 to $1631 in 2050 (Figure 5.35). To put this last figure in perspective, this world average per capita demand for chemicals in 2050 will equal that in the USA in 2010, so we can imagine how many chemical products will be around in our world.

The difference in per capita demand for chemicals between the ADV and BRIC economies will be reduced significantly, especially thanks to India and China that will enjoy an extraordinary growth during the next four decades.

India is expected to multiply its per capita chemical demand by almost 30 times by 2050, while China and Russia will multiply theirs by 6 times and Brazil by 4 times. On the other hand the REST economies, despite multiplying their per capita demand by 4 times, will still remain at half of the world average per capita demand.

By 2050 the USA after doubling its per capita demand from 2010, is expected to have the highest per capita demand in chemicals in the world with US$ 3495

288 | 5 The Chemical Industry by 2050

(a)

	2010	2020	2030	2040	2050
US	1696	2,017	2390	2904	3495
EU	1297	1458	1736	2124	2631
JAPAN	1593	1522	1870	2208	2648
CANADA	1290	1284	1482	1782	2158
BRAZIL	513	749	1051	1512	2181
RUSSIA	296	814	1090	1398	1675
INDIA	63	260	495	987	1803
CHINA	570	1014	1545	2217	3180
World	456	700	918	1222	1631
REST	238	465	572	707	862

(b)

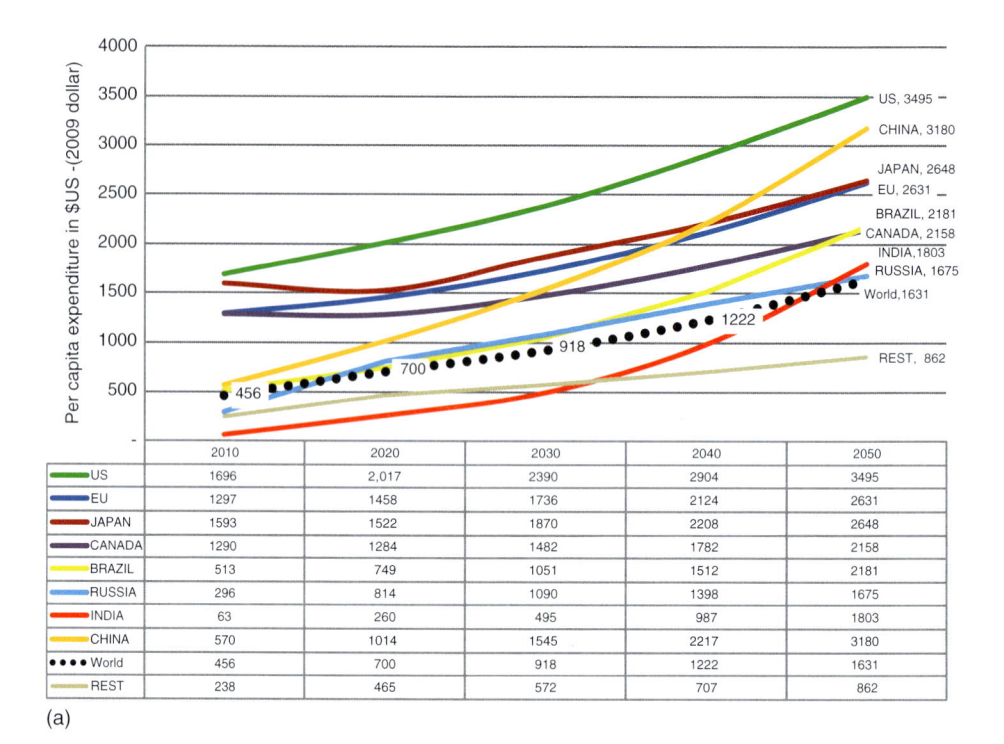

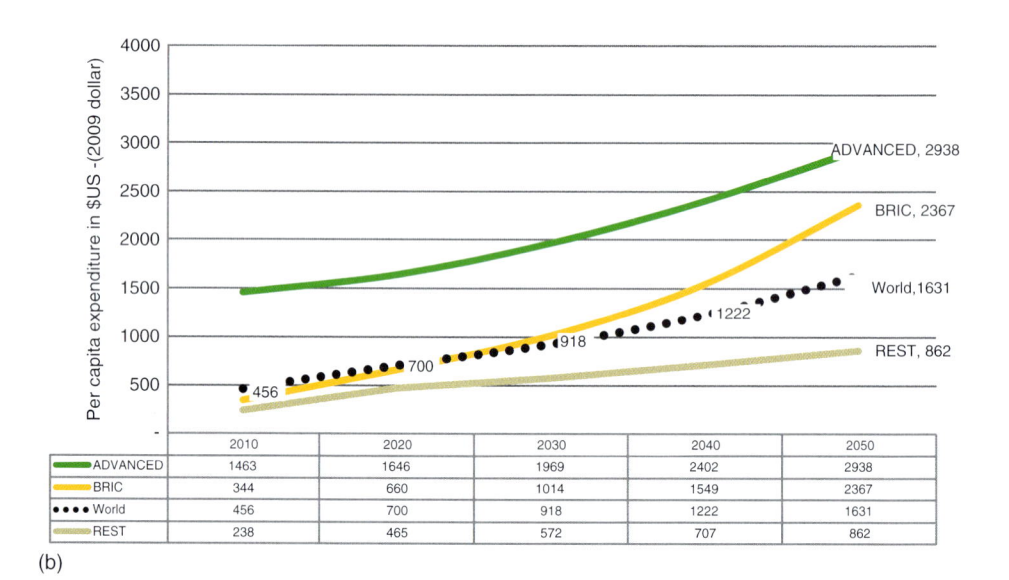

	2010	2020	2030	2040	2050
ADVANCED	1463	1646	1969	2402	2938
BRIC	344	660	1014	1549	2367
World	456	700	918	1222	1631
REST	238	465	572	707	862

Figure 5.35 World average per capita expenditure on chemicals 2010–2050 (modified BAU scenario). (a) By country, (b) by area.

annually. China will be second with US$ 3180 annually after surpassing Japan and Europe by 2040. Japan and Europe will have the third and fourth highest per capita demand with US$ 2648 and US$ 2631, respectively.

Brazil is expected to have a per capita demand for chemicals of US$ 2181 annually, similar to that of Canada with US$ 2158 and slightly higher than India and Russia. India on the other hand is expected to enjoy a formidable growth in chemical demand, increasing its per capita demand from the low figure of US$ 63 in 2010 to a staggering US$ 1803 by 2050. Russia with US$ 1675 of chemicals demand annually will be close to the world average and similar to the level of per capita demand of the USA in 2010.

The massive expected growth in population and our world economy will serve to raise the world per capita demand for chemicals to unprecedented levels, indeed by 2050, the world per capita demand will be equal to that of the USA in 2010 with US$ 1631 annually.

5.6.2.2 Pharmaceuticals

Under the modified BAU scenario, the world average per capita demand for pharmaceuticals is expected to grow dramatically, increasing from US$ 128 in 2010 to US$ 427 in 2050 (Figure 5.36). This figure will be close to the pharmaceutical per capita demand in Europe in 2010, but still far from the per capita demand of countries like the USA in 2010.

Despite a massive growth in per capita demand for pharmaceuticals globally, there will still be a large difference between the ADV and the BRIC economies as well as between the USA and the rest of the world.

The fact that our projections are based on the per capita levels in 2010, explains somehow the large differences for the USA versus the rest of the world. The USA with an expected GDP per capita of almost US$ 97 000 by 2050 and the highest per capita demand for pharmaceuticals in the world in 2010, came out with a projected per capita demand for pharmaceuticals of US$ 2330 by 2050.

India with the lowest per capita demand of pharmaceuticals in 2010 is expected to have a gigantic increase in per capita demand, from US$ 11 annually in 2010 to US$ 283 by 2050; however India will still remain at the bottom of the BRIC economies and below the world average per capita demand for pharmaceuticals.

Among the BRIC, Russia is expected to have the highest per capita demand for pharmaceuticals with US$ 777 annually, followed by Brazil and China both with US$ 591.

Looking at the projection from this modified BAU scenario, we feel conservatively confident on those for the BRIC and REST economies, after all it might not be very unrealistic to envision a situation where the BRICs consume by 2050 a similar amount of pharmaceuticals as the ADV economies in 2010.

Perhaps we should feel more cautious about some of the projections for the ADV economies, and more particularly for the USA. Although it seems quite logical to see a further increase in pharmaceuticals demand as the GDP per capita rises, the reality is that the USA will be having levels of per capita demand in pharmaceuticals never seen before.

290 | *5 The Chemical Industry by 2050*

(a)

(MODIFIED "BAU" Scenario)

	2010	2020	2030	2040	2050
US	1035	1230	1502	1881	2330
EU	486	548	681	868	1119
JAPAN	557	561	712	869	1075
CANADA	704	687	829	1040	1313
BRAZIL	103	153	238	376	591
RUSSIA	99	267	407	586	777
INDIA	11	45	86	162	283
CHINA	28	92	189	342	591
WORLD AVG.	128	179	236	317	427
REST	44	90	111	137	167

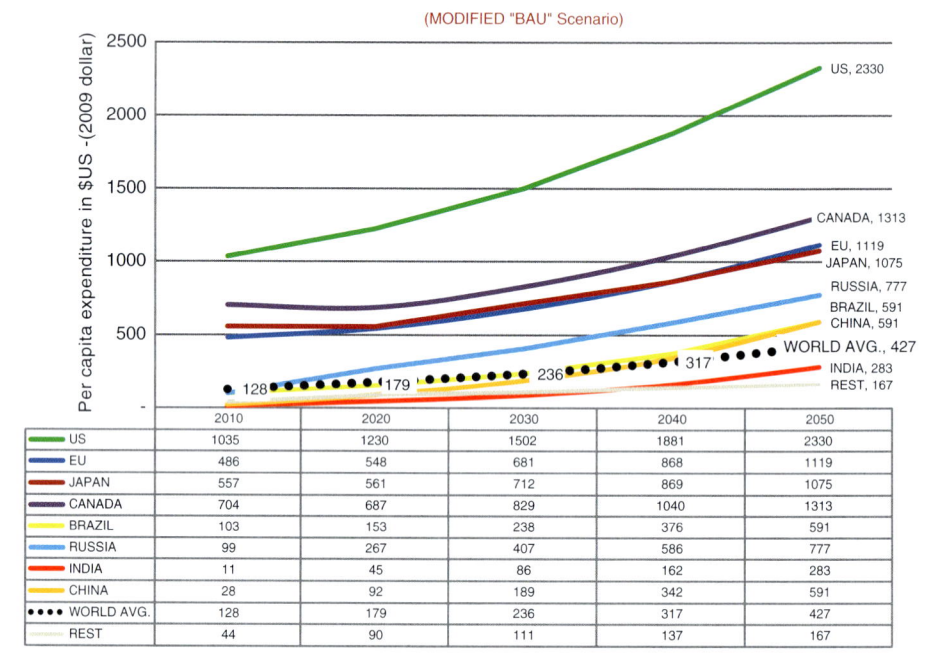

(b)

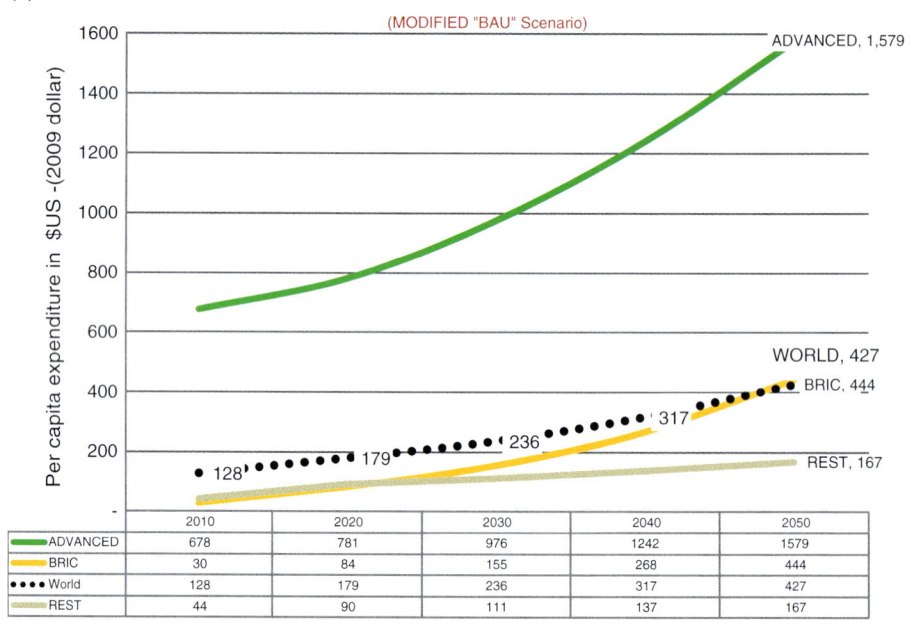

(MODIFIED "BAU" Scenario)

	2010	2020	2030	2040	2050
ADVANCED	678	781	976	1242	1579
BRIC	30	84	155	268	444
World	128	179	236	317	427
REST	44	90	111	137	167

Figure 5.36 World average per capita expenditure on pharmaceuticals 2010–2050 (modified BAU scenario). (a) By country, (b) by area.

Globally, the ADV and BRIC economies, with the exception of India, are expected to be above the world average, while the REST group is expected to be below. However, the fact the REST group is composed of a very large and diverse number of countries makes us believe further updates of this chapter should go deeper into this, as large opportunities remain to be seen.

5.6.3
Companies – Changes in Global Sales Rankings and Company Structures

The massive changes expected in the structure of the chemical and pharmaceutical markets, with a much larger role and sales from the BRIC economies, will trigger similar changes in the way companies operate and sell.

In 2010, 46% of the chemical sales, excluding pharmaceuticals, occurred in the ADV, 31% in the BRIC, and US$ 23 in the REST. However, when looking at the top 10 chemical companies, 8 were based in the ADV economies. The only two exceptions were Sinopec based in China and SABIC based in the Middle East (Figure 5.37).

Considering that, according to the modified BAU scenario for 2050, the BRIC economies will become the largest chemical region by sales with 51% of the world sales, followed by the REST with 28% and the ADV by 21%; we could expect a reshuffle among the top chemical companies in the world with a much larger presence from chemical companies from the BRIC and the REST areas.

This gradual shift from the ADV to the BRIC and REST economies could imply several changes for the chemical companies. On the one hand we could expect that the chemical companies from these areas might become much bigger than they are today, either thanks to simple organic growth or thanks to M&A activities. In that sense a large number of trans-national M&A activities, where the BRIC companies could start acquiring their chemical counterparts around the world could be something quite logical to expect. All these activities might serve to start positioning the chemical companies from the BRIC and the REST economies at the top of the global chemical rankings.

On the other hand, the current distribution of sales of the global chemical companies might also start to shift from the ADV to the BRIC and REST economies. For instance, if we consider BASF, the largest chemical company in the world, we estimate that in 2010, BASF had more than 70% of its sales and 80% of its employees in the ADV economies (Figure 5.38). However, by 2050 we could question if that ratio will remain constant especially if BASF wants to remain the largest chemical company in the world. Indeed, we could foresee a gradual relocation of sales and employees for all the large chemical companies, like BASF, DOW, LBI, INEOS, DUPONT, or others from the ADV economies to the BRIC and REST Economies, especially for companies like these that could be as large as US$ 250 billion by 2050.

The gradual shift of the chemical market to the BRIC and REST economies might trigger several questions in the chemical companies, especially for the more global ones. Questions like where their major sales center, R&D centers or even

292 | 5 The Chemical Industry by 2050

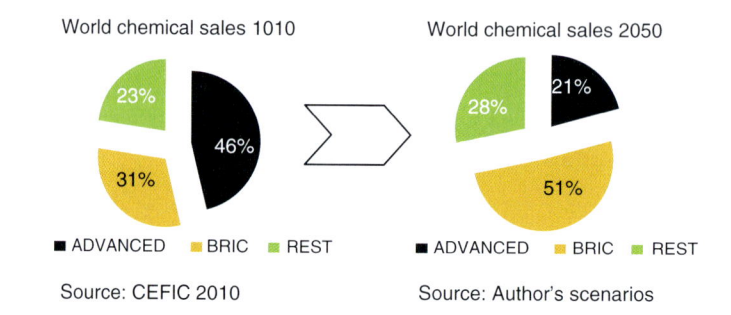

World chemical sales 1010
- ADVANCED: 46%
- BRIC: 31%
- REST: 23%

Source: CEFIC 2010

World chemical sales 2050
- ADVANCED: 21%
- BRIC: 51%
- REST: 28%

Source: Author's scenarios

TOP 10 - CHEMICAL INDUSTRY - excluding pharmaceuticals

$ US billion — 2010

RANK	COMPANY	REVENUE	% on the MARK	OPERATING PROFIT	R&D	EMPLOYEES	COUNTRY	AREA
Top1	BASF	84.7	3%	10.8	2.0	109,140	GERMANY	ADVANCED
Top2	DOW CHEMICAL	53.6	2%	2.8	1.7	49,505	USA	ADVANCED
Top3	EXXON	53.6	2%	3.4			USA	ADVANCED
Top4	SINOPEC	48.7	2%	2.3		65,623	CHINA	BRIC
Top5	LYONDELBASELL	41.1	1%	2.9	0.2	14,000	GERMANY	ADVANCED
Top6	SABIC	40.5	1%	10.1	0.2	33,000	SAUDI ARABIA	REST
Top7	SHELL	39.6	1%				UK	ADVANCED
Top8	MITSUBISHI CHEMICAL	38.2	1%	2.7	1.6	53,882	JAPAN	ADVANCED
Top9	INEOS	34.5	1%	1.5		13,682	UK / SWITZERLAN	ADVANCED
Top10	DUPONT	31.5	1%	3.	1.6	60,000	USA	ADVANCED
	TOP 10	466.0	15%	40.2	7.1	398,832		
	TOTAL MARKET	3121.0						

Source: ICIS Top 100 chemical companies in 2010

Figure 5.37 Top chemical markets and companies 2010–2050. Source: ICIS top 100 chemical companies in 2010.

global headquarters should be located would become more frequent. Indeed we should expect a gradual transition of the industry and its companies from the ADV to the BRIC and REST economies; and that transition might not be only on their operations, or sales, but also in the culture, way to do business and even on their global headquarters.

A similar but less pervasive trend could be expected for the pharmaceutical industry. The high level of technology and R&D content of the pharmaceutical industry might prevent radical transformation in the location and origin of the pharmaceutical companies, however, some changes will certainly occur.

In 2010, 75% of the pharmaceutical sales were in the ADV economies, however, in 2050 and according to our BAU scenario, only 43% of the pharmaceutical sales will take place in the ADV economies (Figure 5.39).

In 2010, similarly to the chemical industry, 8 out of the top 10 world pharmaceutical companies by revenue were based in the ADV economies, 2 in the REST and none in the BRIC. Those based on the REST tend to be in countries like Switzerland.

It is critical to notice that even when we look at the top 50 pharmaceutical companies in the world, just 4 companies are in the REST group and none in

5.6 Feature 5: Industry Structure | 293

Source: BASF Facts & Figures 2011

BASF global sales in 2010

- 55%
- 21%
- 6%
- 18%

EU
NAA
South america, Africa, Middle east pacific

BASF employees in 2010

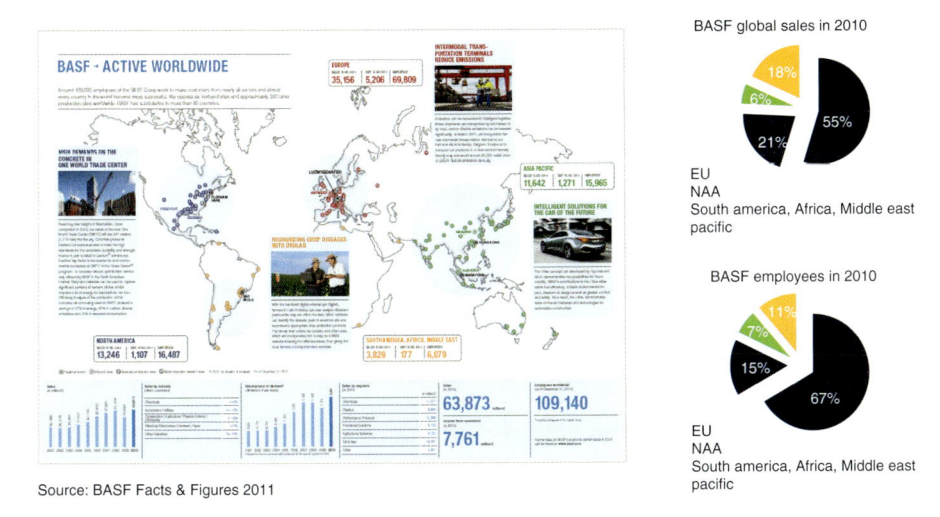

- 67%
- 15%
- 7%
- 11%

EU
NAA
South america, Africa, Middle east pacific

Figure 5.38 BASF global sales and employees in 2010. Source: BASF Facts @ Figures 2011.

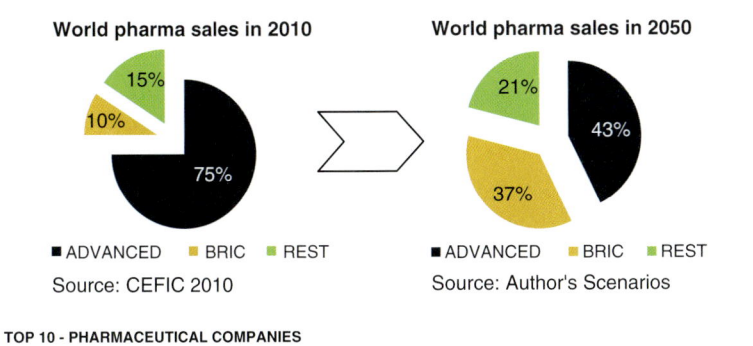

World pharma sales in 2010

- 75% ADVANCED
- 10% BRIC
- 15% REST

Source: CEFIC 2010

World pharma sales in 2050

- 43% ADVANCED
- 37% BRIC
- 21% REST

Source: Author's Scenarios

TOP 10 - PHARMACEUTICAL COMPANIES
$ US Billion   2010

RANK	COMPANY	REVENUE	% MARKET	EBITDA	% Revenue	R&D	% on Revenue	EMPLOYEES	COUNTRY	AREA
Top1	PFIZER	67.8	8%	6.1	9.0%	9.5	14.0%	116,500	USA	ADVANCED
Top2	JOHNSON & JOHN	61.6	7%	13.3	21.6%	6.8	11.0%	115,500	USA	ADVANCED
Top3	NOVARTIS	50.6	6%	9.8	19.4%	9.1	18.0%	99,834	SWIZTERLAN	REST
Top4	ROCHE Holding AG	46.8	5%	9.2	19.7%	9.3	19.9%	81,507	SWIZTERLAN	REST
Top5	GLAXOSMITHKLIN	45.7	5%	6.2	13.6%	6.2	13.6%	98,854	UK	ADVANCED
Top6	MERCK & CO INC	44.1	5%	7.9	17.9%	8.4	19.0%	100,000	USA	ADVANCED
Top7	SANOFI AVENTIS	41.2	5%	8.5	20.6%	-	0.0%	104,86	FRANCE	ADVANCED
Top8	ABBOT LABORATO	35.2	4%	4.6	13.1%	3.7	10.5%	73,000	USA	ADVANCED
Top9	ASTRAZENECA	33.3	4%	8.1	24.3%	5.3	15.9%	62,700	UK	ADVANCED
Top10	BAYER AG	22.2	3%	4.0	18.0%	3.0	13.5%	108,800	GERMANY	ADVANCED
	TOP 10	426.3	49%	73.7	17.3%	58.3	15.1%	852,762.0		
	TOTAL MARKET	875.0								

Source: Several. corporate income statements and IMAP pharmaceutical & biotech industry Report 2011

Figure 5.39 Top pharmaceutical markets and companies 2010–2050. Source: Several, Corporate Income Statements and IMAP Pharmaceutical & Biotech Industry Report 2011.

Financial Review
Pfizer Inc. and Subsidiary Companies

PFIZER Global Sales in 2010: USA 60%, Rest of the World 40%

Revenues by Segment and Geographic Area

Worldwide revenues by operating segment, business unit and geographic area follow:

(MILLION OF DOLLAR)	WORLDWIDE 2011(a),(b)	WORLDWIDE 2010(b)	WORLDWIDE 2009(b)	U.S. 2011(a),(b)	U.S. 2010(b)	U.S. 2009(b)	INTERNATIONAL 2011(a),(b)	INTERNATIONAL 2010(b)	INTERNATIONAL 2009(b)	% CHANGE WORLDWIDE 11/10	% CHANGE WORLDWIDE 10/09	% CHANGE U.S. 11/10	% CHANGE U.S. 10/09	% CHANGE INTERNATIONAL 11/10	% CHANGE INTERNATIONAL 10/09
Biopharmaceutical revenues: Primary care operating segment	$22670	$23328	$22576	$12819	$13536	$13045	$9851	$9792	$9531	(3)	3	(5)	4	1	3
Specialty care	15245	15021	7414	6870	7419	3853	8375	7602	3561	1	103	(7)	93	10	113
oncology	1323	1414	1511	391	506	456	932	908	1055	(6)	(6)	(23)	11	3	(14)
SC&O operating segment	16568	16435	8925	7261	7925	4309	9307	8510	4616	1	84	(8)	84	9	84
Emerging market established products	9295 9214	8662 10098	6157 7790	— 3627	— 4501	— 2656	9295 5587	8662 5597	6157 5134	7 (9)	41 30	— (19)	— 69	7 —	41 9
EP&EM operting segment	18509 57747	18760 58523	13947 45448	3627 23707	4501 25962	2656 20010	14882 34040	14259 32561	11291 25438	(1) (1)	35 29	(19) (9)	69 30	4 5	26 28
Other product revenues: Animal health	4184	3575	2764	1648	1382	1106	2536	2193	1658	17	29	19	25	16	32
Consumer healthcare	3057	2772	494	1490	1408	331	1567	1364	163	10	*	6	*	15	*
AH&CH operating segment	7241	6347	3258	3138	2790	1437	4103	3557	1821	14	95	12	94	15	95
Nutrition operating segment	2138	1867	191	—	—	—	2138	1867	191	15	*	—	—	15	*
Pfizer centersource(c)	299	320	372	88	103	93	211	217	279	(7)	(14)	(15)	11	(3)	(22)
Total revenues	$67425	$67057	$49269	$26933	$28855	$21540	$40492	$38202	$27729	1	36	(7)	34	6	38

(a) 2011 includes revenues from legacy king U.S. operations for 11 months and from legacy king international operations for ten months, commencing on the King acquisition date, January 31, 2011
(b) Legacy Wyeth revenues are included for a full year in each of 2011 and 2010. 2009 includes revenues from legacy Wyeth products commencing on the Wythe acquisition date, October 15, 2009.
(c) Our contract manufacturing and bulk pharmaceutical chemical sale organization.
* Calculation not meaningful.

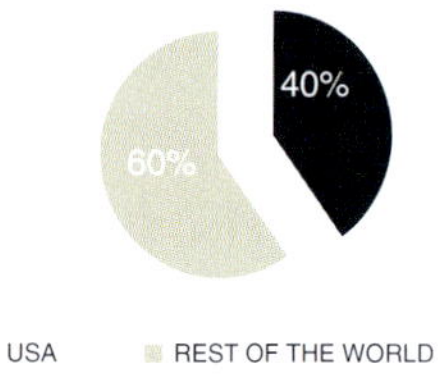

Figure 5.40 Pfizer financial review.

the BRIC. Considering the huge growth the pharmaceutical industry is expected to have in the BRIC economies, we wonder what impact this could have on the pharmaceutical companies and the industry structure.

On the one hand, the huge technological barriers to entering this market might prevent a large number of companies from entering it despite the expected massive growth. Similarly, for existing pharmaceutical companies in the BRIC economies it might be difficult to take full advantage of their local growing markets unless they upgrade their technology. Therefore, with some logical questions and concerns on how local companies in the BRIC could benefit from the spectacular growth the pharmaceutical industry will experience during the next decades, we can expect the current leaders of the industry to gradually expand their sales into these new markets.

Taking as an example Pfizer, as the world's largest pharmaceutical company, and looking at the 2011 Financial Report (Figure 5.40) we can observe that the US economy plays a huge role in the company sales. Indeed, Pfizer does not disclose data for any other individual country, but the USA alone accounted for 60% of Pfizer world sales in 2010.

Considering the amount of growth the pharmaceutical industry is expected to have in the BRIC and the REST economies, we believe important changes will also occur in the structure of pharmaceutical companies. They will extend aggressively their sales into these territories while new players will emerge from them too.

5.7
Feature 6: Social Awareness

The expected massive increase in chemical and pharmaceutical production globally, in combination with their critical capacity to enhance living standards, improve

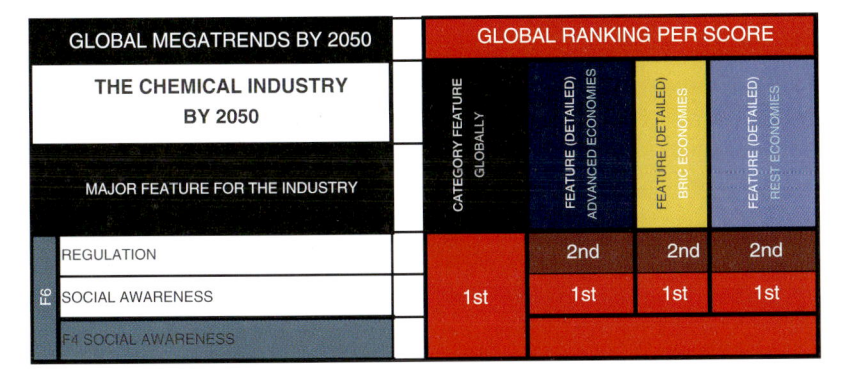

GLOBAL MEGATRENDS BY 2050	GLOBAL RANKING PER SCORE			
THE CHEMICAL INDUSTRY BY 2050 / MAJOR FEATURE FOR THE INDUSTRY	CATEGORY FEATURE GLOBALLY	FEATURE (DETAILED) ADVANCED ECONOMIES	FEATURE (DETAILED) BRIC ECONOMIES	FEATURE (DETAILED) REST ECONOMIES
REGULATION		2nd	2nd	2nd
SOCIAL AWARENESS	1st	1st	1st	1st
F4 SOCIAL AWARENESS				

Figure 5.41 Impact on regulation and social awareness.

quality of life, and reduce greenhouse emissions, will serve to increase significantly the role and visibility of the industry in society (Figure 5.41).

In a more transparent, interconnected, and global world the chemical industry is poised to significantly increase its role in society, and with that society's overall awareness of it. We expect this to be one of the features with highest impact during the next decades.

Society will turn to the industry, asking for solutions and innovation; solutions that will enable all the positive aspects the upcoming economic growth will bring, while minimizing all negative aspects deriving from energy scarcity and climate change. This higher social awareness will result in a much higher interaction of the industry with society and governments, bringing a unique and historical opportunity for the industry to explain its success, how it works and its past and future achievements.

The industry will need to learn to live with much higher levels of transparency and external communication, while working more closely with governments, universities, and society. The expectations will be very high, and the visibility will be total. Under this kind of scenario we expect society's awareness of the chemical industry to increase to the good old plastics times of the 1960s and 1970s, when working for Plastics was one of the most appealing industries to work for. In that sense the capacity of the industry to attract some of the best talent in the world will remain paramount, as the challenges, opportunities, and rewards for the industry will be unprecedented.

On the other hand, all the massive growth, in combination with the clear need to address significant industry challenges in almost every feature of the industry, from feedstock to energy, emissions, and so on, will imply massive increase in regulation. Regulation that will become more and more stringent in content and much more global in nature.

Under this scenario, we expect an industry much more regulated than today, but, more importantly, on a much more global level. A world where regulation will not only become more global but also where non-compliance would become more transparent to society and governments; and penalties much more burdensome for companies and industries.

Appendix – Climate Change

World CO$_2$ Emissions – Different Scenarios

The CO$_2$ emissions in the different scenarios described below are shown as absolute values in Figure 5.A1 and on a percentage basis in Figure 5.A2.

- **2050 – "No Efficiency Scenario"** – If the world continues to release CO$_2$ with the same intensity as in 2010, the expected massive economic growth will create a huge amount of emissions, up to 214 Gt. This is a non-realistic scenario, but it serves to illustrate the need to keep maximizing carbon productivity.
- **2050 – "BAU Scenario"** – If the world achieves the same carbon productivity as that observed during 1970–2010, the world CO$_2$ emissions will increase to 94 Gt annually. This figure is not sustainable for our Earth, provoking irreversible damage to our climate and way of living.
- **2050 – "Sustainable Scenario"** – according to the OCDE/IEA, a leading authority on this subject, if the world is to avoid climate change, it will need to limit its current CO$_2$ emissions to 14 Gt annually.

For this scenario we took the 14 Gt of CO$_2$ emissions as reference and distributed the emissions reduction by area in a simple mathematical, homogenous, and equitable manner. With acknowledgment of the direct correlation between economic growth and CO$_2$ emissions, and the fact that still today different countries and areas remain at different stages in their economic development, if we wanted to maximize the global economic growth while reducing emissions, we assumed it fair and logical to start reducing further where the per capita emissions are higher, while allowing higher per capita emissions where economic growth still needs to occur. In the long run all countries should aim for the same level of emissions on GDP, below 2%, but in the short term we could not treat all countries equally due to their different starting points.

Under these premises, and to see how challenging the results to address climate change and CO$_2$ emissions reductions might be, we set different efficiency ratios for the different areas and countries. For the ADV economies CO$_2$ emissions were limited to 2% of their total GDP by 2050. For the BRIC the average was around 5% with differences across the group – Russia 3%, China and Brazil 4%, and India 6%. For the REST group the average for the group was 8%.

- In all scenarios the *ADV economies are expected to reduce their share of the world CO$_2$ emissions* compared to the BRIC and REST.
- *China and India are expected to become the largest emitters of CO$_2$*, followed by the USA and Europe.
- The REST may account for 25–40% of the emissions, depending on the scenario used.

The CO$_2$ emissions on a per capita basis are shown in Figure 5.A3.

- In all scenarios the *level of CO$_2$ emissions per capita will remain well above the targeted 4000 g of CO$_2$ per capita per day*

Appendix – Climate Change | 297

(a)

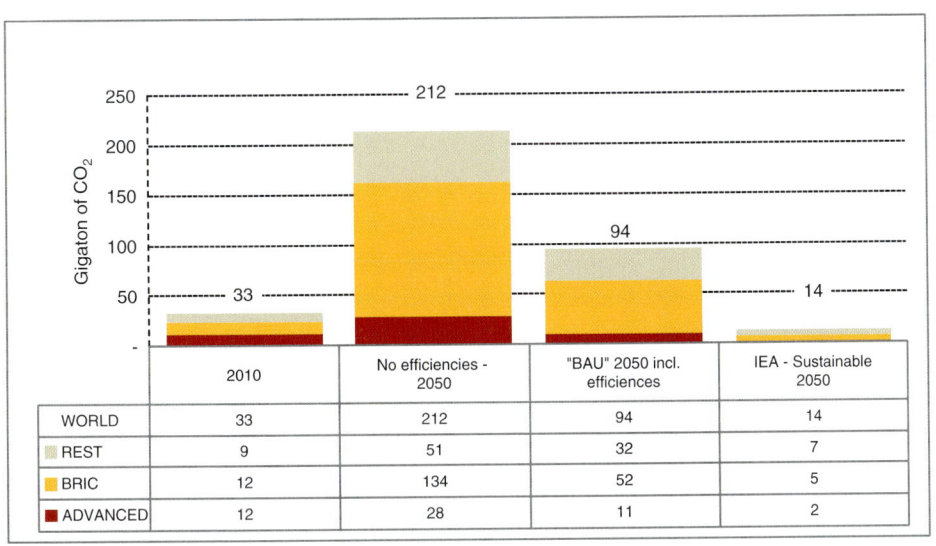

	2010	No efficiencies - 2050	"BAU" 2050 incl. efficiencies	IEA - Sustainable 2050
WORLD	33	212	94	14
REST	9	51	32	7
BRIC	12	134	52	5
ADVANCED	12	28	11	2

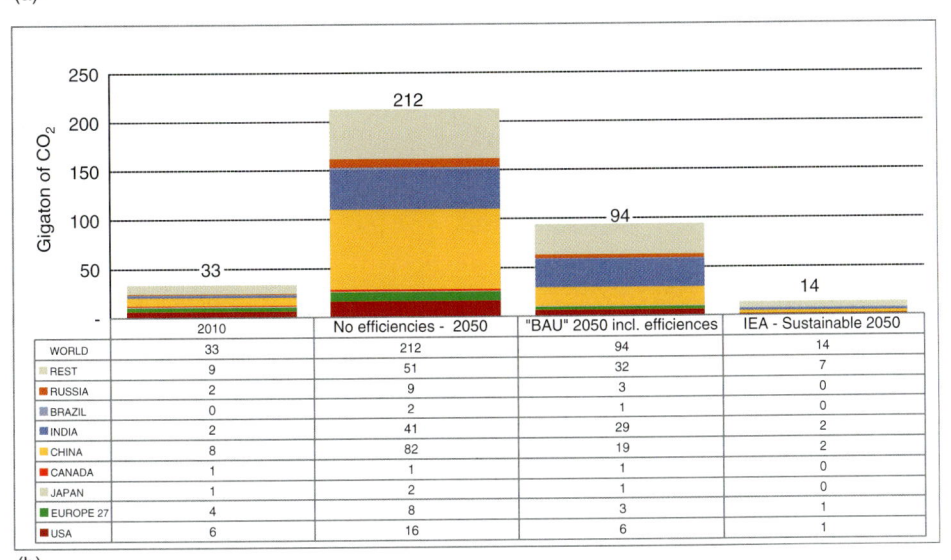

(b)

	2010	No efficiencies - 2050	"BAU" 2050 incl. efficiencies	IEA - Sustainable 2050
WORLD	33	212	94	14
REST	9	51	32	7
RUSSIA	2	9	3	0
BRAZIL	0	2	1	0
INDIA	2	41	29	2
CHINA	8	82	19	2
CANADA	1	1	1	0
JAPAN	1	2	1	0
EUROPE 27	4	8	3	1
USA	6	16	6	1

Figure 5.A1 World CO_2 emissions in Gt with the studied scenarios: (a) by area (b) by country.

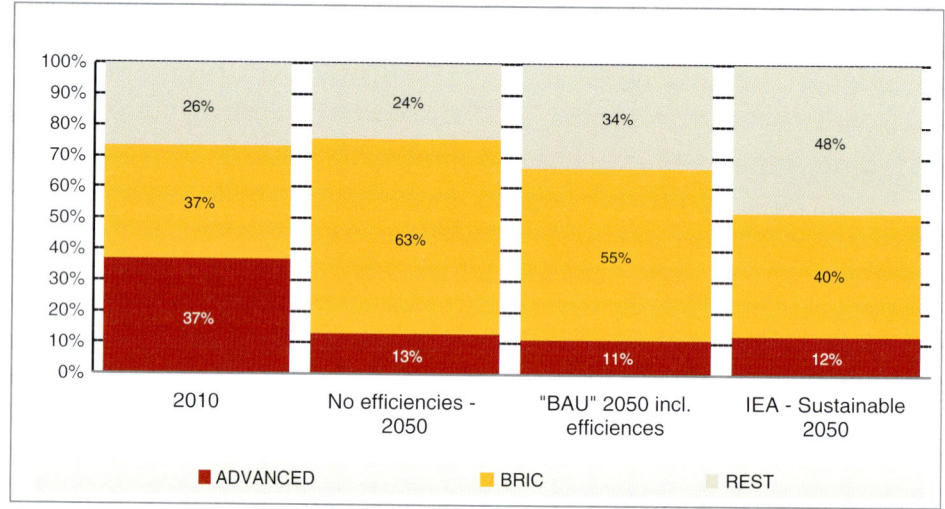

298 | *5 The Chemical Industry by 2050*

(a)

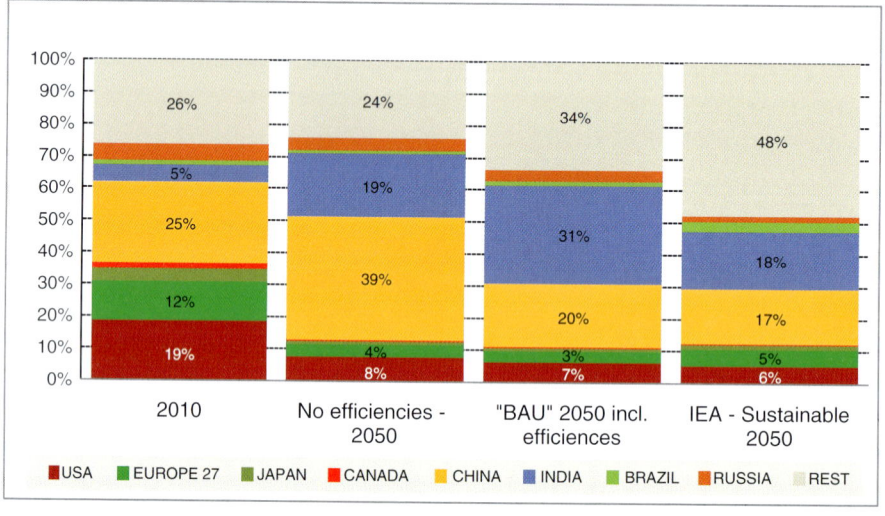

(b)

Figure 5.A2 World CO_2 emissions, expressed as a percentage of world total, with the studied scenarios: (a) by area (b) by country.

Appendix — Climate Change | 299

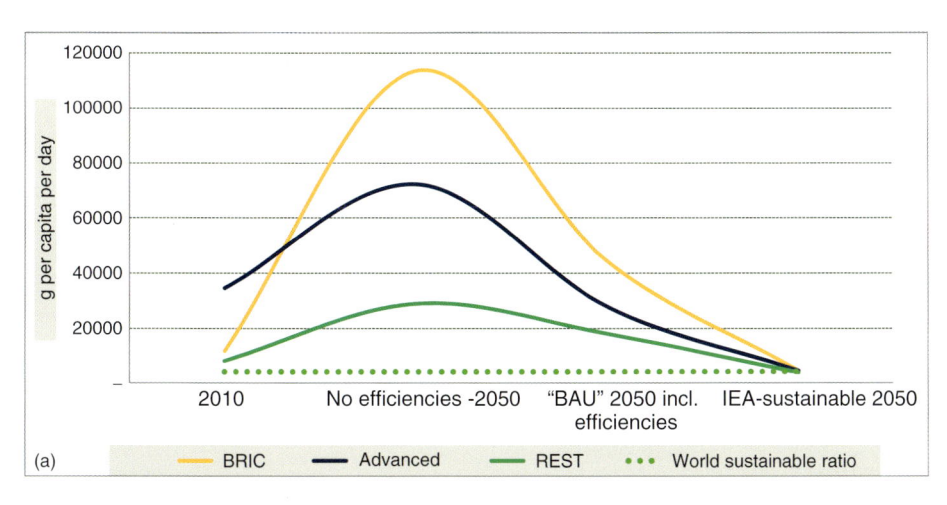

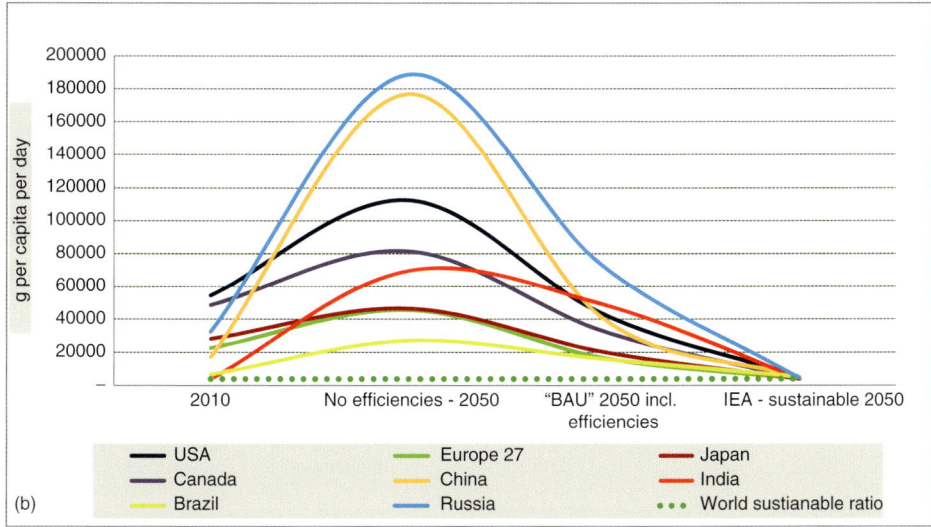

Figure 5.A3 World CO_2 emissions on a per capita per day basis with the studied scenarios: (a) by area (b) by country.

- *Achieving 4000 g of CO_2 emissions per capita per day might be challenging for all countries*, especially for the BRIC and ADV economies.

Further analysis on the sustainable scenario is illustrated in Figure 5.A4.
Figure 5.A5 shows further analysis of absolute values of CO_2 emissions with the BAU scenario.

- *The BRIC and REST economies will account for the largest portion of CO_2 emissions globally*, led by an explosive growth in emissions in China and India.
- *The ADV group despite doubling their economy will reduce CO_2 emissions from 12 to 11 Gt by 2050.* The facts that the ADV will experience a lower economic growth than the BRIC and REST, and that they also have the highest efficiency ratio for CO_2 emissions reductions versus GDP explain that result.

Figure 5.A6 shows efficiency ratios for the period 1970–2010, measured as CO_2 emissions versus GDP.

- These ratios might be conservative, as technological progress might enable further reductions, however, to be factual we deemed it prudent to use these for the BAU scenario. On the other hand, a carbon productivity of 50% in 40 years might be a reasonable target.

Figure 5.A7 shows further analysis of daily per capita emissions with the BAU scenario.

- Under the BAU scenario our world CO_2 emissions will be well above the sustainable level of 4000 g.
- The ADV economies will experience decreases in emissions while the BRIC and REST will increase during the first decades before starting to decrease, however, all will be above sustainable levels.

Figure 5.A8 illustrates the transition from 2010 CO_2 emissions to the 2050 sustainable scenario.
Figure 5.A9 illustrates the transition from the 2050 BAU scenario to the 2050 sustainable scenario.

Appendix — Climate Change | 301

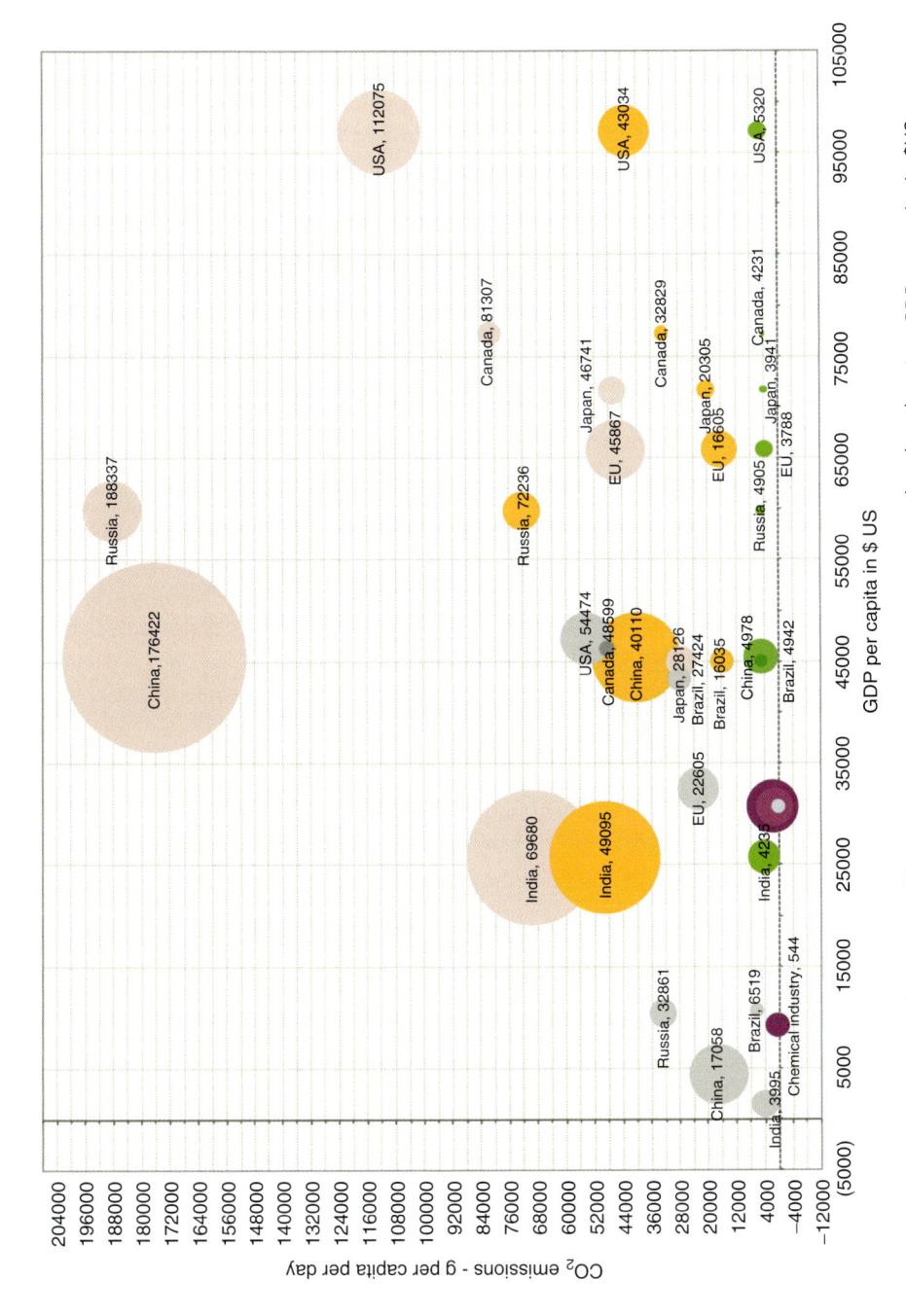

Figure 5.A4 Further analysis of the sustainable scenario CO_2 emissions per capita per day plotted against GDP per capita in $US.

302 | *5 The Chemical Industry by 2050*

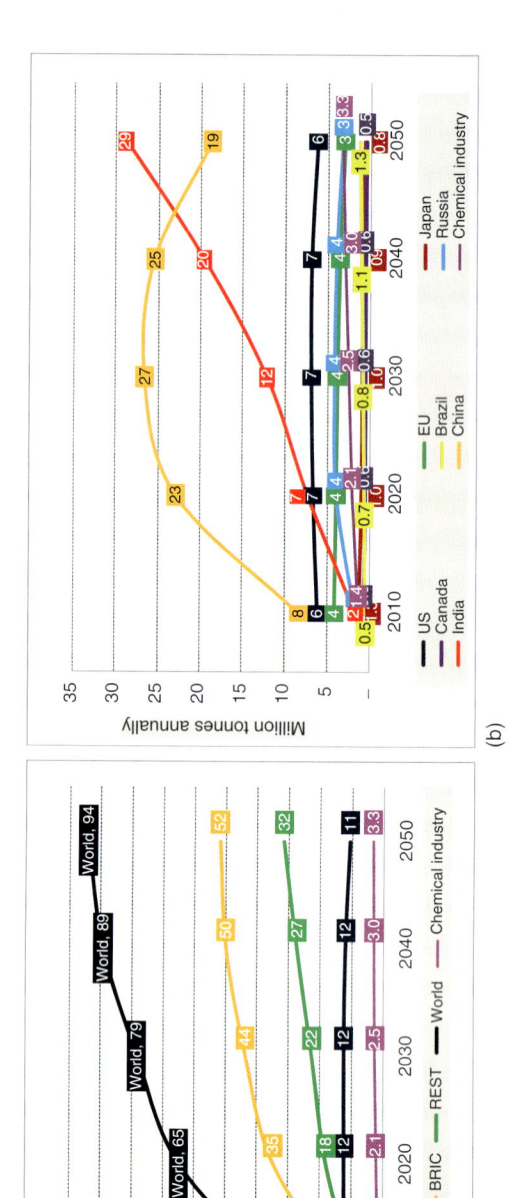

Figure 5.A5 Further analysis of the BAU scenario from 2010 to 2050: (a) by area, (b) by country.

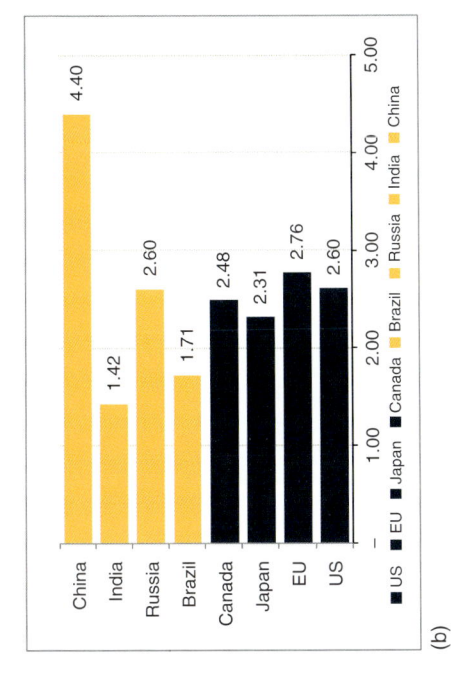

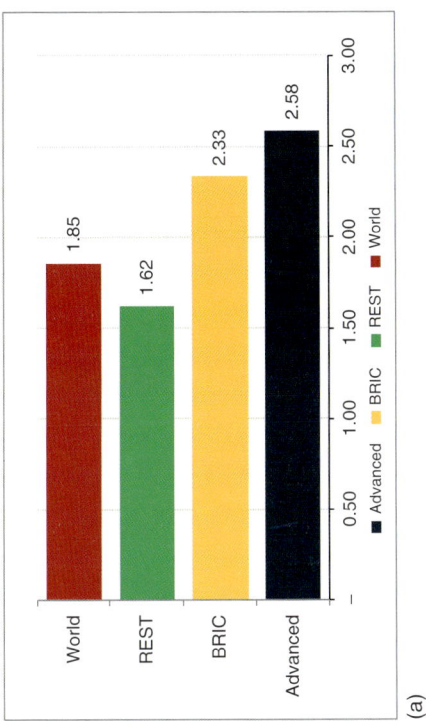

Figure 5.A6 Efficiency ratios measured as CO_2 emissions versus GDP for the period 1970–2010: (a) by area, (b) by country.

304 | 5 *The Chemical Industry by 2050*

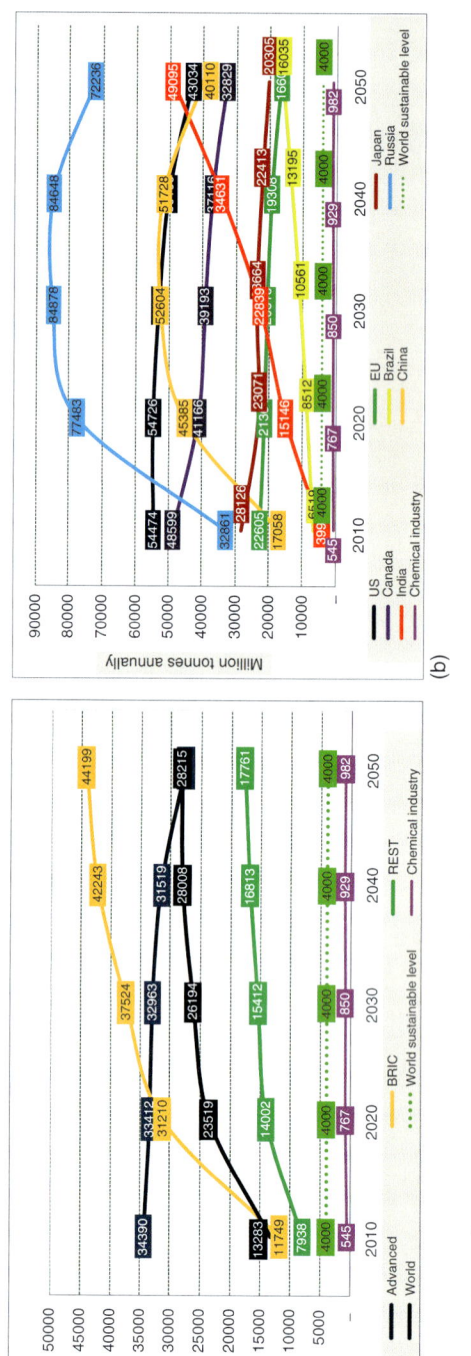

Figure 5.A7 Daily per capita CO_2 emissions by 2050 with the BAU scenario: (a) by area, (b) by country.

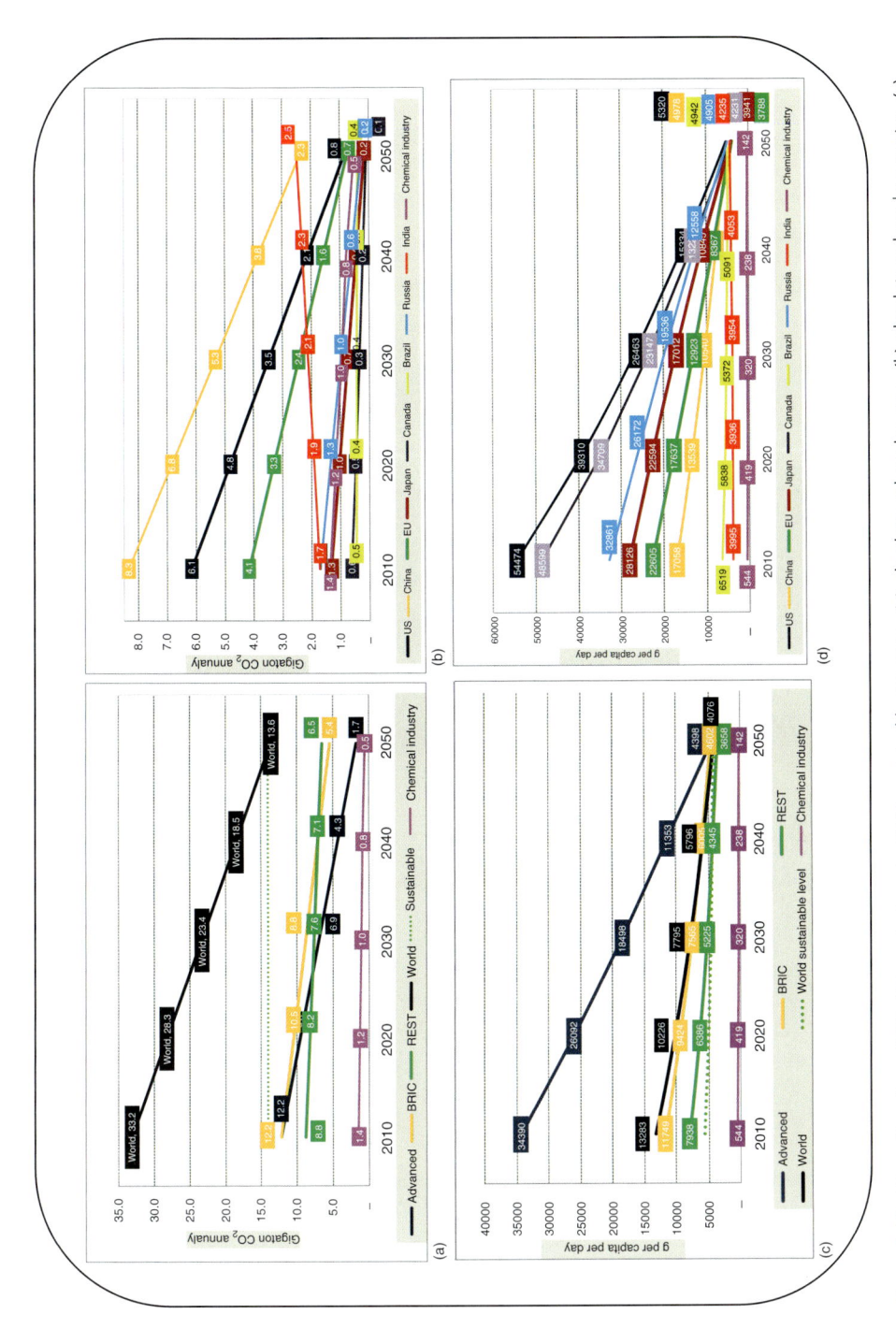

Figure 5.A8 Transition from 2010 CO$_2$ emissions to 2050 sustainable scenario. (a) absolute values by area, (b) absolute values by country, (c) daily per capita values by area, (d) daily per capita values by country.

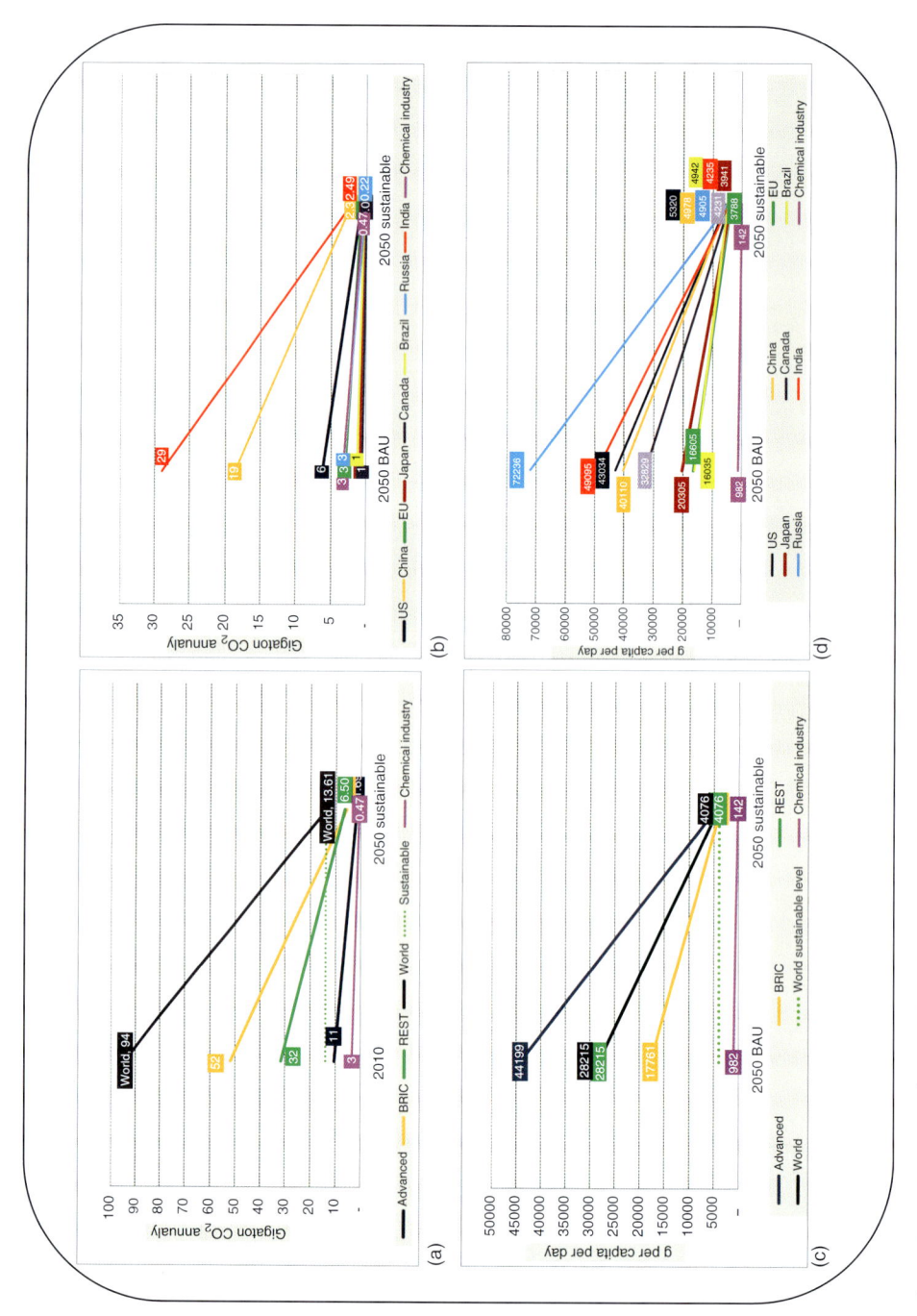

Figure 5.A9 Transition from 2050 BAU scenario to 2050 sustainable scenario. (a) absolute values by area, (b) absolute values by country, (c) daily per capita values by area, (d) daily per capita values by country.

6
Conclusion

A man prepared has half fought the battle.

Don Miguel de Cervantes – Don Quixote

During the next decades the world is poised to witness a period with the longest positive transformation in human history. By 2050 the world is expected to increase its population by an impressive 34% and its average life expectancy by an equally impressive 12%, while the world economy is expected to triple from the 2010 levels.

In a world with more than 9 billion people and a world economy with an estimated GDP of US$ 280 trillion (2009 dollar), the world will have the potential to host one of the largest, wealthiest, wisest, and healthiest societies in human history. The world average life expectancy will increase from 67 years in 2010 to 75 years, with millions of people having the potential to live more than 100 years by 2050, especially in the ADV economies. During this transition millions of people will be free from poverty forever, and the world will have with the highest average GDP per capita in our history, close to US$ 30 675 (2009 Dollar). Information technology and computational progress will have the potential to accelerate and change most of the aspects of our life, the way we work, live, and communicate. The amount of change will be enormous and the speed of it will be spectacular. By 2025 the computational power of one computer could be as large as that of the human brain, while by 2050 the computational power of one computer could be equal to the brain power of the whole world population.

During this transition the chemical industry is not only expected to face one of the most fascinating and transformational periods in its most recent history, but it will be "called into action", playing a vital role for our society, enabling social and economical progress while addressing resource scarcity and climate change. The need to address many challenging, conflicting, and urgent requests, such as massive chemicals demand, changing feedstock, markets, and technologies, resource scarcity, energy reductions, and emissions reductions, will stretch the industry to levels never seen before.

In a world poised to live with just 4000 grams (g) of CO_2 per capita per day, from the current 13 283 g in 2010 and the projected 28 215 g in 2050, the requirements in carbon productivity and energy efficiency will be enormous and unprecedented in human history. Different parts of the world and different industries will be confronted with different levels of carbon productivity but in all cases the challenge

The Future of the Chemical Industry by 2050, First Edition. Rafael Cayuela Valencia.
© 2013 Wiley-VCH Verlag GmbH & Co. KGaA. Published 2013 by Wiley-VCH Verlag GmbH & Co. KGaA.

will be enormous. A world average 90% reduction in CO_2 emissions compared to 2010 will present the largest single challenge and opportunity for our society and the chemical industry.

Climate change and all its embedded consequences will have the potential to trigger the third industrial revolution – the ultimate and lasting one, the sustainable revolution. The first and second industrial revolutions enabled mankind to decouple from its own limitations, unleashing physical work and allowing mass production, higher productivities, technological progress, mobility, globalization, and the longest period of wealth creation. The third revolution will enable human progress to decouple from its own limitation, decoupling forever economic growth and progress from its resources and emissions. During the next decades and century the world is expected to see vast and unprecedented growth; it is up to this generation to learn how to live and progress in a sustainable manner.

The challenge is certainly enormous but the opportunity is not only humongous but historical. The obligation to reduce emissions and avoid climate change is clear and undeniable; its social value is simply invaluable and beyond question, and its economic value is simple overwhelming.

By 2050, the world is expected to release 94 Gigaton (Gt) of CO_2 under the BAU scenario and just 14 Gt under the sustainable scenario. The economic value to reduce these emissions from the BAU to the sustainable scenario could fluctuate between an overwhelming US$ 1040 trillion or 15% of the cumulative GDP from 2010 to 2050, assuming a price of US$ 13 or €10 per ton of CO_2, and US$ 2080 trillion or 30% of the world cumulative GDP assuming a price of US$ 26 or €20.

The economics and opportunities behind emissions reductions and climate change can be so overwhelming that only when placing them in the right context and in the light of what we have at stake can they become not only compelling but also reassuring. The chance to host the largest, wealthiest, healthiest, and wisest society in human history, and the obligation to leave a better and sustainable world for the next generations is not only a tremendous incentive and responsibility but also an enormous obligation and privilege that this generation cannot and simply must not miss.

The required changes are so dramatic and the terms are so short, that the opportunities and their value will be simply humongous. The need to decouple economic growth and prosperity from energy consumption and emissions is not only a must, but also a fascinating challenge and a formidable business opportunity.

The need to increase fuel efficiency to the maximum level – improvements in fuel efficiency account for 24% of the expected emissions reduction under the sustainable scenario – will create massive technological challenges but also tremendous opportunities across all industries. The need to increase the world energy demand, while migrating from fossil fuels to renewable sources and natural gas is not only a tremendously complex and highly expensive exercise but also an invaluable business opportunity. Under the sustainable scenario for 2050 the world will need to double its energy demand, increasing by 21% the use of renewable energy and by 34% the use of natural gas, while reducing its current crude oil and coal demand from 2010 levels by 27 and 18%, respectively.

Finally the need to further reduce emissions around the world will also imply the construction of more than 2500 CO_2 capture and storage units globally; investments that could amount to US$ 100 trillion by 2050.

In this global context of a vast amount of challenges and transformations; the chemical industry will not remain indifferent, indeed it will be subject to one of the most complex and radical transitions. The chemical industry as an industry of industries, a key enabler of emissions reductions and better quality of life, will have the dual task of accomplishing its own transformation while enabling the transformation and emissions reduction in many other industries.

During the next decades the chemical industry will be poised to experience one of its most severe and complete transformations, with radical changes in almost all aspects of the industry, from its feedstock, to its products, emissions, economics, markets, technologies, players, customers, and even in its ways of working, innovating and relating to other industries, governments, and society. The transformation will be gradual, but the impacts will be structural and pervasive.

During the next decades the chemical industry is poised to triple in size from US$ 3.9 trillion sales in 2010 to 18.7 trillion by 2050. The chemical side of the industry is expected to grow from 3.1 to 14.9 trillion, while the pharmaceutical industry is expected to grow from US$ 875 into 3.9 trillion during the same time frame.

The world annual per capita consumption of pharmaceuticals is expected to grow from US$ 128 in 2010 to US$ 427 by 2050. However, and despite this massive growth, the world per capita demand in 2050 will still be below the per capita demand of the ADV economies in 2010, US$ 678. The still large differences in per capita demand, especially in the REST group with just US$ 167 and some of the BRIC economies, such India with expected US$ 283 per capita demand, make us believe that further growth will be possible after 2050.

On the chemical side, the world per capita demand is expected to grow from US$ 456 in 2010 to US$ 1631 by 2050; a figure that is equivalent to the chemical per capita demand in the USA in 2010. Under this scenario chemical demand will grow strongly in all regions but especially in the BRIC and REST economies. The BRIC and REST economies are expected to add more than US$ 6.5 and 3.5 trillion of chemical demand, respectively, by 2050. The ADV economies, with much larger and maturer chemical markets will still add more than US$ 1.5 trillion of chemical demand. Despite all the growth the per capita demand in the REST economies will be just half of the world per capita demand, signaling significant growth potential for the remaining part of the century.

All segments of the chemical industry will be impacted, and massive growth would be expected across all segments and most of the regions.

The petrochemical global demand, as one of the largest and most significant segments of the industry, is also expected to grow massively during the next decades. Indeed Ethylene demand could grow by up to 400% from the current 120 Million tons (Mt) in 2010 to 500–600 Mt by 2050, depending on the scenario used. The large differences in ethylene per capita demand in 2010, with just 3, 10, and 14 kg for India, China, and the REST respectively, versus the current 44 kg in Europe and 77 kg in the USA, present a tremendous growth opportunity for ethylene demand globally.

According to the modified BAU scenario, the chemical industry is expected to add more than 540 world class steam crackers (800 kt ethylene capacity) around the world by 2050, representing a tremendous investment opportunity for the industry and worth up to US$ 1 trillion during the next decades. Expansions and additions of crackers are expected in all areas and most of the economies, but especially in the REST and BRIC where the bulk of the petrochemical demand is expected.

The REST countries (including the Middle East), China and India alone are expected to account for more that 80% of the ethylene expansions by 2050, with 230, 130, and 90 new "world class" steam crackers, respectively. Among the ADV economies, the USA and Europe, on the back of growing economies and populations and shale gas, will account for most of the projected ethylene demand growth in this area. The USA is expected to add up to 45 new world class crackers by 2050, while Europe is also expected to add another 12–15 crackers, depending on the scenario used.

The massive growth in chemicals and pharmaceuticals in the BRIC and REST economies will serve to change the structure of the market, confirming the gradual market shift from the ADV economies to these areas. That shift will imply the appearance of major new customers and suppliers in the BRIC and REST market. In 2010 two of the top 10 chemical companies were based in the BRIC and REST economies, while not even one of the top 50 pharmaceutical companies was based in the BRIC or REST economies. By 2050 the BRIC and REST economies will account for more than 75% of the world chemical demand and 58% of the pharmaceutical demand. In this context, should we expect a reshuffle at the top of the world chemical and pharmaceutical industry?

This reshuffle will be accompanied by a spectacular growth in the top world chemicals and pharmaceuticals companies, as well as the appearance of new megacorporations in the REST and BRIC economies. The top chemical and pharmaceutical companies, on the back of the massive growth, will experience sizes and revenues never seen before. Companies like BASF, DOW, Exxon Chemicals, Sinopec, LiondelBasell, SABIC, Pfizer, Johnson & Johnson, Novartis, and Roche, among others, will have the chance to double and triple in size. Could we imagine the opportunities, challenges, and responsibilities attached to companies with several hundreds of billion dollars in annual revenue?

This massive growth in chemicals demands will not come alone, and significant transformations in energy and feedstock demand, chemical products, markets structures, and customers will need to be complemented with the urgent need to address climate change and reduce greenhouse emissions.

The unique capacity of the industry to reduce emissions is certainly placing the industry in an extraordinary position to support emissions reductions and play a crucial role in the fight against climate change. The chemical industry has distinguished itself for being at the forefront of innovation and technology during the last centuries, enabling vast improvements in living standards and human progress around the world.

During the next decades the chemical industry will be called into action, a call for the industry to excel in what it does best – technology and innovation. Its unique capacity to enable emissions reductions across all industries will stimulate the technological aspect of the industry to its highest levels, producing changes all over the value chain, but more importantly in the way the industry works and relates to its customers, suppliers, governments, and citizens.

According to our calculations, in 2005 the whole chemical industry generated around 1.2 Gt of CO_2, however, the total net greenhouse emissions abatement allowed by the chemical industry in 2005 reached 6 Gt of CO_2 equivalent. *In other words for almost 1 unit of CO_2 released by the chemical industry, 5 units of CO_2 equivalent were reduced.*

Despite the enormous contributions of the chemical industry to the world emission reductions, the chemical industry itself will need to reduce its own emissions significantly. This will not be an easy or inexpensive exercise.

According to our estimations the chemical industry generated around 1.36 Gt of CO_2, equivalent to 545 g of CO_2 per capita per day in 2010. By 2050 under our BAU scenario the industry could generate up to 3.3 Gt of CO_2, equivalent to 982 g of CO_2 per capita per day. However, under the sustainable scenario, where the world can only generate 14 Gt of CO_2 by 2050 and citizens will need to live with a severe diet of just 4000 g of CO_2 per capita per day, *the chemical industry will need to reduce its emissions to just 0.47 Gt, equivalent to just 142 g of CO_2 per capita per day.*

Under these assumptions the chemical industry will need to completely decouple its emissions from its production. Indeed the industry will need to be able to quadruple its production, while reducing its CO_2 emissions by 65% versus the 2010 level and by 85% versus the 2050 BAU level. This is certainly a formidable task, which will require tremendous focus, coordination, and investment.

As of January 2013 the emissions from the European petrochemical industry will be traded under the European Emission Trading Scheme. Similar schemes have started and will be starting around the world, building additional pressure on the industry to reduce emissions. If we assume that in the next decades the whole chemical industry will be subject to Emission Trading Schemes and the price for CO_2 emissions will be at least US$ 13 per unit of CO_2, *the potential value for the chemical industry to reduce from 3.28 Gt of CO_2 emission projected under the BAU scenario to the projected 0.47 Gt of CO_2 projected under the sustainable scenario could amount to a staggering figure of $36 trillion.*

The global need to reduce emissions, ensure energy supply, and reduce fossil fuel consumption, in combination with a massive growth in chemical demand and much better cracker economics for light feedstock globally, will trigger a massive expansion of natural gas and shale gas demand. Shale gas is going to play a pivotal role not only in the world energy mix but also in the chemical industry feedstock mix.

The USA has already gone through an impressive transformation to light cracking, presenting a bright outlook for the chemical industry. China, as a country with the highest expected demand for chemicals and petrochemicals by 2050, the largest world reserves of shale gas, and the highest world contribution of emissions already today, is expected to develop its shale gas resources aggressively during the next decades. Similarly, but for different reasons, Europe will have tremendous incentives to develop shale gas during the next decades. The need for Europe to ensure energy supply, address climate change, and restore the competitiveness of the still growing European petrochemical *industry will serve to catalyze a more complex transition to light cracking.*

That transition to light cracking will be certainly accompanied by a transition to bio-feedstocks. This will take longer that the transition to shale gas, but their growth will be solid and continuous. Biomass or bio-ethanol from sugar, corn, or beet will become more and more popular. Ethylene from sugar is already available and although the economics might not be fully compelling, these will need to be repositioned based on the upcoming need to reduce CO_2 emissions drastically. According to the latest information from BRASKEM, the Brazilian chemical giant, ethylene from sugar can capture or fix up to 2.5 tons of CO_2 per ton of ethylene. Considering the global needs for the chemical industry to reduce its own emissions, and under the assumption that the full cost for CO_2 emissions will be soon captured in the price and cost for chemical production, bio-feedstocks will have a bright future in the next decades.

In terms of its products, climate change will provide perhaps the single most significant business opportunities for the chemical industry, especially for the companies able to lead in this sustainable revolution, enabling other industries to cut further and further their own emissions. As we observed in the business case on the European Tire Labeling, more and more companies and industries will be legally required to create products that comply with already very aggressive emission targets. Upcoming legislation will not only become more and more global, but their targets will also become more and more stringent; after all a tenfold global carbon productivity increase during the next decades opens the door for significant emission reductions.

On the positive side, climate change can create a tremendous business opportunity. Proper Legislation, as the one presented in during business case (European Tire labelling), can drive emissions reductions, and innovation but also a unique "triple win" situation among citizens, governments and companies. Situations where stringent emissions targets will allow governments to cut emissions, citizens will be able to enjoy better products (tires) and less polluted environments, and the industry players, mainly tire, chemical and synthetic rubber producers, will enjoy better prices and margins. On the negative side, companies not able to cope with these reductions will remain uncompetitive, losing market share, reducing benefits, and at serious risk of being out of the market.

Proper emission reductions legislation has the extraordinary capacity to improve market transparency, allowing better producers and products to increase market share and benefits, while creating a virtuoso paradigm, where emissions reductions drive innovation and vice versa. Climate change remains as a tremendous challenge and opportunity for the chemical industry, where the best products and companies will keep gaining market share globally and the worst will gradually fade out.

Therefore, the companies able to lead this third "sustainable" revolution and go beyond the requested limits will become the winners of the future. Emissions reductions for the chemical industry, and life extension and enhancements of the quality of life for the pharmaceutical industry, appear as the largest and most lucrative business opportunities.

Among all the challenges and opportunities the chemical industry will need to face during the next decades, climate change appears as the most pervasive and radical one; massive growth in chemical and pharmaceutical demand as the most exciting and alluring one; and technological collaboration and convergence as the most transformational one. The tremendous increase in computational progress, in combination with the need to reduce dramatically emissions across all industries, will trigger a massive technological convergence during the next decades. The need to accomplish massive emissions reductions across the whole value chain will accelerate that process. The fact that none of the players of the value chain alone, even when they apply their latest technologies, will be able to meet today's targets on emissions reductions, *will foster a major change in technological progress, a change in the way the chemical industry and industries in general work together. A change toward "collaboration and convergence."*

As the different members of the value chain will become aware of this, they will need to learn how to collaborate and work together, and that might be by far one of the major technological progresses for the industry. This realization will be a tremendous shock in some value chains, especially those where confrontation and trade secrecy rather than collaboration has been the leading pattern. The chemical industry, as an industry of industries, could play a critical role in building trust and expertise around the value chain. Radical technologies will need to be developed, but these will no longer be invented by one company or part of the value chain but collectively.

Computational progress will accelerate the trend fostering collaboration and technological convergence. Today we tend to analyze and value companies and industries in isolation; in the future we may need also to value as their capacity to work with other companies and industries. Using as an example our business case on the European Tire Labeling, could you imagine what kind of technologies could be developed if a tire producer worked together to reduce CO_2 emissions with an automotive producer, a synthetic rubber producer, a natural rubber producer, a filler producer, a steel cord producer, Google, Apple, or Facebook? Could we envision the total value these companies collectively could bring together, versus working individually?

Could we envision what sort of new pharmaceuticals could result when different industries and companies converge together? Could we imagine what kind of medicines could be created if Pfizer worked together with BASF, Dow, Apple, Google, and Nike? Could we envision what sort of technological progress could be delivered when converging players across the whole value chain and across several industries work together?

That technological "collaboration and convergence" across industries and companies will be at the core of the third industrial revolution toward sustainability, and with it the technological relevance of the chemical industry is poised to increase to levels never seen before.

Under this depicted scenario we could expect massive attention to be focused on the technological and innovative side of the industry as well as on its major products and applications, raising the public profile of the chemical industry. As the chemical industry learns to work together with other players in the value chain and other industries, the significant increase in the role of governments and citizens will also open the door for further transparency for the industry. The massive economic growth expected for the chemical industry will trigger the creation of large corporate mega-economies. In a global context, where governments will become bigger and citizens more informed and connected, large multinationals will be subject to further scrutiny.

The fact that the chemical industry is in a leading position to enable sustainable economic growth, better living standards and significant emission reductions will foster the openness of the industry toward society. The chemical industry has been enabling better living standards for centuries, however, it has not always been recognized for it. Climate change brings the largest single challenge and opportunity for the industry, the opportunity to enable a sustainable world for us and the generations to come. Under this scenario the industry will need to work closer with governments and citizens to better explain its challenges, aspirations, and successes. The world will place high demands on the chemical industry, eager for solutions and impatient to build on successes, the chemical industry must be at the forefront.

During the next decades the world will need more and better chemical and pharmaceutical products. The chemical industry enabled the first and second industrial revolution, this time the industry will need to enable the third revolution, the "Sustainable" one. The world chemical industry will be called into action, and the chemical industry will respond as it always did, doing what it does best through technology and innovation.

This book is not intended to provide an ultimate and conclusive view on the future of the chemical industry, but rather a set of conservative scenarios for further industry discussion and debate; a solid framework for understanding the basics of the industry and the major upcoming megatrends. This book is not intended to be the end of the discussion, but rather the beginning. Indeed this book might not have all the answers that you were looking for, but I hope it will have some of the questions that you were looking to hear. After all, it is the whole industry that will decide its own future.

By 2050 the world has the potential to host the *"largest, wealthiest, wisest, and healthiest society in human history"*, and with that opportunity the responsibility will be equally high. The chemical industry enabled the first and second revolution, improving living standards and quality of life during the centuries, this time it is again called to play a critical role. Climate change, resources and greenhouse emission reductions appear to be the major and last roadblock in order to attain a sustainable world and achieve this spectacular vision of our world. *The challenge is enormous, the responsibility is humongous and the rewards beyond question. The chemical industry will soon be called into action and with that call it is expected to be at the forefront of this revolution. The stakes are high, very high, but this is an opportunity this society cannot deny to the upcoming generations. The chemical industry will lead into this revolution, and the world will overcome climate change, solving this massive challenge with technology and innovation as it always did; we just hope this time we will not forget it too soon.*

Appendix

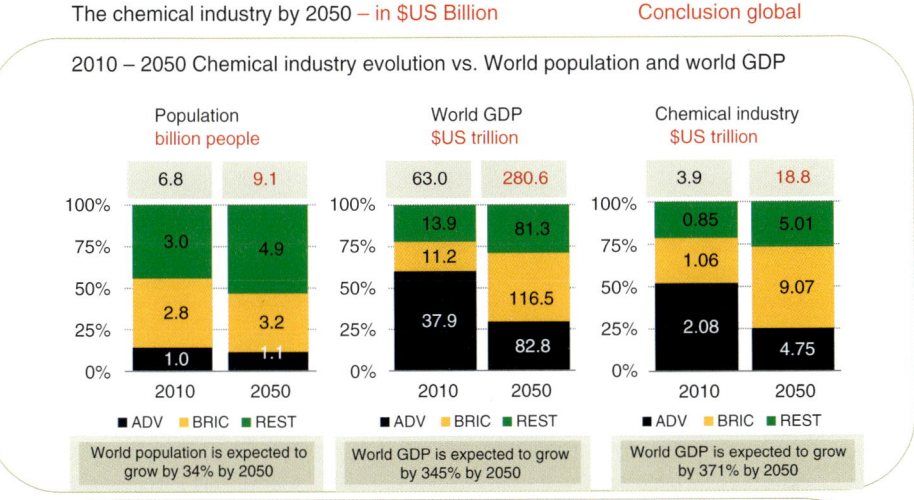

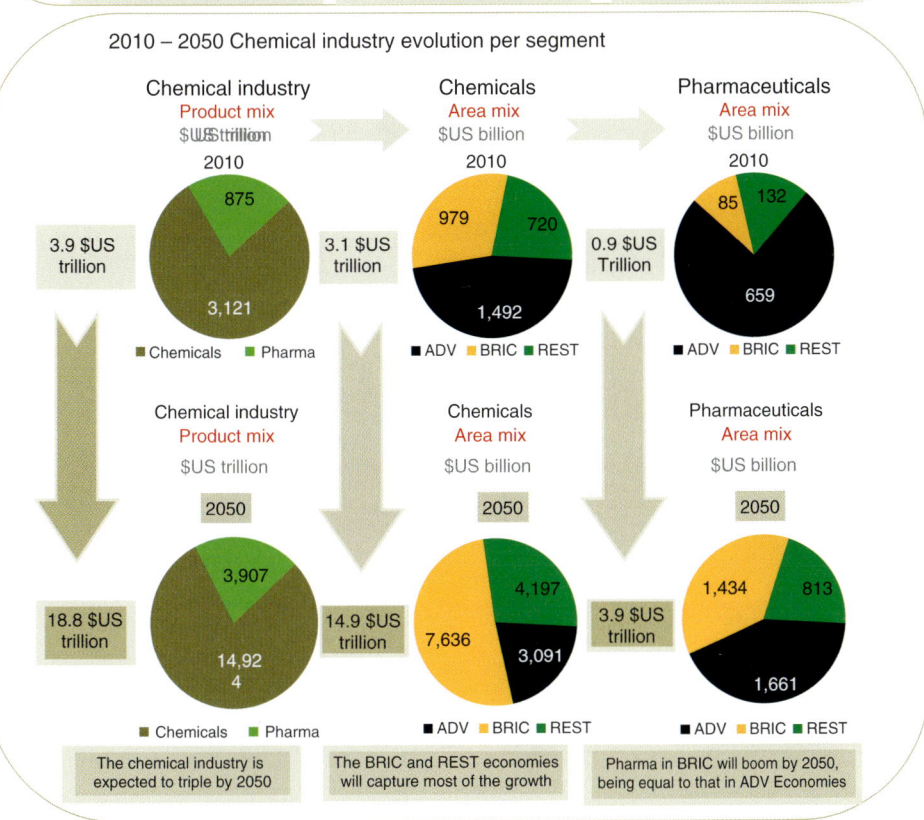

The chemical industry by 2050 – in $US Billion — Conclusion global

2010 – 2050 Chemical industry evolution vs. World population and world GDP

Population — billion people

	2010	2050
Total	6.8	9.1
REST	3.0	4.9
BRIC	2.8	3.2
ADV	1.0	1.1

World population is expected to grow by 34% by 2050

World GDP — $US trillion

	2010	2050
Total	63.0	280.6
REST	13.9	81.3
BRIC	11.2	116.5
ADV	37.9	82.8

World GDP is expected to grow by 345% by 2050

Chemical industry — $US trillion

	2010	2050
Total	3.9	18.8
REST	0.85	5.01
BRIC	1.06	9.07
ADV	2.08	4.75

World GDP is expected to grow by 371% by 2050

2010 – 2050 Chemical industry evolution per segment

Chemical industry — Product mix — $US trillion — 2010 (3.9 $US trillion)
- Chemicals: 3,121
- Pharma: 875

Chemicals — Area mix — $US billion — 2010 (3.1 $US trillion)
- ADV: 1,492
- BRIC: 979
- REST: 720

Pharmaceuticals — Area mix — $US billion — 2010 (0.9 $US Trillion)
- ADV: 659
- BRIC: 85
- REST: 132

Chemical industry — Product mix — $US trillion — 2050 (18.8 $US trillion)
- Chemicals: 14,924
- Pharma: 3,907

Chemicals — Area mix — $US billion — 2050 (14.9 $US trillion)
- ADV: 3,091
- BRIC: 7,636
- REST: 4,197

Pharmaceuticals — Area mix — $US billion — 2050 (3.9 $US trillion)
- ADV: 1,661
- BRIC: 1,434
- REST: 813

The chemical industry is expected to triple by 2050

The BRIC and REST economies will capture most of the growth

Pharma in BRIC will boom by 2050, being equal to that in ADV Economies

6 Conclusion

The chemical industry by 2050 – in %   Conclusion global

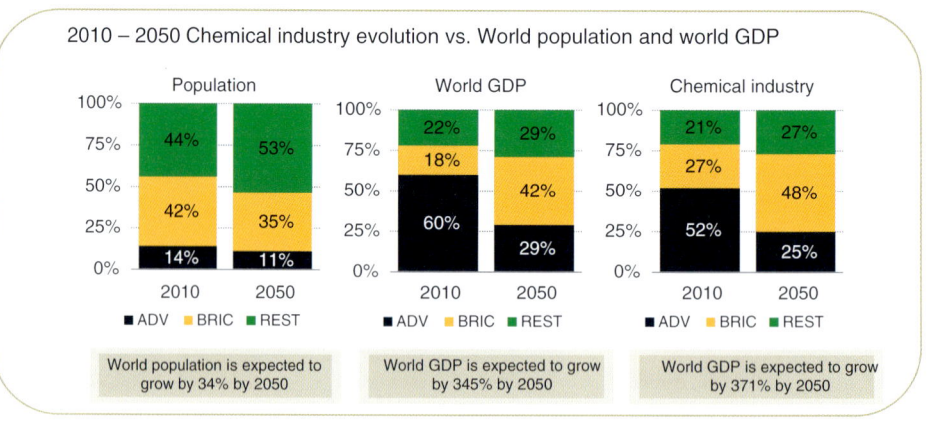

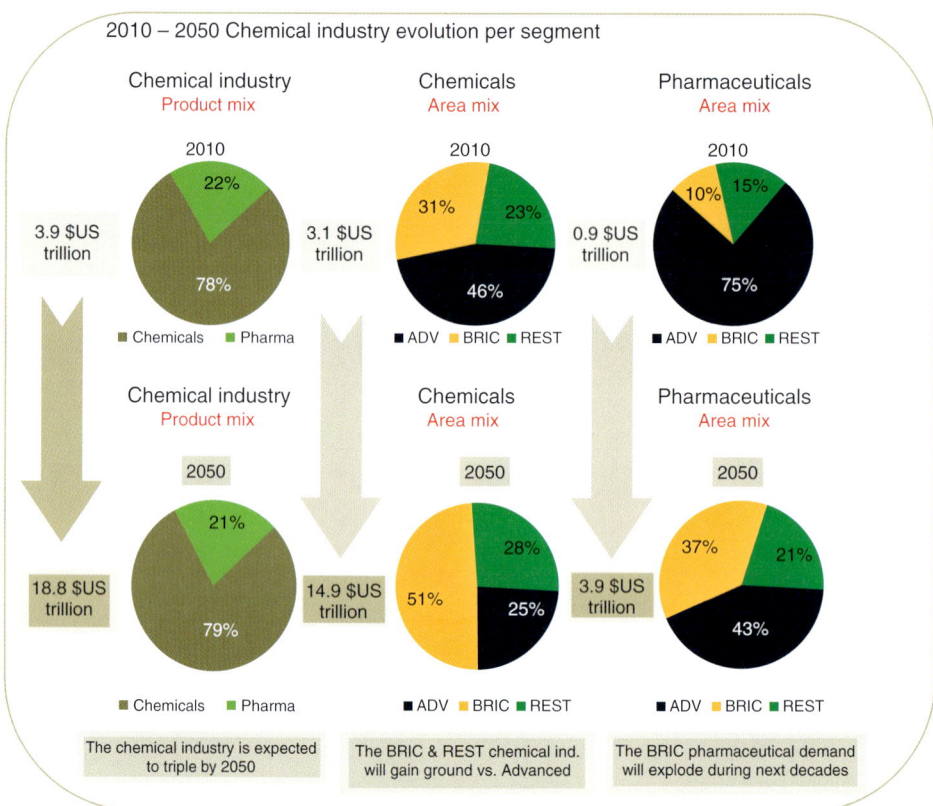

2010 – 2050 Chemical industry evolution vs. World population and world GDP

Population (2010 → 2050)
- ADV: 14% → 11%
- BRIC: 42% → 35%
- REST: 44% → 53%

World population is expected to grow by 34% by 2050

World GDP (2010 → 2050)
- ADV: 60% → 29%
- BRIC: 18% → 42%
- REST: 22% → 29%

World GDP is expected to grow by 345% by 2050

Chemical industry (2010 → 2050)
- ADV: 52% → 25%
- BRIC: 27% → 48%
- REST: 21% → 27%

World GDP is expected to grow by 371% by 2050

2010 – 2050 Chemical industry evolution per segment

Chemical industry – Product mix – 2010: 3.9 $US trillion
- Chemicals: 78%
- Pharma: 22%

Chemicals – Area mix – 2010: 3.1 $US trillion
- ADV: 46%
- BRIC: 31%
- REST: 23%

Pharmaceuticals – Area mix – 2010: 0.9 $US trillion
- ADV: 75%
- BRIC: 10%
- REST: 15%

Chemical industry – Product mix – 2050: 18.8 $US trillion
- Chemicals: 79%
- Pharma: 21%

Chemicals – Area mix – 2050: 14.9 $US trillion
- ADV: 25%
- BRIC: 51%
- REST: 28%

Pharmaceuticals – Area mix – 2050: 3.9 $US trillion
- ADV: 43%
- BRIC: 37%
- REST: 21%

The chemical industry is expected to triple by 2050

The BRIC & REST chemical ind. will gain ground vs. Advanced

The BRIC pharmaceutical demand will explode during next decades

6 Conclusion

Chemical industry evolution 2010–2050 – Growth   Conclusion global

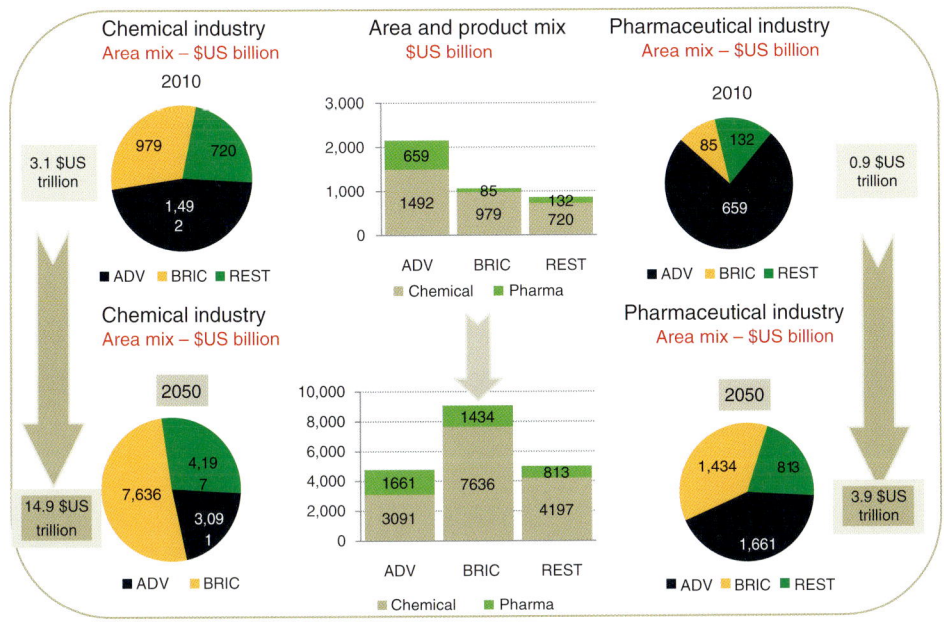

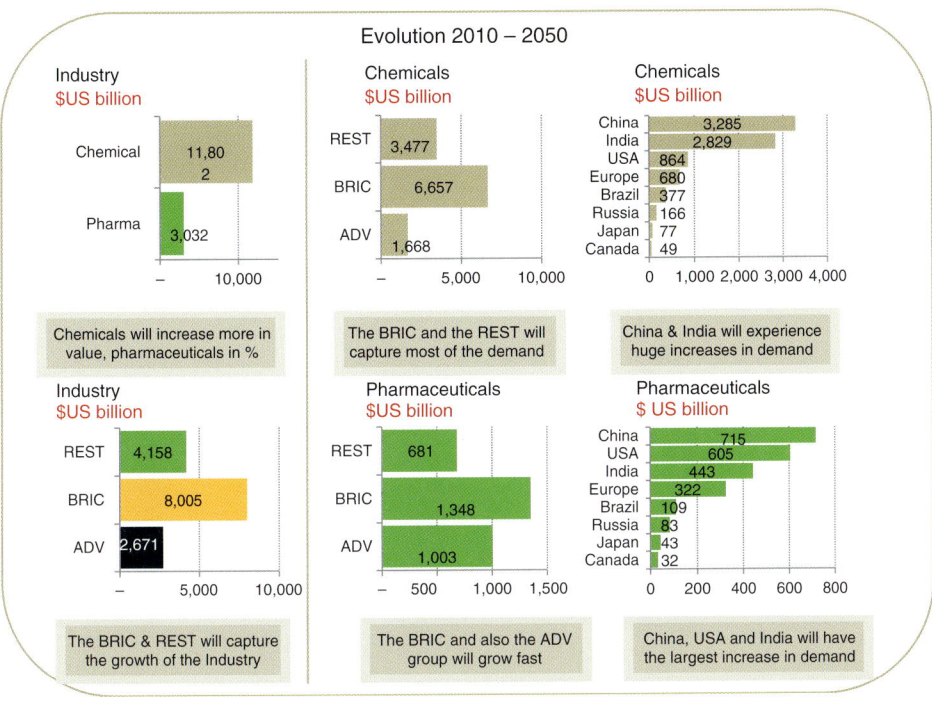

Chemical industry
Area mix – $US billion
2010: 3.1 $US trillion
- ADV: 1,492
- BRIC: 979
- REST: 720

Area and product mix
$US billion
- ADV: Chemical 1,492; Pharma 659
- BRIC: Chemical 979; Pharma 85
- REST: Chemical 720; Pharma 132

Pharmaceutical industry
Area mix – $US billion
2010: 0.9 $US trillion
- ADV: 659
- BRIC: 85
- REST: 132

Chemical industry
Area mix – $US billion
2050: 14.9 $US trillion
- ADV: 3,091
- BRIC: 7,636
- REST: 4,197

Area and product mix 2050
- ADV: Chemical 3,091; Pharma 1,661
- BRIC: Chemical 7,636; Pharma 1,434
- REST: Chemical 4,197; Pharma 813

Pharmaceutical industry
Area mix – $US billion
2050: 3.9 $US trillion
- ADV: 1,661
- BRIC: 1,434
- REST: 813

Evolution 2010 – 2050

Industry $US billion (2010)
- Chemical: 11,802
- Pharma: 3,032

Chemicals will increase more in value, pharmaceuticals in %

Chemicals $US billion
- REST: 3,477
- BRIC: 6,657
- ADV: 1,668

The BRIC and the REST will capture most of the demand

Chemicals $US billion
- China: 3,285
- India: 2,829
- USA: 864
- Europe: 680
- Brazil: 377
- Russia: 166
- Japan: 77
- Canada: 49

China & India will experience huge increases in demand

Industry $US billion
- REST: 4,158
- BRIC: 8,005
- ADV: 2,671

The BRIC & REST will capture the growth of the Industry

Pharmaceuticals $US billion
- REST: 681
- BRIC: 1,348
- ADV: 1,003

The BRIC and also the ADV group will grow fast

Pharmaceuticals $US billion
- China: 715
- USA: 605
- India: 443
- Europe: 322
- Brazil: 109
- Russia: 83
- Japan: 43
- Canada: 32

China, USA and India will have the largest increase in demand

Index

a
American Chemical Council (ACC)   97
Arab Embargo oil shock   65
artificial silk. See rayon, from wood fibers

b
brine electrolysis (chlorine)   188–189
business as usual (BAU)   2–3, 14, 15, 143, 144, 146–155, 247–250, 258, 260–264, 308, 310
– scenario   144

c
carbon dioxide   161
– capture and storage (CCS)   158
– and carbon productivity challenge   111
– emissions   308, 311
– – cumulative economics   107–108
– – per GDP   158
– – reduction in tire and automotive industry   113–129
– – world   268–277, 297–307
– steam cracker   245
– variations   101
caustic soda   188
celluloid (1870)   187–188
climate change   308
– carbon dioxide emissions reduction in tire and automotive industry   113–129
– revolution against (2010–2050)   194
corporate mega-economies (CME)   13, 47, 49–50
crude oil
– demand growth, by 2030   76
– production cost, globally   73
– projections, BAU scenario   153–154
Cuadrilla Resources   98
current status, with 2010 data   163

– economic relevance   163–166
– feedstock and energy   171–172
– industry relevance and profitability   166–168
– industry structure and companies   173–174, 176–184
– major features   197
– – global megatrends   199
– major sectors and products   173
– recent history excluding pharmaceuticals   186
– – brine electrolysis (chlorine)   188–189
– – celluloid (1870)   187–188
– – industrial revolution and inorganic chemistry (1750–1850)   187
– – internationalization (1960s)   194
– – plastics demand explosives (1950s)   191–194
– – polyamide nylon (DuPont)   190
– – rayon from wood fibers (1880)   188
– – revolution against climate change (2010–2050)   194
– – steam cracker (1910–1920)   190
– – styrene cracking   190
– – synthetic dyes from coal for textiles, and chlorine bleach   187
– – synthetic fertilizers   189–190
– – synthetic rubber (1930s)   191
– safety   184–186
– technological relevance   166–168
– upcoming trends   195–197

d
demise/sleeping period, of energy   69
demographics   18
– area and age distribution   18
– – change in age distribution   18

The Future of the Chemical Industry by 2050, First Edition. Rafael Cayuela Valencia.
© 2013 Wiley-VCH Verlag GmbH & Co. KGaA. Published 2013 by Wiley-VCH Verlag GmbH & Co. KGaA.

e

economic megatrends   29–39
- foreign direct investment (FDI)   40–42

emission unit allowances (EUA)   276–277

energy demand
- by area   79
- and feedstock, in 2006   171
- world   81
- – by 2030   76, 78
- – 2050, BAU scenario   151–152

energy–heat intensity   64

Energy Information and Administration (EIA) US   6, 155

energy megatrends   61
- and economics   70–71
- lessons   69
- life cycle   69
- oil peak   71–80
- recent developments   80, 82–98
- recent transitions   63–68
- shocks as valuable source of information   70
- success criteria   69–70
- transitions in life   70

engagement rules, new   60

European Emission Trading System (EU ETS)   240, 275–276, 311

European Policy Evaluation Consortium (EPEC)
- combined tire performance versus rolling resistance and wet grip   12

European tire labeling   119

f

Facebook   50–52
- users, by country   53

foreign direct investment (FDI)   40–42

g

global megatrends. See megatrends, global
global temperatures   84
governments, significance of   54–61
greenhouse gases   102, 115–129
- annual emissions, by sector   103
- efficiency of different transport modes   116
- emissions abatement in 2005   281–283
- passenger vehicle   117

gross domestic product (GDP)   55–56
- evolution   36
- governments and stimulus as percentage   57

h

horizontal drilling   87, 129
hydraulic fracturing/fracking   89–130

i

industrial revolution and inorganic chemistry (1750–1850)   187

Intergovernmental Panel on Climate Change (IPCC)   100, 104, 139

International Council of Chemical Associations –(ICCA)   265, 281

International Energy Agency (IEA)   8, 15, 18, 144, 150, 155–157
- Blue Map scenario   157, 159

internationalization (1960s)   194

International Monetary Fund (IMF)   43–46
- and Directors   45

international nuclear and radiology event scale   82

international organizations   44, 46
Internet use, global   50
IS/LM economic model   59

l

light cracking, transition to   312
long term cycles   236–241

m

magic tire triangle   122
Mauna Loa Observatory   13
megacities   27–29
megatrends, global   11–16
- climate change 99–105   139–140
- – carbon dioxide emissions reduction in tire and automotive industry   113–129
- economic   29–39
- – foreign direct investment (FDI)   40–42
- energy   61–62
- – and economics   70–71
- – lessons   69
- – life cycle   69
- – oil peak   71–80
- – recent developments   80, 82–98
- – recent transitions   63–68
- – shocks as valuable source of information   70
- – success criteria   69–70
- – transitions in life   70
- and hierarchy   12
- impact assessment   201–212
- – in ADV economies   214–216

– – in BRIC economies   212, 216–218
– – global ranking of impacts   210
– – major results   212–214
– – matrix   205
– – in REST economies   218–220
– political   42–43
– – BRIC economies   43–61
– social   16
– – demographics   18–25
– – population growth   16–18
– – urbanization   25–29
– wild cards   129–131
– – information technology and singularity   132–138
– – political   131
– – social   131–132
– – technological   132
– – transportation   132
Moore's law   133–135

n

National Oceanic and Atmospheric Administration (NOAA)   139
nuclear energy, and aftermath of Fukushima   80–84
nylon   190

o

oil production cost curve   73

p

paradox of thrift   58
per capita car ownership   78
per capita demand for chemicals, in $US   229–236
plastics demand explosives (1950s)   191–194
– plastics and products applications   192–193
political megatrends   42–43
– BRIC economies   43–61
polyamide nylon(DuPont)   190
population growth   16–18
Price WaterhouseCoopers (PWC)   6, 160, 226
– long-term economic growth model   6–7

r

rayon from wood fibers (1880)   188
real GDP growth   5
rolling resistance   122, 124
– mastering, into tire   124
– state of the art   121

s

shale gas   241
– of Europe, comparison with USA   98
– geology   91
– reserves, location map   93
– US ethylene expansions based on   92
singularity   15
– and information technology   132–138
– projections   137
social awareness   295–296
social media   143
social megatrends   16
– demographics   18–25
– population growth   16–18
– urbanization   25–29
social networks   50–53
– top   51
steam cracker (1910–1920)   190
Stern Review on the Economics of Climate Change   104–105
styrene cracking   190
synthetic dyes from coal for textiles, and chlorine bleach   187
synthetic fertilizers   189–190
synthetic rubber (1930s)   191

t

Tencent   51
tire labelling initiatives   118
Transnational Index (TNI)   49

u

UN Conference for Trade and Development (UNCTAD)   49
urbanization   25–27
– megacities   27–29
USA long term evolution energy mix   64

v

viscose   188
vision, by 2050   141–144, 221–224
– BAU scenario   146–155
– climate change   264–268, 297–307
– – global emission trading systems in operation   277–283
– – world carbon dioxide emissions historical and future scenarios   277
– feedstocks   245–247
– – alternatives   254–256
– – BAU scenario   247–250
– – simulation   250–254
– industry structure   283
– – chemical market   284

vision, by 2050 (contd.)
- - chemicals and per capita demand 287–289
- - company structures and global sales ranking changes 288, 291–294
- - pharmaceutical market 284–287
- - pharmaceuticals and per capita demand 287–291
- methodology 144–145
- outputs and products 256–258
- - global ethylene market and BAU scenario 258, 260–264
- relevance of industry 225–226
- - economic relevance 226–236
- - profitability 241–245
- - technological relevance 236–241
- social awareness 295–296
- sustainable scenario 155–160

w
World Bank Presidents 45
World Economic Forum 61
world FDI evolution, by major countries and areas 41
world GDP, by 2050 35
world life expectancy 22–23
- by 2050 21
world nuclear demand 84
world oil demand
- historical 66
- and reserves 72
world overview (1965–2010) 100
world status (2010) 145–146
World energy supply and demand 143–149